ENERGY DECISIONS AND THE ENVIRONMENT:
A GUIDE TO THE USE OF MULTICRITERIA METHODS

INTERNATIONAL SERIES IN OPERATIONS RESEARCH & MANAGEMENT SCIENCE

Frederick S. Hillier, Series Editor
Stanford University

Saigal, R. / *LINEAR PROGRAMMING: A Modern Integrated Analysis*

Nagurney, A. & Zhang, D. / *PROJECTED DYNAMICAL SYSTEMS AND VARIATIONAL INEQUALITIES WITH APPLICATIONS*

Padberg, M. & Rijal, M. / *LOCATION, SCHEDULING, DESIGN AND INTEGER PROGRAMMING*

Vanderbei, R. / *LINEAR PROGRAMMING: Foundations and Extensions*

Jaiswal, N.K. / *MILITARY OPERATIONS RESEARCH: Quantitative Decision Making*

Gal, T. & Greenberg, H. / *ADVANCES IN SENSITIVITY ANALYSIS AND PARAMETRIC PROGRAMMING*

Prabhu, N.U. / *FOUNDATIONS OF QUEUEING THEORY*

Fang, S.-C., Rajasekera, J.R. & Tsao, H.-S.J. / *ENTROPY OPTIMIZATION AND MATHEMATICAL PROGRAMMING*

Yu, G. / *OPERATIONS RESEARCH IN THE AIRLINE INDUSTRY*

Ho, T.-H. & Tang, C. S. / *PRODUCT VARIETY MANAGEMENT*

El-Taha, M. & Stidham , S. / *SAMPLE-PATH ANALYSIS OF QUEUEING SYSTEMS*

Miettinen, K. M. / *NONLINEAR MULTIOBJECTIVE OPTIMIZATION*

Chao, H. & Huntington, H. G. / *DESIGNING COMPETITIVE ELECTRICITY MARKETS*

Weglarz, J. / *PROJECT SCHEDULING: Recent Models, Algorithms & Applications*

Sahin, I. & Polatoglu, H. / *QUALITY, WARRANTY AND PREVENTIVE MAINTENANCE*

Tavares, L. V. / *ADVANCED MODELS FOR PROJECT MANAGEMENT*

Tayur, S., Ganeshan, R. & Magazine, M. / *QUANTITATIVE MODELING FOR SUPPLY CHAIN MANAGEMENT*

Weyant, J./ *ENERGY AND ENVIRONMENTAL POLICY MODELING*

Shanthikumar, J.G. & Sumita, U./*APPLIED PROBABILITY AND STOCHASTIC PROCESSES*

Liu, B. & Esogbue, A.O. / *DECISION CRITERIA AND OPTIMAL INVENTORY PROCESSES*

Gal, Stewart & Hanne/ *MULTICRITERIA DECISION MAKING: Advances in MCDM Models, Algorithms, Theory, and Applications*

Fox, B. L./ *STRATEGIES FOR QUASI-MONTE CARLO*

Hall, R.W. / *HANDBOOK OF TRANSPORTATION SCIENCE*

Grassman, W.K./ *COMPUTATIONAL PROBABILITY*

Pomerol & Barba-Romero / *MULTICRITERION DECISION IN MANAGEMENT*

Axsäter / *INVENTORY CONTROL*

Wolkowicz, Saigal & Vandenberghe/ *HANDBOOK OF SEMIDEFINITE PROGRAMMING Theory, Algorithms, and Applications*

ENERGY DECISIONS AND THE ENVIRONMENT:
A GUIDE TO THE USE OF MULTICRITERIA METHODS

Benjamin F. Hobbs
The Johns Hopkins University

Peter Meier
International Development and Energy Associates

SPRINGER-SCIENCE+BUSINESS MEDIA, LLC

Library of Congress Cataloging-in-Publication
Hobbs, B.F.
Energy decisions and the environment : a guide to the use of multicriteria methods/
Benjamin F. Hobbs, Peter Meier.
p. cm. –(International series in operations research & management science ; 28)
Includes bibliographical references and index.
ISBN 978-1-4613-7017-8 ISBN 978-1-4615-4477-7 (eBook)
DOI 10.1007/978-1-4615-4477-7
1. Energy development—Environmental aspects—Decision making. 2. Electric power plants—Environmental aspects—Decision making. 3. Multiple criteria decision making.
I. Meier, Peter, 1946- II. Title. III. Series

TD195.E49 H63 2000
333.79'14—dc21 00-041225

Originally published by Kluwer Academic Publishers in 2000. Second Printing 2003.
Softcover reprint of the hardcover 1st edition 2000

This printing is a digital duplication of the original edition.

Printed on acid-free paper.

CONTENTS

ACKNOWLEDGMENTS

We have many debts, but first and foremost to our long-standing and long-suffering collaborators, with whom we have applied the techniques described in this book to practical utility problems, and who co-authored some of the Chapters: Donna LeClair of BC Hydro helped with Chapter 8 (the BC Hydro case study) and Graham Horn of Planit Management is co-author of Chapter 9 (the BC Gas case study). Allen Wilson of Seattle City Light and Lee Czerniawski of Centerior Energy (now at Carolina Power and Light) also made critical contributions to our earlier studies of Seattle City Light (Chapter 10) and Centerior Energy (now FirstEnergy). Professor Arudpragasam, former Chairman of the Sri Lanka Central Environment Authority, helped with Annex 2.2.

Thanks also go to Wing Chen at Seattle City Light; Carlton Buford, Art Caune, and Rob Martinko at Centerior Energy; Eugene Stakhiv of the US Army Corps of Engineers; Eric Hirst and Stan Hadley of Oak Ridge National Laboratory; George Edgar of the Wisconsin Energy Conservation Corporation; Michael Talbot; Ian Wigington, Sharon McCarthy, and Phil Murray at BC Gas Utility Ltd.; Gina Beim of Pile Dynamics, Inc.; Richard Anderson of The Johns Hopkins University; and Vira Chankong of Case Western Reserve University. They and others have helped the case studies with their data, insights, and good cheer, and have contributed to the ideas presented in this book. The first author is grateful for the early encouragement he received from Jerome Dobson and the late Al Voelker of Oak Ridge National Laboratory. We also thank Paulin Aszkenasy for copy editing, and our families for their patience and moral support.

Portions of Chapter 9 are reprinted from *Energy Policy*, Vol. 25(3), B.F. Hobbs and G.T.F. Horn, "Building Public Confidence in Energy Planning: A Multimethod MCDM Approach to Demand-Side Planning at BC Gas," pp. 357-375, with permission from Elsevier Science. Portions of Chapter 10 are based on "Multicriteria Methods for Resource Planning: An Experimental Comparison" by B.F. Hobbs and P. Meier which appeared in *IEEE Transactions on Power Systems*, Vol. 9(4), pp. 1811-1817, Nov. 1994, Copyright 1994 IEEE.

Funding for portions of the work described here was provided to us by the Martin Marietta Corporation (under contract to the US Department of Energy), BC Gas, Ltd., BC Hydro, the National Science Foundation (under award ECS 9696014), Seattle City Light, the US Nuclear Regulatory Commission, the US Army Corps of Engineers Institute for Water Resources, and the US Environmental Protection Agency (STAR grant R825150). However, despite the assistance received from others, the opinions expressed here, and the responsibility for the contents of this book, are solely those of the writers.

1 INTRODUCTION

Nothing is more difficult, and therefore more precious, than to be able to decide
NAPOLEON I, Maxims (1804-1915)

Alternatives, and particularly desirable alternatives, grow only on imaginary trees
SAUL BELLOW, Dangling Man (1944)

Tradeoffs: it is a cliché that they have to be made in nearly all personal, business, and governmental problems. People, firms, and government agencies have many objectives, goals, criteria, attributes, or performance indices that they use to judge possible courses of actions. The problem is that it is frustratingly rare that all of these *desiderata* can be met or maximized by one alternative. Instead, some options will be good in some criteria, and other alternatives will do better in others. Choosing an alternative means that priorities must be set; accomplishing some goals means sacrificing others. Setting priorities is hard for an individual to do, and even more difficult when there are many people involved in a decision, each with their unique perspective and values. Confronting a decision consists of understanding how well different options might perform on the objectives, and the uncertainties involved; of eliciting the values and priorities of the stakeholders in the decision, and perhaps helping the stakeholders to form those values; of applying those values to the options, resulting in recommendations; and of clarifying how values shaped the recommendations.

Tradeoffs among conflicting objectives lie at the heart of many energy planning and policy problems, and that is often why those problems are so difficult. Tradeoffs are particularly glaring in the energy sector because of the large environmental, social, and economic costs often involved in providing reliable and convenient energy supplies. The question we address in this book is that it is not *whether* tradeoffs should be made, but *how* they should be made. Our thesis is that a systematic approach to examining tradeoffs and expressing value judgments can yield critical insights, facilitate constructive discussion among conflicting interests, and effectively build consensus.

The purpose of the book is to present the principles and tools of a field called *multiple criteria decision making* (MCDM), and to illustrate through actual applications how MCDM methods can help bring insight to problems and facilitate agreement among diverse stakeholders.[1] We emphasize the use of these techniques in dealing with environmental dimensions of electric and gas sector policy and planning problems. However, the methods are applicable to all decisions involving multiple objectives, not merely those concerning energy/environment tradeoffs. Our goal is to demonstrate the usefulness of these methods to those responsible for energy system operation, planning, and policy decisions, and to put practical tools into their hands.

Another goal of this book is to describe experiments in which energy decision-makers compared the usefulness, results and validity of several different MCDM methods. We hope that these experiments, together with our reviews of a wide range of applications, will inspire MCDM researchers to develop better methods and to test them against real problems.

In case the reader is unconvinced of the ubiquity of multiple, conflicting objectives, we describe a number of energy decisions involving several criteria in Section 1.1. Then, in Section 1.2, we summarize how formal MCDM methods can assist and improve decision-making processes. In Section 1.3 we discuss as an alternative to multicriteria-based procedures the use of economic benefit/cost analysis and describe the circumstances under which either that methodology or MCDM would be preferred. Section 1.4 concludes this introduction with an overview of the remaining chapters.

1.1 TRADEOFFS IN UTILITY OPERATIONS, PLANNING, AND POLICY DECISIONS

We summarize a wide range of multicriteria energy sector problems in this section. The time scale for decisions extends from minutes (for generation system dispatch) to decades (for research and development), and the geographic scale varies from local (distribution reliability and local integrated resource planning) to national and international (defining environmental policies and sustainability). What all the problems have in common is that there are many alternatives to choose from, and that there are important tradeoffs among economic, technical, environmental, and social objectives.

Many are applications that are emerging as the gas and electric utility industries are restructured around the globe. These applications show that the move away from integrated planning by a utility monopoly does *not* mean that environmental and social concerns are no longer important or no longer need to be considered. The forums for debating these issues have merely changed, in some cases for the good.

Several of the following applications are elaborated upon later in this book, where we show, step by step, how MCDM methods can be used to handle them systematically and with insight.

When is a power system operating in an emergency condition, and what loads should be shed first and when? "Satisfactory" power system performance involves voltage, frequency, and reserves for withstanding contingencies, among other considerations. Weighting and combining these into a single index summarizing the state of a system is, at its heart, a multicriteria problem, and systematic methods for accomplishing this task have been proposed (Ekel *et al.*, 1998; Gandibleux, 1999; Khare and Christie, 1997; Teo and Jiang, 1997).

What should be the reliability of a transmission and distribution system? Investments can lower the probability of unacceptable voltages or interruptions, but cannot eliminate them. Ramirez-Rosado and Adams (1991), as well as Matos (1999), consider these tradeoffs explicitly in their multi-objective algorithms for distribution system design. Chen (1998) and Venkatesh *et al.* (1999) use MCDM methods to evaluate capacitor locations in networks.

To what extent should a power generating system deviate from least-cost operation in order to lower emissions of NO_x, SO_2, or CO_2? This is the emissions dispatch and unit commitment problem. To answer this question, it is important to understand what emissions reductions can be accomplished at what cost, and the relative priority to be given to emission reductions (Liu and Huang, 1998; Talaq *et al.*, 1994). Lowering emissions can also affect other objectives; if, for instance, fuel consumption changes, jobs in mining or other sectors could be affected (Heslin and Hobbs, 1989). MCDM methods are also used to trade off system security, losses, and fuel cost objectives (e.g., Chattopadhyah and Momoh, 1999; Nangia *et al.*, 1998).

What investments should be made in pollution control or impact mitigation at a particular power plant? Costs must be traded off against the environmental improvements that would be obtained -- though an improvement in one environmental attribute may well be accompanied by a deterioration in another (as in the case of a flue gas desulfurization (FGD) system that reduces SO_x air emissions but increases solid waste). Alternative impact mitigation measures may also differ in attributes such as permanence, how they affect the survival of different organisms at different life stages, and worker and public safety (USACOE and Montgomery Watson Associates, 1996). Such decisions may be particularly difficult in developing countries, where investments in disease prevention, for instance, might be a considerably more cost-effective way to improve public health than investments in power plant pollution mitigation (Meier and Munasinghe, 1994).

Where should new generation facilities be located and transmission lines be routed? Early legal interpretations of the US National Environmental Policy Act of 1970 (NEPA) recognized that a careful and systematic balancing of environmental, economic, and other criteria is needed in energy facility location decisions. This led to the development and application of a wide range of MCDM-based siting and routing methods (Hobbs, 1979; Rowe *et al.*, 1979; Keeney, 1980). Although recent technological changes mean that generators are smaller and have fewer impacts, siting issues will remain important for developing countries with high load growth; for utilities which increasingly use small-scale generation to delay the need for transmission upgrades (Weinberg *et al.*, 1993; Lenssen, 1996); and in situations where sharpened competition increases the demand for transmission facilities.

Which rate structures should be adopted for regulated utilities? MCDM methods can be used to evaluate tradeoffs between administrative ease, economic efficiency, equity, and the environment in cases when significant environmental costs have not been internalized (Karni *et al.*, 1992; Megdahl, 1994; Woo *et al.*, 1994).

What mix of different energy supplies and demand-side management (DSM) programs should be chosen by a gas or electric utility? Hirst and Goldman (1991) argue that a distinguishing characteristic of integrated resource planning (IRP) is its explicit recognition of multiple objectives. Tradeoffs occur among various economic criteria (rates, value to customers, total resource cost, financial performance), environmental impacts (air and water emissions, resource consumption, land use, solid waste, wilderness), and social effects (regional economic development, maintenance of traditional cultures, employment). Energy suppliers and regulators have often used MCDM methods to understand tradeoffs (Andrews, 1992a; Crousillat *et al.*, 1993) and to quantify and apply value judgments -- e.g., in fuel choice (Keeney and Sicherman, 1983), cogeneration and generation plant design (Balestieri and Correia, 1997; Goumas *et al.*, 1999; Yokoyama *et al.*, 1997), resource bidding systems (Kahn *et al.*, 1989; Putta, 1990; Staschus, 1991), generation and demand-side planning (Mills *et al.*, 1996; Pan and Rahman, 1998; Kalika and Frant, 1999; Georgopoulou *et al.*, 1998), and demand-side management (DSM) program screening (Vine and Harris, 1990). In a restructured industry, MCDM methods could help governments and stakeholders evaluate the tradeoffs involved in setting up systems for "portfolio standards" or tradable credits for renewable energy or energy conservation (Rader and Norgaard, 1996). Bulk purchasers can use MCDM methods to decide whether they are willing to pay more for greener sources of power (Maloney, 1996).

How should "systems benefits charges" or other access fees designed to support social and environmental programs be spent? In many places, deregulation is eliminating the local utility monopolies which formerly were responsible for IRP. Because competitive utilities will have strong motivations to curtail expenditures on social and environmental programs, new or revamped public policies may be needed to ensure that society's objectives are appropriately reflected in energy production and consumption decisions. To maintain subsidies for DSM, renewable energy, and low-income customers, some jurisdictions are using access fees to fund such programs (Cavanagh, 1996). Allocating these funds among the many alternative programs will involve the same economic, social, and environmental criteria as traditional IRP.

What other means should governments adopt to maintain social and environmental programs traditionally sponsored by utility monopolies? Besides systems benefits charges, these policies could include: tax breaks and other subsidies; tightened energy efficiency standards for buildings and appliances; government procurement policies; and market transformation measures, such as the US refrigerator "golden carrot" competition and US EPA's "Green Lights" program (Tonn and

Schweitzer, 1996; IEA, 1996). The design of these policies again requires consideration of the whole range of criteria involved in IRP.

How does the environmental performance of different energy producers compare, and how is this performance improving over time? Corporations in many industries have developed environmental indices to complement traditional financial accounting measures for use by investors, regulators, and other external parties (e.g., NORTEL, 1996). Energy providers are developing their own indices (Miakisz, 1995). Regulators may use such indices in performance-based ratemaking (PBR) systems for adjusting rates (Biewald, 1996). Further, environmentally conscious customers and local distributors could refer to these indices when selecting power suppliers; so-called "green marketing" is popular among power companies that compete for retail customers in the US (Moskovitz, 1993). To define such an index, value judgments must be made about the relative importance of various types of impacts, evidence of corporate commitment, and implementation of procedures for pollution prevention and environmental stewardship.

The past environmental performance of other entities, ranging from individual assets to entire nations, can also be compared in this manner. Athanassopoulos *et al.* (1999) compare UK power plants using environmental and economic attributes, while Rogers (1999) argues for improved environmental quality indices in national development planning.

How should an energy sector be restructured? Decisions about privatization, regulatory reform, deregulation, vertical and horizontal break-ups and mergers, auction design, and creation of new institutions such as independent system operators require balancing of a range of objectives (Hirst and Tonn, 1996; Tonn and Schweitzer, 1996). Among them are economic efficiency, national self-sufficiency, institutional practicality, financial viability, product and service differentiation, and market power and other equity concerns. Environmental impacts, too, can be important; fears of increased NO_x emissions were a major objection to the US Federal Energy Regulatory Commission's 1996 rulemakings on transmission access. MCDM methods could be used to evaluate restructuring policies, just as they have been used for evaluating other national energy policies (e.g., Hamalainen, 1988; Jones *et al.*, 1990; Jungermann, 1985; Keeney *et al.*, 1987; Xie and Kuby, 1997).

How ambitious should environmental goals be, and how should they be implemented? Hirst (1996) argues that policies directly addressing environmental performance will be more effective than IRP in promoting environmental quality, a point also made by critics of emission "adders" and other traditional IRP means of considering external costs (Andrews, 1992b). Thus, as restructuring proceeds, this question will be most efficiently addressed when formulating general environmental policies, rather than in individual IRP processes (Jaccard, 1995). Tradeoffs between environmental, economic, and social objectives will be front and center when debating the stringency, effectiveness, costs, and distributional impacts of environmental standards, emissions taxes, or tradable allowance systems (e.g., Anandalingam, 1987; Stam *et al.*, 1992). Such tradeoffs become most difficult to evaluate in integrated assessments of greenhouse gas policies, because of their potentially profound consequences, long time horizons, and North-South equity issues (Bell *et al.*, 2000; Ramanathan, 1999).

How much research and development is desirable, and where should it be directed? R&D planning involves making informed guesses about what might be learned from R&D, and how that will affect the flow of benefits to individuals, firms, and society. These benefits can be multidimensional, including lowered costs, new energy uses, improved reliability, and decreased environmental and social impacts. Choosing an R&D project portfolio is explicitly recognized by private and public research organizations to be an MCDM problem (e.g., GRI, 1994; Golabi *et al.*, 1981; Kagazyo *et al.*, 1997; Perlack *et al.*, 1995; Stewart, 1991).

What is energy and economic sustainability? Many view the concept as being objectively definable based on the physical reality of limited resources. However, others argue that the concept itself is value laden and involves tradeoffs among conflicting criteria (Farrell and Hart, 1998; McDaniels, 1994; Hersch, 1999). Among these criteria are environmental degradation, rate of resource consumption, resource allocation and income distribution, and more traditional economic criteria. Evaluating the sustainability of alternative plans and policies is inherently an MCDM process.

Despite its length, this list is by no means exhaustive. These sample problems lay the groundwork for the next section, in which we summarize what formal MCDM methods can do to help managers, planners, policy makers, and stakeholders facing a multicriteria decision problem.

1.2 HOW CAN MULTICRITERIA DECISION-MAKING METHODS HELP?

The intent of MCDM methods is to improve the quality of decisions involving multiple criteria by making choices more explicit, rational, and efficient. MCDM methods have six basic functions that support this overall goal:

1. *To structure the decision process.* An MCDM approach helps analysts and decision makers think systematically about the problem by providing a logical framework for defining alternatives, comparing their performance on important objectives, and considering different viewpoints (Thomas and Samson, 1986).
2. *To display tradeoffs among criteria.* An example of a tradeoff is: "a decrease of 10,000 tons in SO_2 emissions would require a cost increase of \$2,000,000." Such displays aim to help planners, regulators, and the public understand the relative advantages and disadvantages of alternatives. Two common approaches include X-Y plots, where individual points represent the performance of alternatives upon two objectives, and value paths, in which a series of vertical bars or scales represent the degree of achievement for each criterion, and a single alternative is represented as a "path" connecting points on those scales (Cohon, 1978).
3. *To help people reflect upon, articulate, and apply value judgments concerning acceptable tradeoffs, resulting in recommendations concerning alternatives* (Stewart, 1992). Recommendations can be of several types: choosing the single best option, construction of a portfolio consisting of several options, definition of two or more alternative strategies representing different viewpoints, and screening out of undesirable options. The object is to help

people understand the implications of their value judgments and to inspire confidence in the soundness of the decision without being unnecessarily difficult. When alternatives have many attributes (10 or more are common in energy planning), psychological research has shown that decision makers are inconsistent in their subjective evaluations of the options. Often, the mind will focus on two or three attributes, ignoring the others, or it will flit inconsistently among the attributes (Shepard, 1964; Slovic and Lichtenstein, 1971). An example of an MCDM method that accomplishes value quantification is the additive value function, in which the total value of an alternative is set equal to the weighted sum of its criteria.

However, when performing this function, it is critical to keep in mind that people usually do *not* know exactly what they want at the outset of a planning or policy formation process. Therefore, it is naive to view this function as simply one of eliciting quantitative preferences that are firmly established in someone's head -- as if value functions were eternal, constant, and well defined. In reality, when confronting difficult and new problems, people usually start out only with general priorities and only vague notions of how much of one attribute they are willing to give up for another (Fischoff *et al.*, 1979). During the course of a decision-making process, people's attitudes will evolve in response to new information, interactions with others, and viewing the problem from different perspectives. Appropriately applied, this function of MCDM allows individuals to explore the implications of alternative value judgments and to build a set of value judgments in which they may have confidence (Gregory *et al.*, 1993).

4. *To help people make more consistent and rational evaluations of risk and uncertainty.* Psychological studies show that people have difficulty being consistent in decisions concerning which risks are acceptable (e.g., Krzysztofowicz and Duckstein, 1984). Some MCDM methods attempt to measure decision makers' attitudes towards risk (e.g., risk averseness) using utility functions and then use those functions to evaluate alternatives (e.g., Keeney and Sicherman, 1983). Other methods for initial screening of alternatives attempt to preserve a range of choices so that if the detailed engineering phase finds one alternative to be infeasible, other feasible plans will still be available (Keeney, 1987).

5. *To facilitate negotiation.* MCDM methods accomplish this in two ways: firstly, by quantifying and communicating the priorities held by different stakeholders, and secondly by moving the discussion away from alternatives and toward fundamental objectives and tradeoffs among those objectives. A focus on values facilitates negotiation because it encourages people to think about their common interests, and to avoid the defensive discussions that often result from each stakeholder anchoring on a preferred alternative (Raiffa, 1982). Untangling whether differences of opinion arise from disagreements over facts or over values is difficult in emotionally charged debates over issues such as nuclear power, but is critical to making progress on negotiations (Hamalainen, 1991). A discussion of values can also help define new alternatives that better satisfy group objectives (Keeney, 1992; Gregory and Keeney, 1994). An approach we have found helpful is to have each member of a group create their own additive value function and then to

rank the alternatives. The group is then assembled to review the results and identify areas of agreement and disagreement. The quantified value functions allow the reasons for disagreement to be made explicit and discussed (Brown, 1984). Consensus cannot of course be guaranteed, but even when it is not achieved, MCDM analysis provides useful documentation of the disagreements.

6. *To document how decisions are made.* By detailing how each of the steps of MCDM analysis has been applied, an agency or firm can communicate the basis of its decision to stakeholders. Indeed, some of the first applications of formal MCDM methods by the US Government were spurred by NEPA, which required agencies to document the rationale for decisions in environmental impact statements.

In this book, we summarize practical tools for accomplishing these functions. The emphasis is on the first, second, third and fifth functions (structuring the decision problem, tradeoff display, quantification of value judgments concerning tradeoffs, and negotiation).[2]

1.3 WHY NOT MONETIZE ALL CRITERIA?

A reasonable question is: instead of using MCDM methods such as those just described, why not just translate all criteria into dollars, and then choose the alternative with the highest net benefits or lowest total cost? (Watson, 1981; Mirasgedis and Diakoulaki, 1997). Indeed, utility commissions in several states have required utilities to monetize the air and water emissions of candidate resources and then combine them with the utility's direct costs in order to determine total societal cost (Putta, 1991; Dodd and Lesser, 1994).[3] As another example, reliability criteria are sometimes monetized to determine optimal levels of reliability or, in the case of the UK, to determine capacity payments to generators. Some methods for monetizing environmental impacts are reviewed in Chapter 2 (Section 2.3).

Explicit monetization has the following strong points:

- it is based upon the well-developed theory of welfare economics;
- it purports to measure values for the entire public (or a random sample of that public), rather than a subset of perhaps unrepresentative stakeholders;
- its results can, in theory, be validated by repeating the study or by using several methods; and
- its values for, say, the benefits of emissions reductions can be established and applied uniformly across decision situations and jurisdictions. This promotes cost-effective environmental improvement: for example, by making sure that alternatives for lowering CO_2 emissions are consistently evaluated.

However, monetization also has disadvantages:

- the basic assumptions of welfare economics are not universally accepted -- for example, that only peoples' preferences matter; that non-human creatures and ecosystems have no inherent rights; and that only net benefits matter, and not the distribution among different stakeholders (see Box 1.1 for examples of similar issues in other fields);

- fundamental value judgments concerning, for example, the worth of a human life are made by analysts and may be buried in the calculations, rather than put in the hands of policy makers and stakeholders where they belong; and
- the most valid methods can be difficult or impossible to apply in practice, leading to the use of more expedient yet less valid methods. An example of the latter is the control cost technique for quantifying environmental values.[4]

Box 1.1: Examples from other fields

The desire to capture complex attributes in a single number for purposes of comparison is not limited to the field of economics. Gould (1996) recounts the controversial history of the attempts to capture the multi-faceted set of human capabilities in the single number of the IQ test -- which, he argues, while clearly superior to 19th Century craniometry, is so demonstrably flawed as to render its use extremely dangerous.

Other examples cited by Medawar (1977) of "the illusion embodied in the ambition to attach a single number valuation to complex quantities" include efforts of demographers to seek causes for trends in a single measure of "reproductive prowess," or the desire of soil scientists to abstract the "quality" of soil as a single number.

Indeed, many of the flaws of using IQ to compare the "merit" of individual humans (or groups of humans) apply equally to the use of dollars as a measure of "merit" of individual projects, for economic value is but one dimension of value. Indeed, while the economic value of those quantities for which markets exist may be adequately objective, contingent valuation methods for assessing economic value of non-market goods are subject to the same shortcomings of IQ testing: since the results are extremely sensitive to the way in which questions are asked, it is almost impossible to decouple the results from the biases of those who pose the questions and evaluate the answers. This is an issue which practitioners of MCDM forget at their peril.

Compared to monetization, MCDM has a number of strengths (Watson, 1981; Jones, 1989). For example:

- learning and understanding by users is emphasized;
- tradeoffs among fundamental concerns are more explicit;
- values are directly obtained from stakeholders;[5] and
- dominated alternatives (those that are no better than some other alternative in at least one criterion) are easily ruled out.

The potential weaknesses of MCDM include:

- information "pollution" (the generation of so much data concerning the performance of alternatives on numerous criteria that the information cannot be digested by stakeholders);
- results that are difficult to repeat or verify because of the cost of assembling and educating stakeholders;

- possible failure of the priorities of stakeholder groups or their representatives to reflect the values of the community at large;
- improper application of MCDM methods, so that the true preferences of the stakeholders are distorted; and inconsistencies across jurisdictions regarding value judgments. For instance, CO_2/cost tradeoffs might be expressed in different ways by different agencies, and consequently be conflicting or difficult to compare.

However, careful design of an MCDM process can avoid many of these potential problems. Indeed, one of the objectives of this book is to assist in this design process.

Thus neither monetization nor MCDM is unambiguously superior to the other, because the two approaches have complementary strengths. For that reason, we believe that the best approach in practice is often a combination of the two methods (Gregory *et al.*, 1993; Lee, 1995b). For instance, weighting in additive value functions involves judgments, either implicit or explicit, of the rate at which the user is willing to trade off criteria. Thus, a particular set of weights for, say, cost and NO_x emissions implies that a ton of NO_x is worth so many dollars. Monetary estimates of damages from NO_x pollution could therefore be used to place a lower bound on the weight of NO_x. We found this type of information to be useful in the BC Gas planning exercise described in Chapter 9. There, information on monetary estimates on various air emissions was provided to stakeholders as a "reality" check or bound upon criteria weights.[6]

In the case of environmental criteria, we believe that monetization should be emphasized where:

- environmental costs are truly compliance costs (e.g., the price of tradable emission rights) and there are no externalities;
- defensible monetary estimates of damages exist; or
- there is a desire for uniformity of value judgments among jurisdictions and decisions, as there should be for, say, greenhouse gases.

But even in such situations, MCDM methods should be used to display tradeoffs and to explore the implications of different value judgments upon the decision.

In contrast, MCDM methods are most useful for analyzing environmental tradeoffs in the following circumstances:

- when fundamental value conflicts exist among stakeholders that should not be swept under the rug by monetization;
- when the decision process involves the public sector (such as review by the government regulators or public investment) or where unique local issues for which value judgments from other jurisdictions would not be directly relevant (e.g., local preferences regarding land use); or
- when dollar estimates do not exist for social costs, or are controversial.

Since such circumstances characterize most energy planning and policy processes, we believe that the MCDM approach should be widely useful when analyzing environment/economy tradeoffs in the energy sector.

1.4 SCOPE OF THE BOOK

This book is divided into two parts, one presenting basic MCDM tools that we have found to be useful, and the other describing a series of instructive case studies. We cannot possibly give an exhaustive review of the methodologies in the MCDM field, which has accumulated a huge literature in the quarter century of its existence.[7] Instead, we choose to focus on a subset of methods we believe have particular potential in helping to solve energy problems. The methods have been selected by the following criteria.

- They are *appropriate* for energy decision processes. That is, they are readily understood and accepted by stakeholders, provide the information needed for decisions in a digestible form, and are practical to apply in the time typically available for analysis.
- They represent *divergent philosophies* of decision making. Using more than one approach therefore helps users view the decision problem from different angles, which, as Chapters 9 and 10 document, can build insight and confidence in the outcome of the process.
- There exist either theoretical reasons or empirical evidence that the methods can be *valid representations* of people's preferences. However, methods that do a good job of modeling some people's preferences may do a poor job for others because different people prefer different ways of thinking about or expressing values (Hobbs *et al.*, 1992). For some methods, however, there is no theoretical or empirical basis for believing that they are as, or more, valid than the methods we do consider. Unless those methods have other advantages, we do not consider them here.[8]

We recommend that when choosing an MCDM method for a particular application, the user keep in mind these criteria of appropriateness, differences in results, and validity.

Part I encompasses Chapters 2 through 7. In Chapter 2 we give a quick definition of the ten basic steps of the multicriteria analysis process, and discuss issues involved in selecting and quantifying decision criteria (Steps 1, 2, and 3). An emphasis is placed upon environmental criteria. We then consider the first few analytical steps in Chapter 3, the display and analysis of tradeoffs (Steps 4, 5, and 6). There, we emphasize the use and interpretation of X-Y plots, including the elimination of alternatives by screening and dominance analysis.

In Chapter 4, we turn to the elicitation and quantification of value judgments by individuals (Steps 7, 8, and 9). Thus, while Chapter 3 presents methods for describing which tradeoffs are *possible*, in Chapter 4 we consider methods for determining which tradeoffs are *desirable*. The three most common tasks in value quantification involve the following manipulations of criteria: value scaling, priority weighting, and combination ("amalgamation") into a single index of option desirability.

Chapter 5 is devoted to the tenth and final step, namely that of using the results of the previous steps to support negotiations among stakeholders with potentially divergent interests. That chapter also discusses how use of more than one MCDM method and resolving their differences can promote insight. Chapter 6 then presents simple numerical applications of the methods as a complement to the more theoretical presentations of the earlier chapters.

Chapter 7 presents a brief review of some applications in the literature, chosen to illustrate the very wide range of utility problems to which MCDM methods have been applied. However, with few exceptions (e.g. PP&L, 1977), MCDM methods have not been widely used in stakeholder or other planning processes involving public participation (Schweitzer *et al.*, 1994). One reason for this lack of use is MCDM's perceived complexity and opaqueness -- a shortcoming that we attempt to remedy in this book.

Part II of the book presents three case studies in depth to illustrate the potential of the methods and the practical difficulties involved in their application. The first two case studies emphasize tradeoff analysis (Steps 4 - 6). Chapter 8 describes the use of tradeoff analysis to facilitate public involvement in power resource planning. Interest group representatives were gathered by BC Hydro, a Canadian electric utility, to make recommendations concerning desirable supply and DSM resource portfolios. Tradeoff analysis helped to narrow the discussion to a few critical issues. The result was a useful clarification of the debate and the achievement of limited, yet surprising consensus on some of those issues. Thus, this chapter also addresses the critical negotiation step, Step 10.

Chapters 9 and 10 are case studies of the use of value quantification for individual and group decision making (Steps 7 through 10). In Chapter 9, BC Gas, a Canadian natural gas utility, used MCDM methods to quantify the values of stakeholders regarding the impact of DSM programs. Each of the stakeholders used three different MCDM methods to screen DSM programs; the conflicts among method results gave them an opportunity to reflect upon their values, and to learn about the tradeoffs. These results were the basis of individual recommendations, which were then reviewed by the group. This multi-method approach enhanced stakeholder confidence in the outcome of the process.

In Chapter 10, we report upon an experimental comparison of MCDM methods for quantifying value judgments. As noted above, methods can differ in appropriateness, results, and validity. We asked personnel and external stakeholders of the American municipal utility Seattle City Light (SCL) to apply several methods for weighting and amalgamating criteria in integrated resource planning. We found significant differences in people's confidence in the methods, and in the results themselves. In fact, rank orders yielded by different methods often differed more than rank orders chosen by different people: a disturbing result. These disagreements reinforce the importance of careful application of MCDM methods, and the value of a multi-method approach for building insight and confidence.

The book ends with the concluding remarks of Chapter 11.

NOTES

[1] The field is also known by several other names, including *multi-objective decision making* and *multi-attribute tradeoff analysis (MATA).* Although some researchers differentiate between multicriteria and multi-objective decision making techniques (primarily by whether discrete alternatives or mathematical programs are involved), we interpret MCDM broadly as encompassing all methods for displaying and evaluating tradeoffs among conflicting criteria.

[2] In the case of the fourth function, we cannot conclude that tools for capturing attitudes towards risk are useful in many circumstances. This is because risks are rarely quantified in energy policy and planning studies (which is to be bemoaned). Further, we have found that typical participants in planning and policy studies do not respond well to the lottery-type questions required to quantify risk attitudes (Keeney and Raiffa, 1976). Regarding the sixth function, documentation, this does not require special techniques, but just a careful reporting of how the steps of the MCDM process have been executed.

[3] For reviews of the types of methods used for the purposes of monetizing environmental benefits and costs, see Freeman (1993) or Munasinghe (1993). Applications of such methods to energy planning and policy are critiqued in Eyre (1997), Krupnick and Burtraw (1998), and Stirling (1998).

[4] In the control cost method, a regulatory or legislative body's monetary evaluation of an environmental impact is considered to be revealed in its past decisions concerning the costs it is willing to impose to lower that impact. The challenges faced in quantifying environmental damages is well illustrated by the results of the ambitious US - EC Fuel Cycle Externalities Project (Lee, 1995a). To quote another study: *"...no clear consensus exists on quantitative estimates of environmental costs of electricity, or on methodologies for making those estimates ... Some critical disagreements over methodology ... mask deeper disputes over values, basic policy goals, and the intended role of environmental cost studies. It is unlikely that these disputes can be resolved by technical analysis or scientific research. Instead, these disagreements are more likely to be successfully addressed through public debates in the policy arena"* (USOTA, 1994).

[5] The monetization method of contingent valuation also elicits judgments directly from people, although from a random sample of the public at large rather than from stakeholders in a collaborative planning context. For a discussion of how contingent valuation might be strengthened by including the value-structuring capabilities of MCDM, see Gregory *et al.* (1993).

[6] In a BC Gas resource planning exercise that preceded ours, stakeholders were asked to trade off CO_2 against other objectives. Because of the complicated nature of the process and the lack of consideration of CO_2 monetary values used by other jurisdictions, average values they chose for CO_2 were several thousand dollars per ton -- two orders of magnitude above values applied elsewhere in North America. Explicit consideration of previous monetary values can prevent such distorted results. However, the presentation of such precedents can also distort the judgment process; psychologists have found that when making judgments, people tend to anchor on suggested values and then fail to adjust them sufficiently (Kahneman *et al.*, 1982). Although we recognize this is a danger, we believe that it is better not to intentionally withhold information from stakeholders; data on value judgments made elsewhere should be disclosed, but the potential for an anchoring bias should also be explicitly discussed and understood.

[7] Some textbooks in the area include Chankong and Haimes (1983), Cohon (1978), Goicoechea *et al.* (1982), Mollaghsemi and Pet-Edwards (1997), Steuer (1986), Yoon and Hwang (1995), and Zeleny (1982). Many articles have surveyed the field; Stewart (1992) is

a particularly stimulating review of many of the methods of Chapter 4. There are also many journals that publish papers on MCDM, and a large number of applications are published in Springer-Verlag's series of proceedings of the biannual International Conference on Multicriteria Decision Making (e.g., Haimes and Chankong, 1985; Haimes and Steuer, 2000). We refer the interested reader to these sources for in-depth examination of theoretical issues, for presentations of methods we have chosen to omit, and for reviews of the wide range of non-energy applications of MCDM.

[8] An acceptable theoretical basis, in our opinion, consists of 1) a set of axioms about human choices that have normative validity (i.e., it is reasonable to assert that rational behavior *ought* to conform to those assumptions) and are, in theory, verifiable; and 2) a set of methods, consistent with the axioms, for inferring the necessary parameters (weights, etc.) from expressed or revealed human preferences, preferably manifested as choices between alternatives. Empirical evidence for validity is a problematic subject, as we hope Chapters 9 and 10 make clear; but, in general, we believe there should be studies with a reasonable sample of people facing a realistic problem which show that a method does as good, or better, a job of representing preferences as alternative methods.

Multi-attribute utility theory is an example of an MCDM methodology with an acceptable theoretical basis (Keeney and Raiffa, 1976), while a number of empirical studies show that additive value functions represent preferences as well as more complicated amalgamation methods (e.g., see reviews by Huber, 1974; Hobbs, 1986; or von Winterfeldt and Edwards, 1986). Therefore, we mention these methods in Chapter 4. An example of a method that we have chosen not to review is fuzzy set theory as applied to MCDM problems (e.g., Beccali *et al.*, 1998; Chedid *et al.*, 1999; De and Hipel, 1987; Ekel *et al.*, 1998; Khare and Christie, 1997; Liu and Huang, 1998). This approach does not satisfy our theoretical or empirical criteria. In particular, it is not founded on axioms or quantification methods that are based on choices, and we are unaware of empirical studies comparing its validity to other techniques. (Similar criticisms are made by Stewart, 1992.) Because, in our opinion, fuzzy set approaches are opaque to typical stakeholders and have no obvious advantages, we have chosen not to include them in this survey.

2 THE APPLICATION OF MCDM METHODS

Let all good things be done decently and in good order
NEW TESTAMENT, 1 Corinthians 14: 40

We owe almost all our knowledge not to those who have agreed, but to those who have differed
CHARLES CALED COLTON, Lacon 2:121 (1825)

In this and the following three chapters, we present a discussion of the sequence of steps involved in applying MCDM methods to a decision problem. Although most of the examples included here are drawn from energy resource selection, these steps are largely independent of the particular application involved, and apply to a wide spectrum of energy applications that range from power plant siting to bidding programs and the assessment of corporate environmental performance.[1] We begin with an overview of the process, and then discuss each of the steps in turn.

The application of MCDM methods involves the following steps:

Problem definition steps:

1. Selection and definition of attributes, say A_i, $i=1, \dots I$, chosen to reflect important planning objectives and/or environmental concerns. System cost, reliability, effects on prices, air quality impacts, or impact on fisheries are examples: in this step we select which of

these will be used in an application, and precisely how they should be defined.

2. Definition of the alternatives to be analyzed.
3. Quantification of the levels A_{ij} of the i attributes estimated for each of the j alternatives. This generally requires the application of a model to predict the impacts. Uncertainty and risk in attribute levels are quantified at this time.

Tradeoff analysis steps:

4. Construction of tradeoff curves.
5. Dominance analysis.

Evaluation steps:

6. Preliminary screening of alternatives.
7. Scaling of attributes, in which the level of an attribute is translated into a measure of value, $v_i(A_{ij})$, (also known as a single attribute value function). This sometimes involves normalization, such as using a scale of zero to one (in which the lowest value of the attribute value function is assigned zero, the highest is assigned one).
8. Selection of weights w_i for each attribute.
9. Determination and application of an amalgamation rule. Such rules combine the weights and value functions into a single overall value or ranking of the available options. Alternatively, a rule can merely reduce the number of options for further consideration to a smaller number of candidates.
10. Resolution of differences between methods, and between and among stakeholders.

While the steps appear conceptually straightforward, in general most applications give relatively little attention to a number of theoretical requirements for the results to be valid. Equally important, even though we have presented these elements in the form of a sequential list, iterations among these steps will be necessary -- as we shall see in the subsequent discussions. The relationships between these steps are indicated in Figure 2.1.

Although we do not identify it as one of the steps in the MCDM process itself, deciding *who* participates may well determine which criteria are chosen for the process (Step 1) and which options should be examined (Step 2). Moreover, since weights (Step 8) reflect the values of the participants, the outcome of the process will necessarily reflect its participation. The case study of Chapter 8 discusses this issue and its practical consequences.

In this and the subsequent three chapters, we review some useful methods for accomplishing each of the ten steps. Their use is illustrated with a simple three-attribute problem in Chapter 6, brief summaries of a range of energy applications in Chapter 7, and in-depth case studies in Chapters 8 to 10.

2.1 ATTRIBUTE SELECTION AND DEFINITION (STEP 1)

We place attribute selection as the first step, prior to alternative definition, because we believe that hard thinking about attributes can then suggest new alternatives that better achieve the objectives. Keeney (1992) and Gregory and Keeney (1994) convincingly argue that expanding the set of options by "value-focused thinking" can therefore lead to better decisions.

Figure 2.1: Steps in an MCDM process

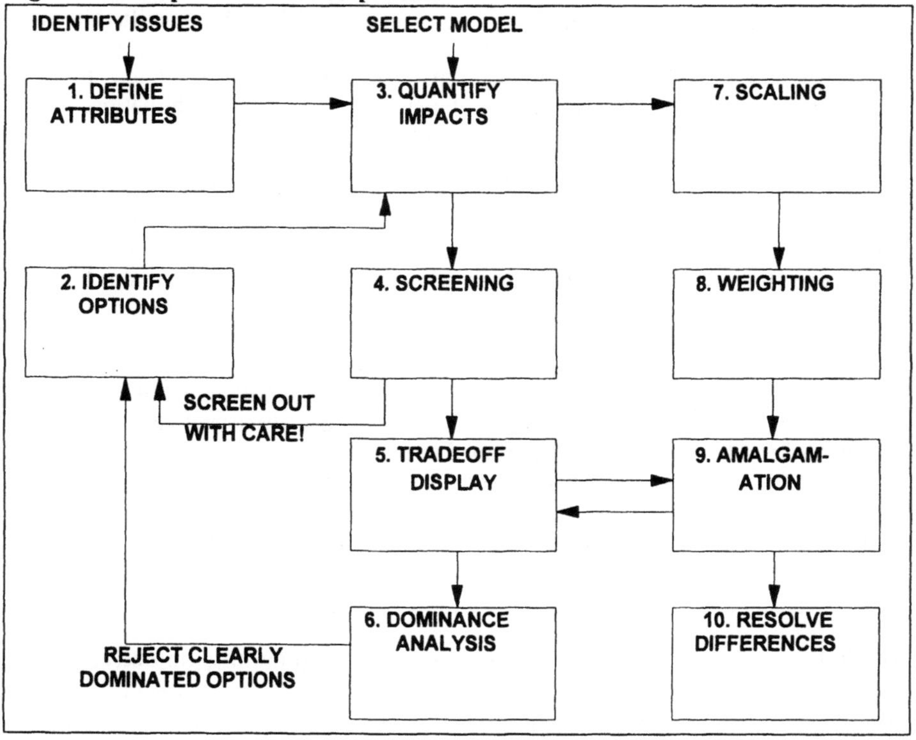

Selecting the issues to be included in a planning or policy study may appear to be relatively straightforward, yet in fact proves to be one of the most difficult, and ultimately critical, tasks. In large part, the criteria that are selected reflect who participates in the process, and whose perceptions about what is important are included. Yet there are many practical difficulties: while attributes should reflect the main criteria that individuals view as important, there is little point in selecting attributes that cannot be quantified, or where quantification involves such high uncertainties that the result may be meaningless.[2] Indeed, in some problems such as climate change mitigation, the uncertainties that plague most of the attributes are so large that use of MCDM or any other decision method may well be problematic (Morgan *et al.*, 1999).[3]

Box 2.1: MCDM case study of Sri Lanka

In the early 1990s the Environment Department of the World Bank undertook a research project to develop ways of incorporating environmental concerns into power sector planning (Meier *et al.*, 1993; Meier and Munasinghe, 1994). Although project specific environmental assessments have long been a routine part of World Bank procedures, it was recognized that many environmental issues needed to be analyzed at a much earlier stage in the project cycle, and that absent consideration of environmental concerns in systems planning, fundamental questions about the need for new generation projects remained unanswered. These questions became all the more important as the World Bank shifted its attention to broader questions of power sector reform, with its power sector lending program in South Asia changing from loans for new power stations to the support of institutional restructuring, tariff reform, T&D rehabilitation, and demand-side management (see, e.g., World Bank 1998b).

In the late 1980s Sri Lanka began to face increasing environmental opposition to new power plants, particularly at the proposed coal-fired power project in Trincomalee on the east coast. As originally proposed, this would have used once-through cooling, discharging into some relatively shallow waters of Trincomalee Bay. At the same time, the development of the remaining major hydro sites also ran into increasing opposition, and several projects were redesigned as run-of-river schemes.

The Sri Lanka study used an MCDM approach, structured along the lines presented in this book. The emphasis was on the quantification of tradeoffs among six environmental attributes and economic cost (discussed below in Section 7.9.2). The methodology developed in the Sri Lankan study has since been adopted for a comprehensive assessment of environmental issues in the Indian power sector, where the urgency of reform and restructuring, and the environmental problems related to plants burning low-grade Indian coals, have received increasing attention (World Bank, 1998a).

Attributes, or measures of goodness, express the planning objectives to be considered in a given problem. Unfortunately, the need for care in the selection of attributes is not very evident from much of the literature of application of multicriteria methods to energy problems. Six main issues need to be addressed in the selection and definition of attributes: boundary setting, double counting, value independence, specification of attribute definition, attribute quantification, and the proliferation of attributes. We discuss each of these in Section 2.1.1. We then examine a special category of attribute, namely that which captures the *risk* associated with an alternative, in Section 2.1.2.

2.1.1 Issues in attribute selection

Setting boundaries: The first task is to draw the boundaries of analysis. Two examples underscore the point made earlier: how the boundaries are drawn depends upon the perceptions of the participants as to what is important. In the first example of Sri Lanka, impacts occurring outside the national boundary were excluded; in the second example of BC Hydro in Canada, impacts outside the national boundary were included.

In our case study of Sri Lanka (Box 2.1), the boundaries were drawn on the basis of relevance to national decision-makers. For example, the environmental impacts associated with the extraction of imported coal (that would arise in Australia or South Africa) were excluded; if in fact Australian policy-makers do decide to include the cost of environmental externalities associated with extraction (say by imposition of rigid land reclamation requirements on the coal mining companies), then that would be reflected in a higher coal price to Sri Lanka.

On the other hand, we did include the question of greenhouse gas emissions. The reason is that Sri Lankan decision-makers will need to deal with greenhouse gas emission reductions as part of the international discussions and agreements on climate change now underway.

Many impacts that are purely local in nature can safely be deferred to the project level environmental assessment stage without running any significant risk that these would ultimately stand in the way of implementing a strategy identified at the planning stage. Thus, again in the Sri Lankan study, in the case of the large coal-fired plant proposed for Trincomalee we chose to exclude interest group objections based on very narrow issues (such as the potential impact of the plant on whale watching). However, we included the concern held by the broader environmental community that discharge of thermal effluents into shallow waters constituted a broad risk to aquatic ecosystems. The emphasis was therefore put on the cost of alternative cooling system options to limit thermal discharges, not the presence or absence of specific species at specific locations -- the latter being the proper concern of the *project* level environmental assessment.

Obviously judgments must still be made, and there may be some impacts that at some projects are purely of local, and at others of national significance. Resettlement at hydro plants is a good example: at some projects, the resettlement impacts may be quite small, and would neither strain the ability of local authorities to manage, nor would they be likely to become a political issue at a national scale. But at other sites, the scale of resettlement may be such that indeed it may require the attention of national decision-makers. The Kukule hydro project originally proposed for central Sri Lanka is a case in point: if the high dam variant had been built, the need to resettle some 11,000 people defines an issue of national importance; on the other hand, the run-of-river variant that was finally selected (and which is now under construction) displaces only 135 people in 27 families.

Boundary issues featured prominently in the BC Hydro Study (Chapter 8), because of the increasing importance of electricity trade with the United States and Alberta. Air emissions associated with electricity imports were therefore included in the corresponding attributes, particularly for greenhouse gas (GHG) emissions. This was relatively straightforward where the type of generation could be readily identified, as in the case of firm contracts from coal-fired plants in Alberta. But in the case of spot market purchases from the United States, lack of information on the type of generation made this more difficult, and a notional generation mix was used as the basis for defining GHG emissions associated with such imports.

Even more vexing was the treatment of exports. For example, suppose surplus hydro were exported to California. Consistency with the treatment of imports would require that a GHG emissions *credit* be given for such exports, since such hydro imports would, at the margin, reduce thermal generation in California. Again lack of information forced us into making some set of assumptions satisfactory to the various stakeholders represented on a consultative committee that guided the BC Hydro study: in this particular case we made the assumption that exports would displace gas-fired generation, with an appropriate adjustment for transmission losses.

As another instance of a situation where boundary setting is difficult, consider the case of a pollutant which is subject to a national cap and a tradable emissions rights system. An example is the case of SO_2 in the United States. At first glance, local SO_2 emissions would seem to be a reasonable attribute. Yet because of the cap, any local increase in SO_2 would ultimately be matched by an equal decrease elsewhere; consequently net national damages from emissions might net out to zero, or even be positive (Hobbs, 1994; Burtraw *et al.*, 1995). This issue is discussed further below.

Double counting: There are at least three types of double counting that need to be considered in an MCDM problem. The first relates to double counting in the attributes: for example, in a power plant siting or IRP problem, there ought not to be a separate attribute for the miles of associated transmission line if there are already other attributes that capture the relevant land use impacts. As another example, including both "monthly bills" and "total resource cost" (as in Maloney, 1996) would double count cost. However, provided one is always cognizant of precisely what impact a particular attribute is supposed to capture, double counting is not too difficult to avoid in practice.[4]

However, the attribute "land use" is one that is frequently encountered, often expressed as "acres", or "acres/MW."[5] Why this should be a separate attribute is generally unclear. The cost of land is (or should be) included in the system cost attribute, as should any costs associated with resettlement and compensation (often an issue in hydro projects). The opportunity costs of lost production are also easily calculated (say, in the case of land in agricultural use). Therefore, it is only the *non-market* values of land that need to be captured in an additional attribute. Yet if there are unique historical, ecological or archaeological values that need to be explicitly considered, then they are surely not captured by an attribute for acres per MW.

A second, more difficult, double counting issue concerns the extent to which those pollutant emissions that are subject to tradable emissions allowances -- such as SO_2 -- ought also to be treated as an attribute (or, in monetization approaches, also assigned some externality value).[6] As noted above, a practical question is whether or not there should be an attribute for SO_2 emissions in an MCDM analysis in the US, given that the total quantity of SO_2 emissions in the US is fixed under the emissions allowance system established under the 1990 Clean Air Amendments, and that the cost of purchasing the necessary allowances is already included in the cost attribute. Therefore, at first glance the answer to the question might seem to be negative.

However, the argument that the emission allowances system has internalized the cost of SO_2 emissions is valid only as long as it does not matter where and when the emissions occur. But the damages SO_2 emissions inflict do depend strongly on location. Emissions upwind of urban areas and sensitive ecosystems, such as New York's Adirondack lakes, are more harmful than releases in areas with predominantly alkaline soils or few people downwind, such as Florida and Nevada.[7]

Elsewhere (Hobbs, 1994; Burtraw *et al.*, 1995), it is shown that the social cost of an additional ton of SO_2 emissions at a particular location X is the sum of two quantities: the price of an allowance, plus the external cost resulting from the emissions at X, minus the external costs avoided because emissions at other locations will decrease. That is, the social cost accounts for how an emissions allowance would otherwise be used, and the resulting damage if it were not consumed at X. Since allowances are an internal cost anyway -- utilities must either buy allowances or forgo selling them -- the appropriate estimate of social cost is just the second of the two quantities. This adder might in fact be negative.[8]

In practice, therefore, whether or not there should be an attribute for SO_2 emissions would depend upon whether a case can be made that there is a differential locational effect *in addition to* the cost of allowances that should have been included in the cost attribute. If such a case can be made, and SO_2 emissions are included as an attribute, it is likely that participants should be repeatedly reminded, particularly during the scaling and weighting elements of the MCDM process, that the attribute refers to only the *differential* impacts.

The third double counting issue is the comparison of means and ends. Attributes -- the ends -- should represent fundamental objectives and concerns. They should not be confused with the *means* for accomplishing those ends. Including types of alternatives as an objective in itself risks double counting if there are already attributes that capture some of these alternatives' advantages and disadvantages.

For instance, inclusion by Hanson *et al.* (1991) of both "SO_2 emissions" and "coal consumed" as attributes in an integrated resource planning (IRP) study, risks penalizing coal twice for the same sin if individuals consider coal's pollution when weighting it. Similarly, in our case study of DSM described in Chapter 9, energy conservation and two types of emissions are all included as attributes. If conservation is valued because it lowers pollution, this advantage will be double counted. However, some participants in that study claimed that the conservation attribute represented only the moral obligation to save non-renewable resources for future generations. One safeguard against double counting in that study was that weight assessors chose a single weight for the entire set of environmental attributes (two emissions and conservation), and then divided it among the constituent attributes. Where double counting is suspected, this "hierarchical" weighting approach can help prevent it (see Section 4.2.3).

Conceptual independence: Attributes should be conceptually distinct. A strict type of conceptual independence is called "preference independence".[9] For example, if CO_2 emissions and cost attributes were preference independent of SO_2 emissions, then one would be willing to spend $15/ton to avoid a ton of CO_2 emissions, regardless of whether SO_2 emissions were low or high. Decision analysis differentiates between statistical independence and preferential independence; the former refers to the correlation structure of the alternatives, the latter to structure of the user's preferences -- a distinction that might be characterized as "facts" versus "values."

Additive value functions (discussed in Chapter 4) assume a certain type of conceptual independence, but statistical independence is irrelevant to the validity of the additive form. For example, whether alternatives with high CO_2 emissions also tend to have high SO_2 emissions is a question of statistical, not preference, independence.

A good illustration of the importance of conceptual independence is illustrated by the poor criteria definition adopted by the New York State Hazardous Waste Facility Siting Board. The Board established "danger of fire and explosion" and "population near site" as separate criteria to be added together (New York Department of Environmental Conservation, 1980). But these attributes are *not* preferentially independent -- a high likelihood of explosion would motivate a user to put a higher weight on the criterion "population near site." One possible approach to avoiding this kind of problem is to define a new attribute as the product of these two. Keeney and Raiffa (1976) describe several approaches to redefining attributes to avoid preference independence problems.

Specificity of attribute definition: The rationale for selecting a particular attribute needs careful consideration. For example, one frequently finds a criterion "solid waste disposal" expressed as tons of waste generated. Precisely why the *quantity* of solid waste should be the best proxy for the impacts associated with solid wastes is unclear, except that it is easy to estimate. Since a large part of the total land area required for a coal-fired power plant may be for solid waste disposal, there may in any event be double counting if there is also a criterion for "land use" (expressed in hectares). Moreover, the *quantity* of solid waste is perhaps primarily related to the *cost* of disposal (higher quantities correlate very well with higher disposal costs): yet again, such costs themselves should be included in the system cost attribute.

In fact the relevant environmental risk associated with ash and scrubber sludge disposal is the risk of toxic leachates contaminating nearby aquifers, which has only an indirect relationship to the quantity of waste produced. Rather, the risk is more directly related to site characteristics such as soil permeability, depth to the groundwater table, and the intensity of rainfall.

The problem is that actual environmental risk may be poorly determined at the long-range planning stage: even if the general locations are known, site studies are unlikely to be available in the sort of detail necessary. Yet choosing the most convenient proxy, or even one with seeming precision, such as volume of solid waste generated, may not in fact be the most appropriate. In this particular case, the *number* of solid waste sites that have to be opened may be a more suitable attribute.[10]

Quantification of attributes: In some sense, obviously, the definition and quantification of attributes go hand in hand: MCDM methods require that attributes be quantified. However, there is the danger that what is difficult to quantify -- yet important -- may get ignored. On the other hand, just because a criterion is easily quantified does not necessarily mean it has anything to do with actual impact -- as noted above in the case of tons of solid waste. Nor is regulatory convenience necessarily related to impact. Thus despite the regulatory focus on (easily calculated and monitored) emissions, weighting air emissions by local area populations is surely a more precise measure of potential health damage than emissions *per se.* Indeed, use of population-weighted changes in ambient concentrations would be even better as an attribute.

There are four ways of dealing with attributes that are difficult to quantify. First, in rare cases, indirect valuation methods can be used without direct impact quantification: for example, some of the so-called "hedonic" pricing methods (Freeman, 1993) can be used to estimate the impact of existing power facilities. By statistically examining real estate prices in the vicinity of a facility, and comparing them to values of property without a facility nearby, hedonic pricing can assess how individuals value the perceived environmental impacts -- without necessarily having to quantify the actual impact such as incremental air pollution (at a fossil plant) or incremental radiation or probability of catastrophic accident (at a nuclear plant).

A second way of dealing with impacts difficult to quantify is to use mitigation scenarios. For example, the health impacts associated with EHV transmission lines may be viewed as significant.[11] Yet there is presently no basis for quantifying that effect. One way to include an assessment of the issue is to define a scenario that eliminates the need for additional EHV lines (by dispersed siting, or reduced reliability), and then asking whether one is willing to incur the incremental costs involved.

The third and most often practiced approach is proposed by Keeney and Nair (1977). Environmental effects that are difficult to quantify, such as some ecological impacts, are described categorically. For instance, several categories can be defined, ranging from "mild/beneficial impact" to "severe impact," accompanied by a detailed textual description of what is meant by each. Then experts or stakeholders can directly construct a value scale for these categories by one of the methods of Step 7 (Chapter 4). Thus, the attribute is directly quantified as a single attribute value function, without first quantifying a physical or otherwise "objective" scale.

Box 2.2: Particulate emissions as a proxy for rural air quality

In Chapter 8 we present a case study of the application of MCDM methods at BC Hydro to help the development of a strategic plan. Central to this process was the participation and work of a Consultative Committee of stakeholders: it played a critical role in the development and evaluation of alternative resource portfolios.

Particulate emissions were added as an attribute mid-way through the process to reflect the strong importance given by some members of the Consultative Committee to reduction of rural particulate emissions from beehive burners (that would be reduced if woodwastes were diverted from such burners and used at power generation projects more easily capable of meeting stringent emission standards). As suggested by the tabulation below, the attribute value for particulate emissions is heavily influenced by emissions from proposed wood waste-fired projects, which are given emission *credits* for particulate reductions due to reductions in beehive burner emissions. Indeed, this gave the seemingly anomalous result that a "green" energy resource portfolio, with no wood-waste plants and with only minimal additional particulate emissions from gas-fired plants, did much worse on this attribute than a portfolio having much higher levels of fossil resources but which also included some woodwaste plants.

2010 Particulate emissions (tons):
(for a scenario with high woodwaste utilization)

coal	0
gas (Burrard)	91
gas (IPPs)	306
woodwaste	-2200
imports	220
exports	-75

Although calculated as tonnes of particulates, the attribute was intended as a proxy for rural health. The Consultative Committee noted that a given change in particulate emissions from a woodwaste plant had a much greater *relative* impact on rural health than in urban areas, because emissions from other sources were much lower. On the other hand, particulate emissions from the urban gas-fired Burrard plant, located near Vancouver, are extremely low, but potentially expose large numbers of individuals: thus when particulates from other sources are taken into account, the cumulative *change* in health risk attributable to Burrard emissions, even though large numbers of individuals might be exposed, is very small. On the other hand, particulate emissions from beehive burners expose small numbers of individuals to very high emissions; and the use of woodwaste at power generating plants would therefore result in a relatively large improvement in ambient air quality and benefit to rural health.

A fourth approach is to define an appropriate proxy attribute, of which we give an example in Box 2.2. However, great care must be exercised in such cases to ensure that the proxy really reflects the objective (and, as noted in Box 2.2, to ensure that there is no double counting). Often it will be necessary to bring in outside experts to ensure scientific validity -- expertise which may often be absent from stakeholder groups and environmental non-governmental organizations (NGOs) that participate in utility planning.[12]

In Annex 2.2 we give an example of how an attribute was developed for biodiversity.[13] This was an objective about whose importance there was general agreement in the Sri Lankan case study, but which heretofore resisted quantification in MCDM studies. However, thanks to the participation of an eminent ecologist, an attribute was derived that balanced the goals of stakeholder interest, scientific rigor, and practicality.

Attribute proliferation: A proliferation of attributes tends to make weighting more difficult. Further, it may introduce a bias simply because one is reluctant to weight any particular attribute as near zero, which results in an under-weighting of truly important attributes (see Chapter 10). Moreover, tradeoffs become difficult to understand and display to decision-makers in comprehensible form if there are too many of them. The guiding principle ought to be that one starts from those impacts which are the most important (such as greenhouse gas emissions, health effects of fossil fuel pollution, the risk of groundwater contamination, the risk posed to aquatic ecosystems from thermal plumes, etc.), and then, if at all possible, one attribute is selected for each of these concerns.[14]

2.1.2 Risk as an attribute

Risk is present when there is randomness or uncertainty associated with an attribute. Everything else being equal, alternatives with more risk are generally less preferred by planners, policy-makers, and the public than more risky options. This subsection gives an overview of methods for quantifying risk as an attribute so that it can be traded off against other attributes when evaluating energy alternatives.

Risks and uncertainties in energy planning and policy problems arise from many sources. Variability in natural processes (e.g., river flows that affect hydropower production or weather conditions that affect electricity demands) is important, and can usually be estimated from historical records. Such records are also useful for equipment reliability analyses. On the other hand, historical data cannot be directly used to quantify institutional, political, and economic uncertainties. Expert judgment might be relied upon to quantify such uncertainties by specifying subjective probability distributions (von Winterfeldt and Edwards, 1986), or plausible ranges of outcomes.

How options with different risks are evaluated depends not only on the probabilities and consequences involved, but also on many other factors (Starr and Whipple, 1980). These include whether risks are 1) borne voluntarily or not; 2) controllable by the individuals facing them; 3) immediate or delayed; 4) disagreed upon by experts; 5) borne by those who also receive the benefits; and 6) familiar or unusual.

Value judgments regarding risks can be quantified in two basic ways. One is to capture attitudes towards risk in the form of utility functions. We examine this approach in Chapter 4. The other approach, described here, is to define one or more separate attributes that characterize the degree of risk associated with each option. For instance, if expected cost is one attribute, then a second attribute representing risk could be the standard deviation of cost, or the probability that cost exceeds some pre-specified level. The more risk averse the decision maker, the more weight would be assigned to the risk attributes.

An example of the risk-attribute approach is portfolio analysis (Zeleny, 1982). Two indices -- the expected value of an attribute, and its standard deviation -- are used. There exist tradeoffs among these. One alternative might have a good expected value (for example, high average net benefits) but a large standard deviation (because of vulnerability to, say, disruption of energy supplies because of political or weather conditions). A different alternative might do less well on an expected value basis, but catastrophic failures might be less likely (because the alternative relies on more reliable yet costly supplies). Portfolio analysis is a standard financial tool for private firms to compare the riskiness of alternative investment portfolios. For instance, Pacific Gas & Electric, a US utility, has used the approach to compare alternative combinations of natural gas supplies and various contracts, including futures and options.

In theory, the standard deviation and expected value are all that are needed to evaluate alternatives under either of the following two restrictive circumstances:

1. Each decision-maker has a quadratic single attribute utility function which takes the form:

$$U_i(A_i) = a + bA_i + cA_i^2$$

   where $U_i(A_i)$ translates the original attribute A_i into a measure of desirability or "utility" (usually on a zero to one scale), such that if a decision-maker prefers risky option X to risky option Y then the expected utility of an option X exceeds that of Y, and vice versa. (Expected utility is defined as the probability weighted average of $U_i(A_i)$. Section 4.1.2 describes the use and estimation of utility functions.) Or:

2. All probability distributions are normal and each decision-maker has an exponential single attribute utility function (as defined in Section 4.1.2).

Neither circumstance is likely to fully hold in practice. In that case, theory suggests that additional information (such as the skew of the distribution) be considered. Nevertheless, the standard deviation of an attribute might still be a convenient and readily understandable means of summarizing the relative risks of competing alternatives.

A problem with applying the standard deviation as an index of risk is that it measures the spread of an alternative's performance around the mean. However, a decision-maker may be most concerned with values that are much worse than the mean and not too worried about those that are better. Several other indices for risk may be more useful in that case. For instance, the level of the attribute which is exceeded, say, just one out of 10 years (10% exceedence probability) is one such index. Such indices are technically known as "fractiles," and are often used to characterize, for example, streamflows and hydro-power production. (They may also be called "scenarios" or "outcomes," but the term "fractile" will be used here because it is more precise.) Several such indices might be used, representing various levels of risk. An extreme case is to define the index as being the worst level that has a non-zero probability (such as Zeleny (1982) proposes in his "Prospect Ranking Vector" method): however, for many uncertainties such as hydropower production or future electricity demands, defining such a level is a difficult and ultimately arbitrary task.

An alternative to fractiles is to define an attribute as the probability of exceeding some threshold. Examples of such thresholds might be the level of cost that would risk bankrupting the firm, or a concentration of pollutant that could not be tolerated by a desirable species of wildlife.

2.2 DEFINE OPTIONS (STEP 2)

Most decision-makers like to have some reasonable number of options to choose from. Which number is reasonable in any set of circumstances may of course vary, but 20 may be too many, and two is likely to be too few. But whatever the number, the most important criterion is that the set selected shows meaningful differences in the type of option and its impacts. An energy resource planning analysis that looks, say, only at five different coal plant choices and a single level of DSM program implementation will satisfy nobody. At the same time, at the outset of an IRP process, one may not know which alternatives are the interesting ones to examine. In short, the definition of the options is likely to be part of an iterative process, rather than something determined irrevocably at the outset. Keeney (1992) presents a number of useful approaches to defining options that are likely to do well under the attributes defined in Step 1.

The problem of proliferation of options is well illustrated by a case study of the application of MCDM at Centerior Energy of Ohio (now FirstEnergy) (Hobbs and Meier, 1994a).[15] The possible combinations of different levels of DSM, generation unit life extension planning, dispatching method, and new generation and purchase options ran into several hundreds. This had to be reduced down to some manageable number. The question is how many, and by what criterion is the reduction process to take place?

Moreover, these plans had to be evaluated in the light of many different uncertainties (Schoemaker, 1991). A number of combinations of external circumstances beyond Centerior's control -- the purchase price of electricity from neighboring utilities, nuclear availability, DSM effectiveness, natural gas prices, to name a few -- all complicated the process.

This of course describes the classical problem of decision-making under uncertainty, for which decision trees and influence diagrams make an excellent representation.[16] But with many chance nodes (representing the outcomes of the uncertain variables) and many decision nodes (which represent the alternative actions that can be taken), such trees often become very complex ("bushy") which may make them difficult for decision-makers and management to absorb.

At Centerior we therefore focused only on a smaller number of "futures" -- in effect some subset of the branches of the decision tree. With management strongly concerned about rate impacts, the futures defined for our case study represented the extremes that would combine into very favorable, or very unfavorable, circumstances for the evolution of rates. Thus, as shown on Table 2.1, Centerior defined "low stress," "reference," and "high stress" futures. These cases corresponded to the likelihood that the utility would require rate relief, which was the dominant concern of Centerior's management (and of its relationship with the regulatory commission).

Table 2.1: Definition of futures in the Centerior Energy case study

	low stress [L]	reference [R]	high stress [H]
Purchase power, $/MWh	21	27	33
Allowance price, $/ton SO_2	444	278	222
CO_2 tax, $/ton CO_2	0	0	22
Nuclear availability, %	80	80	60
DSM effectiveness, %	100	100	50
Load growth, %/year	3	1.5	0
Gas prices, cents/mmBTU	178	355	533

The process of defining such futures (or scenarios) can stimulate thinking about what options might "work" in different futures, or be robust across these futures. This can lead to the addition of more plans to the initial list (Keeney, 1992; Gregory and Keeney, 1994).

There is probably some danger of leaving option selection entirely to technical analysts: if the analysis is to be representative of the issues likely to come before subsequent regulatory proceedings, it is better that options of interest to likely intervenor groups be included in the analysis, however unpalatable that may be, or however improbable that may be viewed by planners. Thus we cannot stress too much the desirability of as broad and as early as possible a public participation in this process -- a point that is hardly new.

Broad-based participation within an organization is also important. For instance, in resource planning several different organizational departments need to participate at this stage, with, at a minimum, representation from the environment, rate, marketing and DSM departments, in addition to those directly involved in the planning process and in the preparation of the regular cycle of submissions to the regulatory commission.

As an example of the results of such an option definition process, in Figure 2.2 we show those branches of the total decision tree that represent the set of 15 options used in the Centerior study (the Reference cases, plus cases ***a*** through ***n***, which are indicated on the far right-hand side of this figure). The options were chosen to be representative of the diversity of resources and economic and environmental outcomes; and each was assessed for each of the three futures (scenarios L, R, and H) identified in Table 2.1. The specific assumptions for these plan options are summarized in Table 2.2.

Figure 2.2: The Centerior study decision tree

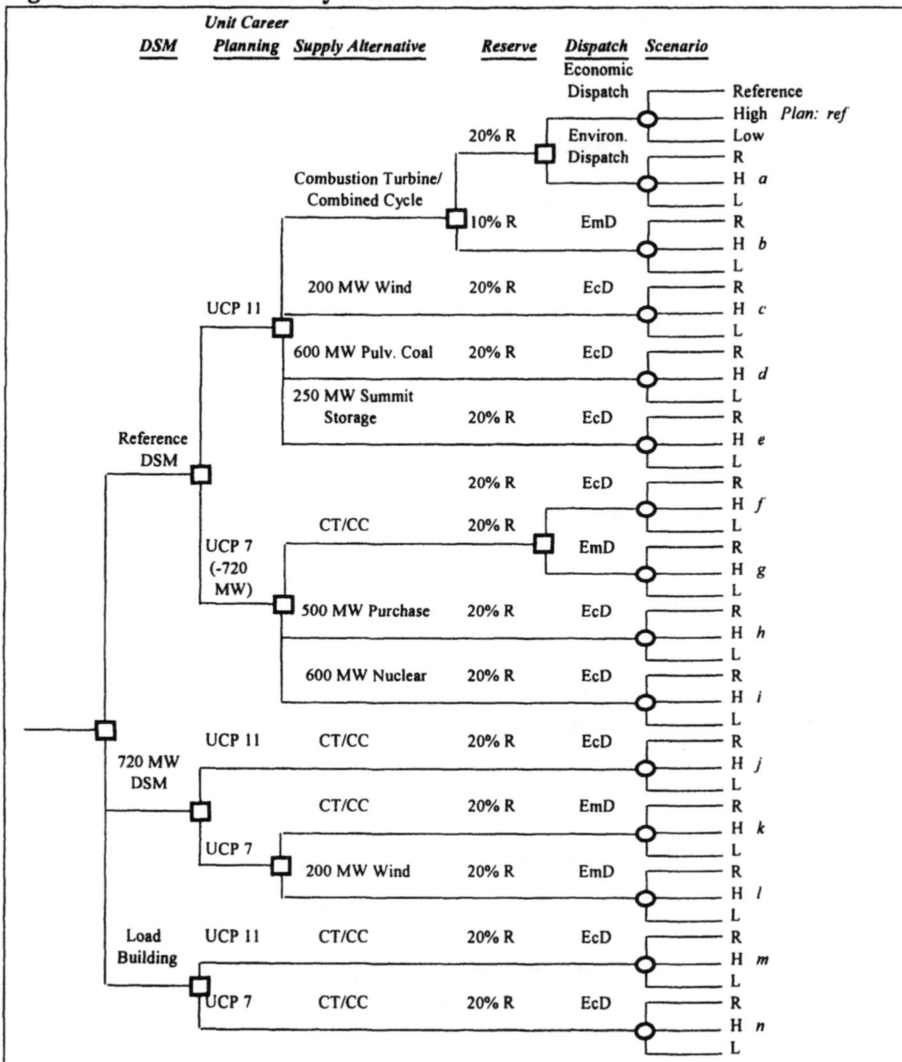

A different approach was taken in the Seattle City Light study (described in Chapter 10), where a mathematical programming model was used to generate a range of resource mixes. This was accomplished by varying the relative weight of CO_2 and utility cost in the model's objective function, and by adding and deleting certain options. For instance, in one run a CO_2 penalty of \$20/ton was applied, while new hydroelectric capacity was prohibited in response to concerns about its impact on fish. In another run we set no penalty and no constraints on resource choice, yielding a least utility-cost plan. The process yielded 36 tentative resource plans that were then narrowed down to 12 using some of the methods described under Step 6 (see Chapter 3).

Table 2.2: Plan options in the Centerior study

Plan	Description	DSM [MW]	UCP[a] [units]	new capacity [type]	reserve margin [%]	dispatch method[c]
ref	Reference	360	11	CT/CC[b]	20	economic
a	Reference + emissions dispatch	360	11	CT/CC	20	emissions
b	Reference with 10% reserve	360	11	CT/CC	10	economic
c	Reference + wind	360	11	200 MW wind	20	economic
d	Reference + coal	360	11	600 MW coal	20	economic
e	Reference + Summit	360	11	200 MW Summit Plant[d]	20	economic
f	Reduced UCP[a]	360	7	CT/CC	20	economic
g	Reduced UCP + emissions dispatch	360	7	CT/CC	20	emissions
h	Reduced UCP with 500 MW purchase	360	7	500 MW purchase	20	economic
i	Reduced UCP with new nuclear	360	7	600 MW nucl.	20	economic
j	Increased DSM	720	11	CT/CC	20	economic
k	Increased DSM, reduced UCP, emissions dispatch	720	7	CT/CC	20	emissions
l	Increased DSM, reduced UCP, wind + emissions dispatch	720	7	200 MW wind	20	emissions
m	Load building	360	11	CT/CC	20	economic
n	Load building + reduced UCP	360	11	CT/CC	20	economic

[a] UCP = unit career planning. The plans with 7 units have 720 MW less of capacity subjected to UCP

[b] CT/CC = combustion turbine/combined cycle

[c] In emissions dispatch, penalty added to CO_2 during dispatch, resulting in lower emissions

[d] Compressed air storage facility

2.3 QUANTIFY IMPACTS (STEP 3)

In this section we discuss four issues associated with quantification of the impacts of attributes: the use of models, economic valuation, the quantification of environmental impacts, and the influence of location.

The use of models: The use of rigorous mathematical models to predict impacts on each of the attributes lies at the heart of the MCDM process; indeed, impacts need to be quantified for each alternative under each future, and often for every year within the planning horizon. Nevertheless, in many studies -- particularly power plant siting -- there has been a tendency to use "expert judgment" to score alternatives. In such procedures, groups of experts directly "score" different alternatives on some scale: this represents a combination of steps 3 and 7. As we shall see in Section 7.9.1 where we review some applications of MCDM to siting, such short cuts may lead to questionable results.[17] In our view, whatever the application, there is no valid alternative to a rigorous calculation and quantification of impacts by application of an appropriate model. Even if all models are subject to certain assumptions and limitations, these can always be spelled out clearly, so that decision-makers and reviewers can understand how results were derived.

Models are in wide use by electric utilities for resource planning, and can be used to estimate total system costs and emissions. Traditionally, their principal focus has been to determine the economic and financial consequences of alternative resource plans. Some are optimization models, that seek the least-cost capacity expansion path subject to exogenously specified demand forecasts and reliability levels (such as EGEAS). Others are simulation models, in which cost, reliability, and rate impacts are calculated for a series of specific options specified by the user (such as MIDAS and ENVIROPLAN).

Table 2.3: Examples of resource planning models used in electric utility MCDM applications

Application	Model	Reference for model	Reference for study
Sri Lanka	ENVIROPLAN	Meier (1999)	Meier and Munasinghe (1994)
Centerior Energy	MIDAS	Farber *et al.* (1988)	Hobbs and Meier (1994a)
Seattle City Light	System specific, in-house model.	Wilson (1989)	Chapter 10, this book
British Columbia Hydro	ENVIROPLAN+ in-house model	Meier (1999)	Chapter 8, this book
Andhra Pradesh, India	IRP Manager + ENVIROPLAN	Meier (1999)	USAID (1995)
Bihar, India	ASPLAN	Analytical Solutions (1997)	Sone Command Area Development Authority (1998)
TVA	EGEAS	Bloom *et al.* (1984)	TVA (1995); Andrews and Govil (1995).
Cincinnati Gas & Electric	PROVIEW/ PROSCREEN	Energy Management Associates (1993)	Bloemer and Knue (1993)

Table 2.3 lists some of the models used in recent electric utility MCDM applications. Economic cost (specified either as present worth or annualized cost), reliability (specified as loss of load probability or the present value of unserved energy), rate impact (specified as average cost per kWh), emissions, and capital requirements are attributes that will most likely be viewed as part of every MCDM application to utility resource planning, and for which these kinds of models are well suited.

But the most important criterion for selecting a techno-economic model is not theoretical rigor, but the degree to which the analysts and decision-makers have confidence in model results. A model that has been in use for some time for resource planning is likely to be the best choice, for almost certainly the decision-makers will ask what confidence the analyst has in being able to predict the impacts of an option. The discussions are likely to be contentious enough without model validity becoming an additional issue.

For example, in MCDM analyses at Seattle City Light (SCL) and Centerior Energy, we used models that had been in use for many years: MIDAS in the case of Centerior, and a system specific, in-house model in the case of SCL. A number of possible shortcomings might have been cited in both models, but everyone involved seemed confident about their use, and their ability to calculate to within reasonable accuracy the tradeoffs that were the principal focus of the exercise. In both cases the participants were confident that the models themselves would not be questioned by the regulatory entity (the City Council in SCL's case) that would review the plan. On the other hand, in the BC Hydro application, we introduced a new model (albeit used and tested elsewhere), in part because none of the participants (including the utility itself) had great confidence that the existing models met the needs of the planning process.

While the absence of a suitable model would be quite rare in energy utility MCDM applications, in situations where no models exist or application is otherwise impractical, then there may be little choice but to rely on expert judgment (of the type discussed above in the case of ecological impacts following the procedure of Keeney and Nair, 1977). Careful documentation of the points of the impact scale is critical. One should avoid vague generalities, and be very specific: rather than saying "high impact" one might say "high probability of extensive damage to coral reefs due to water temperature increase greater than 2° C over present ambient temperatures," and so forth.

In most of the MCDM applications conducted by the writers, the need for "what if" analysis has always arisen.[18] The ability to test the many ideas and inputs from stakeholder groups and consultative committees that often participate in utility planning is a key part of confidence building necessary for successful planning. Explicit modeling allows the user to test the attractiveness of different mitigation approaches, technology choices, or other concerns of the participants in the planning process (Andrews, 1992a).

Economic valuation: As pointed out in Chapter 1, in any general situation there will be certain environmental impacts for which economic valuation poses few difficulties, and/or where application of well-established procedures is unlikely to be controversial. So-called use values in particular pose relatively few difficulties in this respect. For example, estimating the opportunity costs of lost agricultural production of inundated land and the costs of rehabilitation and resettlement at a hydro plant, or estimating the benefits of a potential reservoir fishery, is generally straightforward, and can be valued by the market value of the lost production.

In some situations, what gets monetized and included in the economic attribute may be dictated by a regulatory body -- as in the case of certain utility regulatory bodies in North America which prescribe monetary values to be used for air emissions. Unfortunately, the procedures used by some Commissions to set values are controversial, particularly those based upon the costs of pollution control rather than avoided damage costs,[19] and may therefore not meet our criterion of "unlikely to be controversial."

Common sense rather than the application of hard and fast rules is the indicated approach. Economic valuation of human life is to be avoided whenever possible, for the simple reason that controversy over the validity of any particular value that may be chosen tends to detract from the more important issue of comparison of alternatives. If indeed the risk to human health that is associated with utility air emissions is an issue, then a health risk attribute might be used, based upon a calculation of the change in population exposure to the ambient concentration of certain pollutants. Such an attribute is more likely to provide the necessary insights into how alternative plans perform on this attribute than attempting to monetize the incremental mortality and morbidity and include the result in the economic attribute.

Imputed values for human health effects may also provide guidance. In the Sri Lankan case study, one of the issues was whether or not flue gas desulfurization (FGD) should be required at proposed coal-fired generating plants (Meier and Munasinghe, 1994). Estimating the value of human life in developing countries is a particularly hazardous venture (especially when conducted by foreign consultants from the United States or Europe), for inevitably there is the danger that one gets embroiled in ethical debates about why the dollar value of a human life in Sri Lanka should be worth only a few percent of that in the United States. We calculated the value of human life necessary to justify the installation of an FGD system to avoid the health effects of SO_x emissions, for which a valuation of about $1.5 million emerged. Common sense dictated that in a situation where local hospitals lacked all modern diagnostic equipment, application of the corresponding money for the purchase of such equipment would provide health benefits at least two orders of magnitude greater than by the installation of FGD equipment. The policy conclusions were thus evident without the result being dependent upon some particular value of human life.

Quantification of environmental impacts: Much more difficult is the quantification of the environmental *impacts*. Most of the techno-economic models mentioned above have the ability to calculate pollutant emissions, and deal with environmental regulations that apply to the *quantity* of pollutants that may be discharged into the environment. Thus emissions of criteria pollutants, greenhouse gases, land use, and quantities of solid wastes are routinely compiled by such models.

However, the extent to which emissions address regulatory convenience rather than actual impacts is an important question. For some residuals, emissions may be much more appropriate than other indicators as rational proxies for actual impacts. Thus, as already noted, tons of solid waste, easily calculated, is probably a rather poor indicator of the extent to which solid waste disposal actually poses an environmental risk. On the other hand, emission of greenhouse gases is an appropriate proxy for the incremental contribution to global warming.[20]

Environmental impacts and location: Impacts depend upon the location and timing of emissions, their transport and transformation in the environment, and the location, exposure, and vulnerability of affected populations and resources. Only in a very few exceptions is the location of emissions unimportant.[21] Even though emissions tend to dominate the regulatory perspective, from the viewpoint of actual impacts it is not emissions themselves that are important, but their fate in the environment. Two things follow from these observations: firstly, that it matters a great deal where emissions occur, and secondly that any quantification and/or monetization of impacts are also likely to be dependent upon location. Most of the monetary valuations of air emissions adopted by US state regulatory commissions thus far, however, are independent of location; as noted earlier, it is entirely unclear why, if the regulatory objective is to minimize impacts, a ton of NO_x emitted from a combustion turbine with a relatively low stack in Boston should be subject to the same externality value as a ton emitted from a plant in the less populated areas of central Massachusetts. The values used for air emissions in some states in the early 1990s are shown in Table 2.4.

Table 2.4: Monetary valuation of air emissions ($/ton)

	CO_2	NO_x	SO_x	particulates
New York	1.1	1,832	832	333
Massachusetts	22	720	1,700	4,400
Nevada	22	6,800	1,560	4,180
California (PG&E)	26	7,105	4,060	2,380

There are also situations where the exact opposite is true, most notably in the case of greenhouse gases: in this case from the perspective of actual impact on the environment it matters not one bit whether a ton of CO_2 is emitted in Manhattan or rural upstate New York, or in Florida or Colorado. The incremental impact on global warming will be the same. Yet different regulatory commissions have chosen to value CO_2 emissions at very different levels (e.g., in Massachusetts at 22 $/ton, in New York at 1.1 $/ton).

The impacts of global warming will be unevenly distributed: for example, Boston and Manhattan will be much more susceptible to the impacts of sea-level rise than Colorado. Yet even if the impacts of global warming were assessed by the Massachusetts regulatory commission as more severe than elsewhere in the country, does it follow that imposing a higher externality value on greenhouse gas emissions in Massachusetts than elsewhere is rational? This can lead to inefficiencies: expensive efforts to limit emissions at one place, when a much greater tonnage could be eliminated at the same cost elsewhere.[22]

A general procedure for quantifying environmental effects: The quantification of environmental impacts might proceed in the following sequence (Rau and Wooten, 1980; Wood, 1995). We use here the example of an air pollutant, but the framework is equally applicable, albeit with some modification, to other types of impacts:

1. *The timing and quantity of emissions.* This can generally be estimated fairly well with the kinds of models in general use by utilities. Models such as MIDAS or EGEAS can be run in a number of alternative dispatching modes to the traditional least-cost dispatch, including dispatch to minimize specific emissions.
2. *The location of emissions.* At the planning stage there may be high uncertainty about where particular new power plants may be located. There do exist models that predict locations of emissions (e.g., Meier, 1979; Cohon *et al.*, 1980). However, a glance at the siting scenarios

projected in the 1970s and 1980s raises a number of questions about the ability to make realistic assumptions for much beyond 10-15 years.[23]

3. *The fate of pollutants in the environment.* For most of the important pollutants, enough is known about atmospheric transport to be able to make fairly good predictions of the changes in ambient concentrations resulting from a given quantum of emissions at a given location.[24] Such models are routinely used in environmental impact statements. Modeling the fate of pollutants in surface waters can also be said to be relatively free of substantial uncertainty, since again many water quality models for rivers, lakes, and coastal waters are generally accepted by the scientific community as valid. More difficult, and more subject to uncertainty, is modeling the fate of pollutants in the subsurface environment.
4. *Exposure of sensitive populations.* To calculate the aggregate dose requires knowledge of how great the affected population will be. Projecting population for local areas over the long time horizons that are typical of energy planning is notoriously difficult, and subject to large uncertainties (Meier, 1972). Equally important, an existing projection may be affected by the facility itself, for which the classic example is nuclear power plants: because of the uneven distribution of the property tax benefits, in some areas of the country nuclear power plants have attracted additional population far beyond the initial projections.
5. *Estimation of the dose-response function.* Epidemiological studies to estimate dose-response functions are subject to great uncertainty because of the need to control for the large number of other factors that affect morbidity and mortality. Indeed, in regulatory proceedings this is likely to be one of the most contentious issues, and it is quite unclear whether commissions and their staffs have the ability to resolve conflicting testimony presented by parties and their witnesses.
6. *Valuation of the response.* Perhaps most controversial is the final step in this sequence, namely the monetary valuation of the impact of the affected populations to the dose in question. The most difficult of all is the economic valuation of human life, subject to such large uncertainties that use of any particular estimate is likely to be so controversial that attention is focused on validity of the valuation, rather than on the tradeoffs that must be made. One of the advantages of MCDM methods, of course, is that this final monetization step need not be taken.

The central difficulty in this sequence is that none of the Steps 3 through 6 can be undertaken unless the location (Step 2) is known. For this reason, while most project-level environmental impact statements normally progress to Step 3, few long-range planning studies progress much beyond the first step.[25]

One approach to the problem of unknown location at the long range planning stage is the use of "surrogate sites." This was pioneered by the nuclear energy center studies conducted by the United States Nuclear Regulatory Commission in the mid 1970s (USNRC, 1977). It was charged by Congress with assessing the feasibility of clustering nuclear power plants into "energy centers" as a way of addressing concerns about safeguards at widely dispersed centers. It was quickly recognized that in order to assess the environmental impacts of such centers, one would have to hypothesize particular sites. Unfortunately, in this specific instance of nuclear energy centers, residents near the selected sites had a hard time accepting the notion that one might study a site without actual plans to proceed to construction.

Advocates of monetary valuation argue that while it may well be true that location is indeed important, in the absence of practical ways to determine which differences in impact can be attributed to different locations, applying an average value to all locations is better than doing nothing at all. Alternatively, applying the worst-case value to all locations is better than making exceptions or imposing lower values at some other locations in the face of very imperfect information about the spatial variation in impacts.

Another way of expressing this is to note that such conclusions in fact tell us something about the risk preferences of utility regulators, and that in turn leads us to a decision analysis framework for examining these issues more scientifically. Rather than always assuming the worst case, in the expectation that more detailed study, or resolving the scientific uncertainties, takes too much time and money, the proper question is whether in fact developing additional information has value. Decision analysis, and its concept of the value of perfect information, provides just such a framework.[26]

Detailed environmental models -- particularly for air and water quality -- are of course routinely applied at the project-level environmental impact stage to translate emissions into estimates of changes in ambient concentrations. In other words, items 1 through 5 in the above list can readily be determined at the project level. But at the long-range planning stage, the necessary site-specific data may simply not be available.

* * * * *

In this chapter, three critical steps that set the stage for understanding and valuing tradeoffs have been described. We have raised several important issues that make the steps of attribute selection, definition of options, and performance quantification difficult. Users of MCDM methods must exercise considerable judgment, as there are few cut and dry recipes for resolving these issues. Compromises must often be made: for instance, the meaningfulness of environmental attributes must be balanced against the need for a timely and not overly cumbersome process.

Next, we turn to the approaches for communicating the information obtained in Step 3, and using it to screen alternatives using the concept of "dominance."

ANNEX 2.1

An example of an MCDM method for evaluating proposals

MCDM methods lie at the heart of many systems for evaluating bids submitted in response to requests for proposals (RFPs) for generation capacity provided by independent power producers (IPPs). A recent RFP issued by a developing country for the supply of diesels on a build-own-operate basis (BOO) illustrates the approach, and some of the pitfalls.

Proposals were evaluated on the basis of nine criteria (described more or less verbatim, in italics), as follows. Our comments on the attributes are noted in each case.

1. Base purchase price

The marks for the base year purchase price, A, will be calculated by the formula

$$A = 300\left[1 + \frac{0.055 - P^*}{0.055}\right]$$

where P is the proposed base year price in US$/kWh, and subject to a maximum of 350 marks.*

Some algebra shows that as a result of the last constraint, there is no benefit to a first-year price below 4.583 cents/kWh.

2. Tariff structure

Project proponents who accept the tariff structure as described under (1) will be given a maximum mark of 50. Alternative structures will not be considered, since such alternatives will undermine the basis of the model power purchase agreement. Therefore zero marks will be assigned for alternative tariff structures.

This is a pointless attribute. What happens if the winner is one who scores zero marks on this attribute? What about some minor adjustment (say to some foreign exchange indexing proposal)? If indeed the intent is to irrevocably screen out those who are not prepared to accept the tariff structure described in the power purchase agreement (PPA), then a simple screen for that acceptance suffices.

3. Levelized tariff

The marks, B, for the proponent's levelized tariff, P, will be determined by the formula[27]

$$B = 300\left[1 + \frac{0.055 - P}{0.055}\right]$$

with a maximum of 350 points, where P is the proponent's tariff, in US$/kWh.

Since the levelized tariff P will be strongly dependent upon the first-year price (attribute 1), there is potential for double counting.

4. Development experience

Project proponents who have achieved financial close in at least one power project (similar to the one proposed) on project finance basis as the main sponsor or co-sponsor will be assigned full 50 marks. No credit will be given to participation in projects which are yet to achieve financial close.

The possible scores are either zero or 50 marks: there is no intermediate marking. This means that an operator who has developed, built and operated a single plant "similar to the one proposed" gets the same marks as one who has developed 20 such plants. "Similar" is not defined. (Similar in size? Similar in fuel? Would experience developing and running a 500 kW micro-hydro plant qualify?)

5. Operation and maintenance (O&M) experience
The project proponents who participate in at least one power project (similar to the one proposed) on project finance basis as the O&M contractor will be assigned full 50 marks. No credit will be given to the project proponent's participation which are to be commissioned.
Again, "similar" is not defined.

6. Engineering, procurement and construction (EPC) experience
The project proponents who participate in at least five power projects (Similar to the one proposed) as the EPC contractor will be assigned the maximum of 25 marks. Pro-rated marks will be given for participation as EPC contractor in a lesser number of projects.

Again, "similar" is not defined.

It is theoretically possible for the winning proposal to be one that scores zero on both criteria 5 and 6. Yet if the pre-qualification -- screening -- were done properly, anyone without the minimum requirement would not have been short-listed. It is highly probable that all bidders will score the same marks on these criteria.[28]

7. Fuel consumption
The marks assigned for fuel consumption, D, will be determined by the formula

$$D = 60\left[1 + \frac{P_{base} - f}{P_{base}}\right]$$

where f is the proponent's fuel cost in cost units/litre, and P_{base} is the utility's "base fuel cost," given as 5.50 cost units/litre for residual fuel oil, 6.10/litre for furnace oil, 12.20/litre for heavy diesel. The marks available under this item are subject to a maximum of 75.

This is a good example of a poor proxy variable that may also double count. If in fact the intent was to reward efficiency, *per se*, why not use the heat rate? As defined, the proposal rewards the unit cost of fuel (on a cost per kWh basis). Why this is of interest is unclear: respondents who have inefficient plants would necessarily also be penalized under the cost attribute -- and hence the attribute double counts.

It is also not clear whether a bidder who uses heavy diesel (say at 12.20 cost units/litre) gets minus 13 points (the result of the application of the formula), or simply zero. Obviously the electricity board wished to strongly discourage all fuels except residual fuel oil.

8. Compliance with power purchase agreement (PPA) (a draft of which was supplied to bidders)
For each exception to clauses in the PPA, one mark will be deducted from the maximum of 50 marks available for this item. Future negotiations on the PPA will be confined to those exceptions and reservations listed by the project proponent in the proposal forms.

This attribute is hopelessly vague. Neither "clause" nor "exception" was defined. A 42-page PPA was provided as an attachment; and, as well recognized by anyone who has been involved in a legal review of an untested PPA, there are likely to be several hundred "exceptions" in such a document. It is not clear whether more than 50 "exceptions" would invoke a penalty (i.e., a subtraction of marks). Even if what was intended here was to penalize proposals for substantive changes to provisions that might take some time to negotiate, how to define "substantive" is highly subjective. There is also some double counting with attribute 2 (acceptance of the tariff structure).

9. Compliance with the implementation timetable
For each month delay from the expected completion date, two marks will be deducted from the maximum of 50 marks available under this item.
This attribute is poorly scaled, since in the case of small diesel projects (which are relatively easy to implement), one may expect few differences in proposed completion dates.

ANNEX 2.2

Defining attributes: Biodiversity in Sri Lanka[29]

Biodiversity was identified as a key environmental concern in our study of Sri Lanka (Meier and Munasinghe, 1994). In this Annex we describe how an attribute for this concern was developed for that study.

Detailed biodiversity information at potential power plant sites is unlikely to be available at the system planning stage. Consequently, the only quantification that appears possible is to derive a probabilistic index that gives the decision-maker information about the likelihood that the detailed environmental impact statement would reveal the presence of an endemic species;[30] would significantly affect ecosystems of high biological diversity; or would affect a habitat already in a marginal condition. Endemicity and biodiversity are not necessarily correlated: an endemic species may be encountered in an area of low biodiversity, and areas of high biodiversity may in fact contain no endemic species. It is certainly true, however, that extinction of an endemic species would very likely constitute a "show-stopper" at the project stage; and it is also true that at least as far as Sri Lanka is concerned, its endemic species are most likely to be encountered in areas of high biodiversity.

A biodiversity index will have several constituent elements. First is the nature of the affected ecosystem itself. Table 2.5 ranks the main ecosystem types encountered in Sri Lanka, and assigns to them a value, x_j, that captures the relative biodiversity value of different habitats. The scale is to be interpreted as a strict ratio scale, (i.e., zero indicates zero amount of the characteristic involved). Note the zero value assigned to tea plantations and rice paddies typical of monocultures.

Table 2.5: Relative biodiversity value

rank	ecosystem	x_j
1	lowland wet evergreen forest	0.98
2	lowland moist evergreen forest	0.98
3	lower montane forest	0.9
4	upper montane forest	0.9
5	riverine forest	0.75
6	dry mixed evergreen forest	0.5
7	villus	0.4
8	mangroves	0.4
9	thorn forest	0.3
10	grasslands	0.3
11	rubber lands	0.2
12	home gardens	0.2
13	salt marshes	0.1
14	sand dunes	0.1
15	coconut	0.01
16	forest plantations	0.01
17	tea plantations	0.00
18	rice paddies	0.00

The second element concerns the *relative* valuation, because the *value* of the area lost is a function of the proportion of the habitat that is lost. This idea is illustrated in Figure 2.3, which shows the estimated non-market value for resident fish, expressed as $ per recreation-day, as a function of the ratio of actual abundance to present abundance. The figure is adapted from a study of the environmental costs and benefits in the US Pacific Northwest (BPA, 1986). The implication is that as habitat or relative abundance decreases, the value of the remaining resource increases. The further implication is that as the habitat declines in size to zero, the value assigned to it tends to infinity (or some extremely high value).

For instance, in the case of Sri Lanka, the point is that the *last* hectare of tropical rain forest would have infinite value, and hence its loss would be unacceptable; whereas if one lost, say, one hectare out of a remaining 10,000 hectares, the value given to that hectare might be quite modest.

Such an approach to valuation of biodiversity is subject to several caveats. First, as noted, ecosystems may require some minimum area for long-term survival, which implies that the value function would tend to infinity as it approaches that minimum value (rather than as it approaches zero as suggested in Figure 2.3).

Figure 2.3: Generic value function for habitat loss

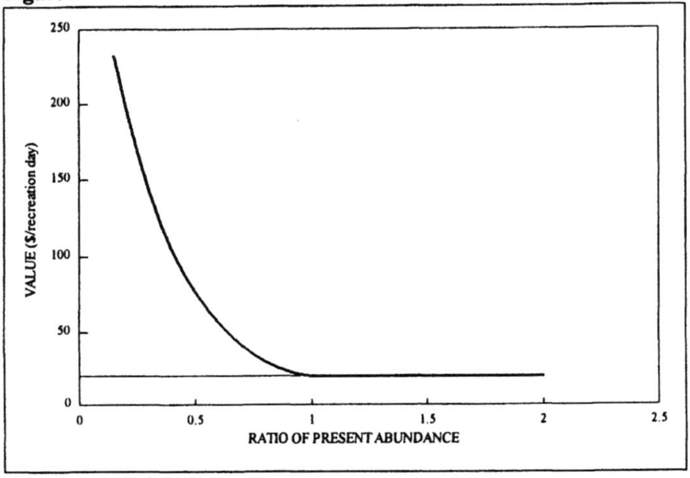

Based on BPA (1986)

Second, and perhaps more importantly, the argument is sometimes made that the value to be ascribed to the loss of habitat associated with some regulatory or governmental decision depends upon whether what remains is *secure*. For example, the cost of the loss of 1 ha of a habitat, if 1,000 ha remains, might be valued as negligible if that remaining habitat is protected from encroachment. On the other hand, if the remaining 1,000 ha is vulnerable to encroachment, then the loss of that 1 ha might be assigned a much higher value.[31]

Such reasoning is not logical, because there is a confusion between costs and benefits. On the cost side, it really does not matter whether the loss is attributable to a power plant, to agricultural development, or even to illicit felling: the loss is the same in all cases. On the other hand, the *benefits* to society of these activities may be quite different.

In any event, for the power sector the difficulties of estimating the appropriate value function proves to be largely moot. Because the fraction of remaining areas of each ecosystem that might be consumed by power sector projects is generally extremely small, we are on the very far right-hand side of the curve of Figure 2.3, where the slope is essentially flat. Consequently, for numerical computation this element can simply be represented by the area lost.

Thus the biodiversity index associated with site i, B_i, is simply defined by Meier and Munasinghe (1994) as

$$B_i = \Sigma H_{ji} x_j$$

where H_{ji} is the ha of ecosystem of type j at site i

x_j is relative biodiversity value of type j (as defined in Table 2.5)

As it turns out, the biodiversity index is strongly correlated with hydro reservoir size, as indicated in Figure 2.4. This is simply a consequence of the fact that in Sri Lanka, all the hydro larger projects inundate relatively significant amounts of natural forest of high biodiversity value: this would not necessarily be the case elsewhere.

Figure 2.4: Biodiversity index and reservoir area

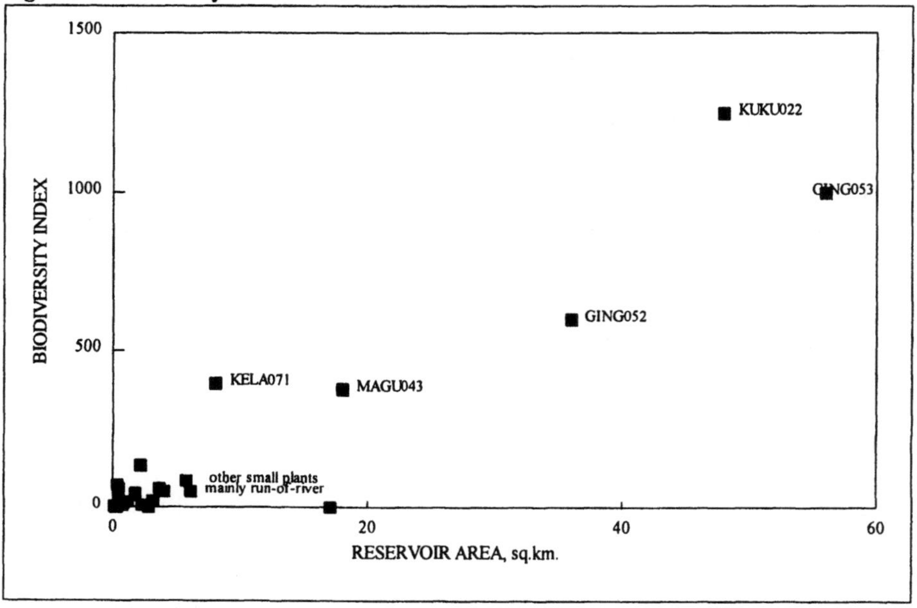

Yet when one excludes the five very large projects with reservoir sizes in excess of 15 km^2, the correlation is quite poor. This reflects the diversity of ecosystem types encountered at the smaller projects (see Figure 2.5).

Figure 2.5: Biodiversity and reservoir area at the smaller hydro projects

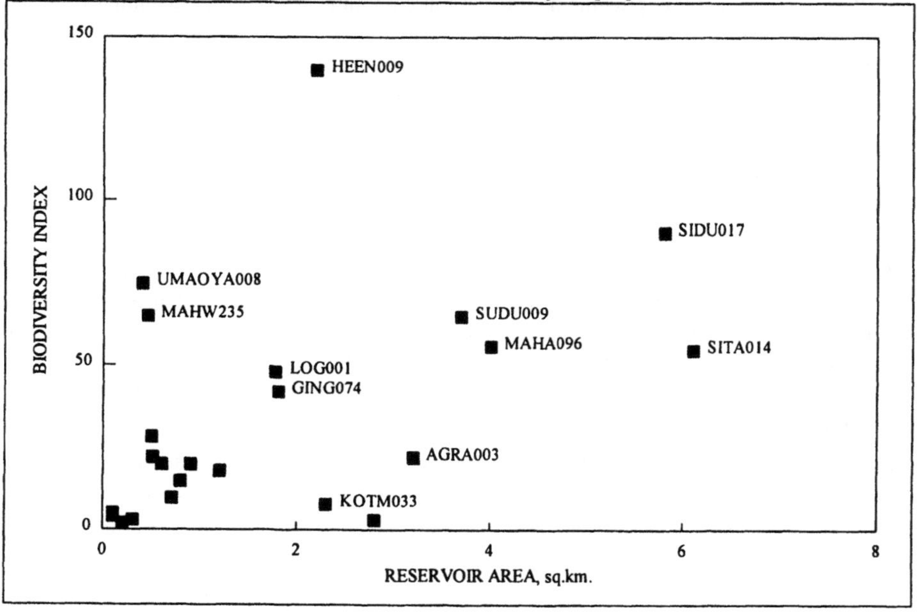

NOTES

[1] A selection of these diverse applications is reviewed in Chapter 7.

[2] As we shall see in Chapter 8, in the BC Hydro application, reducing over 60 initial "objectives" to 11 attributes actually used in the MCDM was a major challenge.

[3] However, Meier and Munasinghe (1999) show how calculation of real option values can aid decision-making in the presence of uncertainties that characterize climate change mitigation.

[4] But if stakeholders insist that all attributes be considered, even if duplicative, double counting can be avoided if attributes are structured as a hierarchy in which weights assigned to general objectives are then allocated among the attributes constituting each objective (Section 4.2.3, below). For instance, one could deal with the double counting inherent in having both "monthly bills" and "total resource cost" as attributes as follows. First, ask each person to assign a weight to the general objective of "economics." Then divide that weight among the bill and resource cost attributes.

[5] For example, acres/MW appears in the rating system of Niagara Mohawk in New York, reviewed below in Section 7.2.

[6] NO_x emissions in non-attainment areas are subject to "offset" provisions that can have much the same effect.

[7] For example, as noted by Joskow (1992), the net damage of an additional ton of SO_2 emitted in Massachusetts might be negative if that ton might otherwise be emitted upwind in Ohio, rather than if emissions, for example, from Massachusetts's coastal Brayton Point plant, are blown out to sea.

[8] The social cost of an incremental ton of SO_2 under the Title IV allowance system equals the increase in control costs that one or more utilities bear in order to free up the one allowance needed to cover that ton, plus the net increase in environmental damage stemming from the rearrangement of emissions. Under a well-functioning allowance system, the former cost equals the price of allowances. The theoretically appropriate adder is then the latter quantity -- which is generally neither zero nor the difference between the gross damage caused by a ton emitted within its jurisdiction and the allowance price. Whether the net external cost is positive or negative depends on two variables: the relative magnitudes of marginal damages of emissions in different locations, and who would buy or sell allowances on the margin. If marginal damages in all locations are equal -- as would be the case for a CO_2 trading system -- then the net cost is zero.

[9] Consider a set of I attributes. A set of $n = 2$ or more attributes $\{A_1, A_2, .., A_n\}$ is preference independent of its complement (i.e., the other attributes $\{A_{n+1}, A_{n+2}, ..., A_I\}$) if the tradeoffs a decision-maker is willing to make among the $\{A_1, A_2, .., A_n\}$ do not depend upon the levels of its complement. Keeney and Raiffa (1976) discuss this property at length, and present easy procedures for checking whether an individual's preferences concerning a set of attributes satisfy that property.

[10] Below (in Step 2) we argue that arbitrary "scoring" of impacts through use of such proxies can be misleading, particularly if more rigorous quantification through modeling is possible.

[11] For a good review of the issues and uncertainties involved in EMF, see, e.g., Florig (1992).

[12] However, as discussed below in Section 7.9.1, participation of "experts" does not necessarily ensure good scaling.

[13] There are a number of practical problems in deriving an appropriate index, as discussed in Annex 2.2.

[14] These points have long been established in the decision sciences literature, and were validated in applications in the 1970s to power plant siting particularly by Keeney and his colleagues (see e.g. Keeney and Raiffa, 1976; Keeney and Nair, 1977). Keeney *et al.* (1981) and Keeney and Sicherman (1983) illustrate a well-founded application of decision theory to utility resource planning (a coal/nuclear choice).

[15] Our MCDM studies of two US utilities, Centerior Energy of Ohio (now FirstEnergy) and Seattle City Light (SCL), were conducted in 1993-1994 as part of a research project sponsored by the US Department of Energy to examine ways of incorporating environmental issues into IRP. Chapter 10 presents some of the findings of the SCL study; case material from the Centerior study is used throughout the book.

[16] There is a growing literature on utility decision-making under uncertainty that is far too extensive to be reviewed here in any detail. Clemen (1996) is an excellent text. Practical examples of such techniques in the energy sector include Hirst and Schweitzer (1989), Hobbs and Maheshwari (1990), Keeney and Sicherman (1983), and Merrill and Wood (1991).

[17] For instance, in the power plant siting example of Section 7.9.1, a comparison was made between judgmental scores for air quality impacts given by "experts" on a scale of 0 to 30, and actual computation of air quality impacts using a dispersion model to estimate population exposure. The model calculations showed that the values of the best and the worst site varied by a factor of 11, whereas the judgment scores showed a variation between the best and worst of less than 3.

[18] Indeed, in the BC Hydro study, the ability to conduct extensive "what if" exercises, and the desire of the participants to examine a very large number of scenarios in the initial stage of the process, emerged as the most important criteria for model selection: the existing in-house models were too detailed (and/or still run on mainframes) for the desired application.

[19] The literature on this point is contentious and growing (e.g., see the articles in the December 1996 issue of Resource and Energy Economics).

[20] In our Centerior case study there was extensive discussion about solid waste and how that concern would be captured in the process. Initially, almost as a matter of routine, estimates of tons of solid waste were generated by the MIDAS model. But after discussion with the participants, the shortcomings of such a quantification became clear. What was deemed to be a much better proxy for the actual environmental impacts, as well as the utility's own transaction costs (the "hassle factor"), was the number of new solid waste disposal sites that had to be opened. (Recall also the earlier discussion in Section 2.1.1 about the specificity of attribute definition.)

[21] GHG emissions, as discussed subsequently, are an important exception, as are some other pollutants -- such as plutonium -- that are so highly toxic that it is reasonable to assert that *any* emission at *any* location must be controlled. But such arguments would be difficult to invoke for most of the pollutants of concern to electric utility resource planning. Even in the case of hydro plants -- where one might accept the premise that *any* hydro plant that eliminates white-water recreation opportunities, or interferes with salmon migration, is unacceptable -- there are certainly *some* locations where this would not be the case.

[22] This is even more true when comparing the cost of GHG emission mitigation efforts among countries, and is the basis for so-called joint implementation approach for GHG emission reduction, under which developed countries, faced with very high costs of GHG emission reductions, would find it cheaper to *pay* developing countries to achieve the same reduction by some other means (e.g., by reforestation). Some have criticized this approach on grounds that this gives the industrialized countries access to the "best" emission reduction projects, foreclosing their use by the developing countries later on. This objection may have some

merit in the present situation, where individual deals may be struck between North American and European utilities and parties in the developing countries; but if and when an efficient worldwide marketplace for emission rights emerges, such an approach is clearly to everyone's mutual benefit, and is certainly more economically efficient from a global perspective. For further discussion, see IPCC (1996).

[23] With the benefit of hindsight, some of the assumptions made in North America in the late 1970s -- for example the world price of oil, or the availability of natural gas for power generation -- appear to lack much foresight. But the correct response to these kinds of uncertainties is not to abandon the attempts to look into the future, but to focus the analysis on the robustness of decisions with respect to the probable uncertainties. Thus, an analyst today might do well to consider scenarios that appear to run against the present conventional wisdom. Any number of plausible events in the Middle East might significantly change the long run world oil price (instability in Saudi Arabia would raise it, resumption of large-scale production in Iraq or a comprehensive Middle East peace settlement may lower it); and notwithstanding the current reluctance of the United States to commit to GHG emission reductions, carbon taxes may yet bring a significant shift to renewable energy for electricity generation.

[24] However, even Step 3 may be quite difficult for heavy metals and toxic hydrocarbons whose fate in the environment is complex.

[25] Critics of the monetization approach argue that going straight to Step 6, without having done detailed studies at Steps 4 and 5, is quite arbitrary.

[26] A good example of such an approach is the application of decision analysis to the identification of the optimal hedging strategy for the global climate change problem by Manne and Richels (1991). The paradigm they use is that of a portfolio of insurance options: what combination of insurance should be bought, if indeed any at all? *"..(W)hat portion goes to R&D to resolve scientific uncertainties? What portion goes to the development of new supply and conservation technologies to reduce abatement costs? And what portion goes to immediate abatement of emissions?"* In particular, they focus on the value of information: how much accuracy is needed in climate modeling and impact assessment? Clearly, with *perfect information*, the best course of action can be charted immediately, and there is no need to hedge bets. Manne and Richels conclude that the need for precautionary near-term emissions reductions is inversely related to the sustained commitment to R&D to develop better climate information (which reduces the need to hedge against an uncertain and potentially hostile future).

[27] The RFP includes the formula to be used for calculating the levelized tariff.

[28] Moreover, a bid bond of $200,000 would also serve to weed out non-serious proposals.

[29] The work described here owes much to the participation in the Sri Lankan case study of Professor K. D. Arudpragasam, Professor of Zoology at the University of Colombo and former Chairman of the Central Environment Authority of Sri Lanka.

[30] A species is said to be endemic to Sri Lanka if it occurs nowhere else in the world.

[31] Although not often put explicitly in these terms, this line of reasoning has in effect been much used by environmentalists to oppose power facilities. For example, the reasoning goes, because coastal wetlands are ecologically valuable, power plants should be sited elsewhere. Yet uncontrolled real estate development poses a much greater threat to coastal wetlands, but since this is harder to control, attention is focused on power plants that make much easier targets for attack.

3 SCREENING AND TRADEOFF ANALYSIS

Someone told me that each equation I included in the book would halve the sales
STEPHEN HAWKING, A Brief History of Time (1988)

We often experience more regret over the part we have left,
than pleasure over the part we have preferred
JOSEPH ROUX, Meditations of a Parish Priest, 5:26 (1866)

In this chapter we turn to the first analytical step in a multi-attribute decision analysis, namely the preparation and analysis of tradeoff plots. Tradeoff plots have three major functions. The first is to serve as a screening device, by identifying those options that do not appear to warrant further attention. The second, related function is to assist in the definition of strategies (combinations of options). The third, and perhaps most important, is to highlight the tradeoffs that must in fact be made by the decision-maker. In its simplest form, a tradeoff graph is just a plot of two attributes, in which the attribute values for the various options under consideration are displayed graphically.

In many MCDM applications in the energy sector, screening, tradeoff, and dominance analyses are the end-products, and the value quantification methods of the next chapter are not needed (e.g., Andrews, 1992a; Andrews and Govil, 1995; Crousillat *et al.*, 1993; Meier and Munasinghe, 1994; Xie and Kuby, 1997; Yokoyama *et al.*, 1997). Often those analyses yield important insights that allow recommendations to be developed without resort to explicit value judgments. For instance, a tradeoff and dominance analysis contributed to the Tennessee Valley Authority's cancellation of its unfinished nuclear plants (TVA, 1995); plans including such plants were generally either dominated or represented clearly unacceptable tradeoffs.

Figure 3.1 illustrates such a tradeoff plot, here for a set of eight hypothetical resource portfolios in a utility planning study: a reference portfolio, (denoted R), and hypothetical portfolios A, B, C, D, E, F and G. Typically, system cost is shown on the y-axis, and other attributes on the x-axis. However, cost versus environment is not the only tradeoff of potential interest to electricity planning, and often there is a need to look at tradeoffs among environmental attributes, or between cost and other financial objectives such as rates.

Figure 3.1: Quadrant definitions in a tradeoff plot

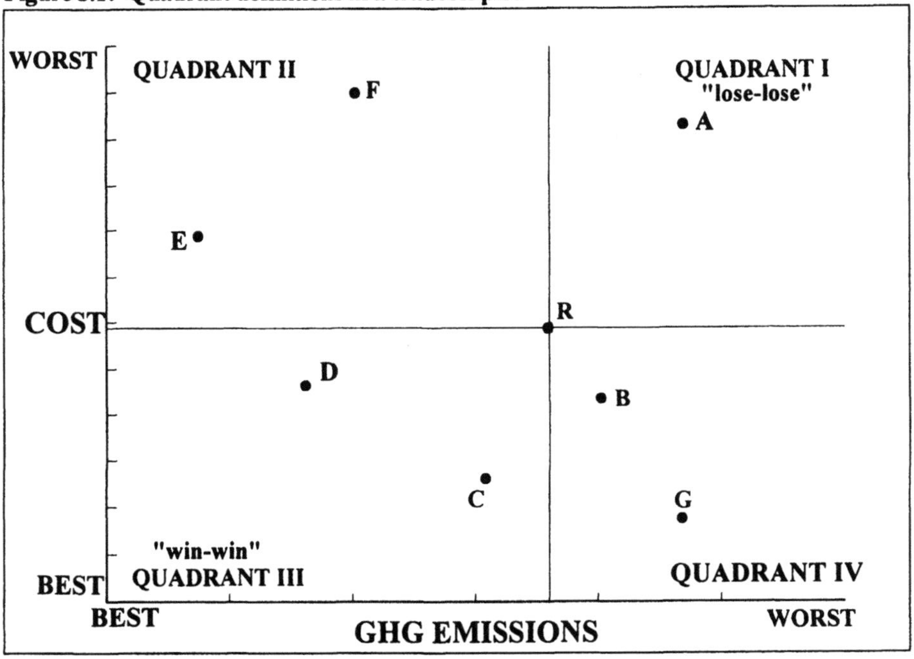

Such a tradeoff plot is divided into quadrants defined by the performance of the reference case. In resource planning, for example, the reference portfolio is generally just the starting-point: as the process unfolds, it is eventually discarded for the preferred portfolio presented in the final plan (see Chapter 8 for a real example).

Alternatives that fall into quadrant I (i.e., to the top right of the reference case), such as A, perform more poorly on both the cost and environmental attribute than the reference case. Such alternatives are sometimes described as "lose-lose." Depending on how far they are from the axes that define the quadrants, such alternatives might be eliminated as being of little further interest. By contrast, alternatives C and D, located in quadrant III, perform better than the reference case on both attributes: such an alternative is often described as "win-win." Alternatives that fall into this quadrant are typically those on which diverse interests can agree, since both the cost and environmental interest are served by choosing either C or D, rather than the reference case R.

Alternatives that fall into quadrants II and IV are those that require *tradeoffs*. For example, alternative E, located in quadrant II, has better (lower) GHG emissions, but increased costs -- say as a result of introducing renewable energy technologies, which reduce GHG emissions by displacing thermal generation, but which increase system costs. In contrast, an alternative (such as G) may also fall into quadrant IV, in which system costs improve, but environmental performance worsens.[1]

Figure 3.2: The tradeoff curve

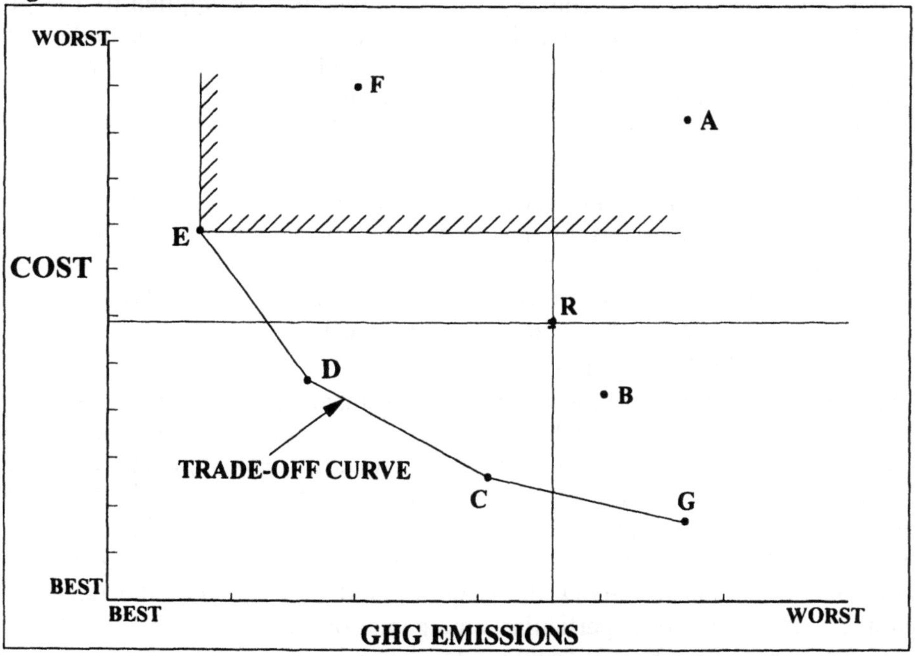

A particularly useful concept is that of *dominance*, and the so-called *tradeoff curve*.[2] Consider the example of Figure 3.2. All of the options in the hatched area (alternatives A and F) are worse in both cost and GHG emissions than option E. Option E is therefore said to *dominate* options A and F. If one eliminates from the plot all options that are dominated in this manner, one is left with a set of options that are *non-dominated* -- in this case the options E, D, C and G. The set of *non-dominated* options is known as the tradeoff curve: intuitively, these are the options that are "closest" to the origin, where the ideal (or zero) value of the attributes occurs.[3] Attention should then be directed to the options that lie on (or near) the tradeoff curve, since these are the ones that diverse interests are most likely to agree upon.

An important methodological issue in MCDM is how to find non-dominated options when the decision problem is a linear program or other optimization problem. Annex 3.1 summarizes two methods for accomplishing this task. Another methodological issue is how to portray tradeoffs; "X-Y plots" such as Figure 3.2 are only one such technique. Others can show more than two attributes at a time, but have their own disadvantages. Annex 3.2 presents some of these other display techniques.

The remainder of this Chapter uses the concepts of tradeoff curves and dominance to explain the ideas and methods involved in Steps 4 - 6.

3.1 SCREENING (STEP 4)

Screening is the first step requiring value judgments. It is applied to a large set of alternatives in order to yield a smaller set that would then be studied in more detail or presented for scrutiny by the people responsible for the final decision. In so doing, screening should accomplish two things:

- Firstly, to eliminate alternatives that are unlikely to be chosen, so that later effort can be focused on the most attractive options;
- Secondly, to give those responsible for the final decision a range of options that emphasizes different attributes. Screening should not "stack the deck" in favor of a particular set of values, nor should it yield a set of options that are essentially similar. The most difficult value judgments should be left for the final decision, not usurped by analysts and buried in the screening process.

The most popular method for screening is exclusionary screening, which we review below. Other approaches, however, can be used to whittle down the number of alternatives to be examined. These include application of the amalgamation methods of Chapter 4 to identify a set of top-ranked options to be analyzed in more detail, rather than just identifying a single best alternative. A problem with this approach, and of exclusionary screening, is that those methods tend to narrow the range of tradeoffs and thus the range of choice. However, one particular amalgamation technique, ELECTRE I, avoids this problem, as it is designed to yield a diverse set of options representing a wide range of tradeoffs. We discuss this in Section 4.3.6.

In exclusionary screening, one or more attributes are used to screen plans by dividing the range of each into two categories: "acceptable" and "unacceptable." These are, of course, value judgments. Then, any alternative that is unacceptable in one or more attributes is dropped. Failure in one attribute is assumed to be as bad as failure in all, which is appropriate in the case of legal requirements (the most common), or where there exist unambiguous physical thresholds (less common, because of scientific uncertainty as to what the precise threshold values might be).

In the example below (Figure 3.3), the area to the right of the cutoff point for the first attribute (A_1) and above the cutoff point for the second attribute

(A_2) is unacceptable: this leaves as the acceptable domain the area *XYZO*. The acceptable solutions will then lie to the northeast of the tradeoff curve, in the pie-shaped slice *WYV*.

For some attributes, there exist natural dividing points between "acceptable" and "unacceptable," often due to legal restrictions. Examples might include presence of endangered species (which might preclude development), and regulatory criteria imposed by the US Nuclear Regulatory Commission (NRC) in the case of nuclear power plant siting (the only acceptable sites being those that strictly conform to the regulatory requirements): here failure in one attribute is indeed as bad as failure in all.

Figure 3.3: Exclusionary screening

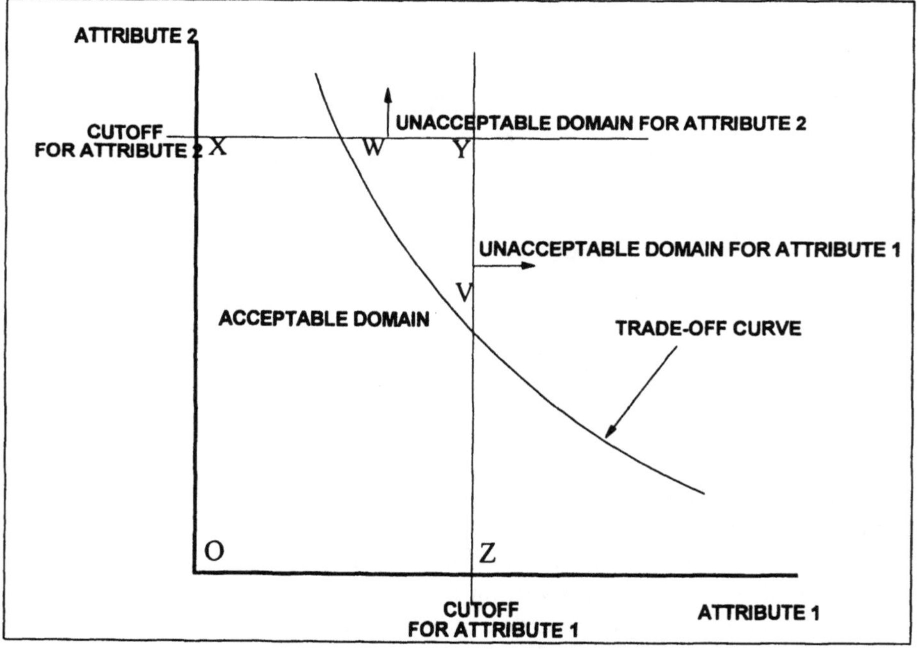

However, analysts are often tempted to treat attributes that lack such natural cut-off points (so-called "discretionary attributes") as exclusionary. For instance, in a power plant siting study it might be decided a site must lie within 10 miles of a river. However, such a cutoff is not a natural one -- why is 9.5 miles acceptable but not 10.5 miles? The danger is that a plan might fail such a criterion by a small amount, yet be so good in other attributes that it would be found best at a later decision stage. More importantly, the violation of a particular constraint can often be mitigated by modifying other features of a selected plan without significantly affecting the benefit-cost ratio or its social desirability. In these cases, exclusionary screening makes tradeoffs that decision-makers would not accept if they could consider them explicitly.

This notion of implied tradeoffs is an important one. Consider Figure 3.4. The axes of the graph represent levels of two different discretionary attributes, both of which are to be minimized. From the point of view of exclusionary screening, plans B and C are "better" than A and D because only B and C are acceptable in both attributes. However, tradeoffs are implied by these cutoffs, as A and D are each better in one attribute than B or C.

Figure 3.4: Ratios of weights implied by cutoffs in exclusionary screening

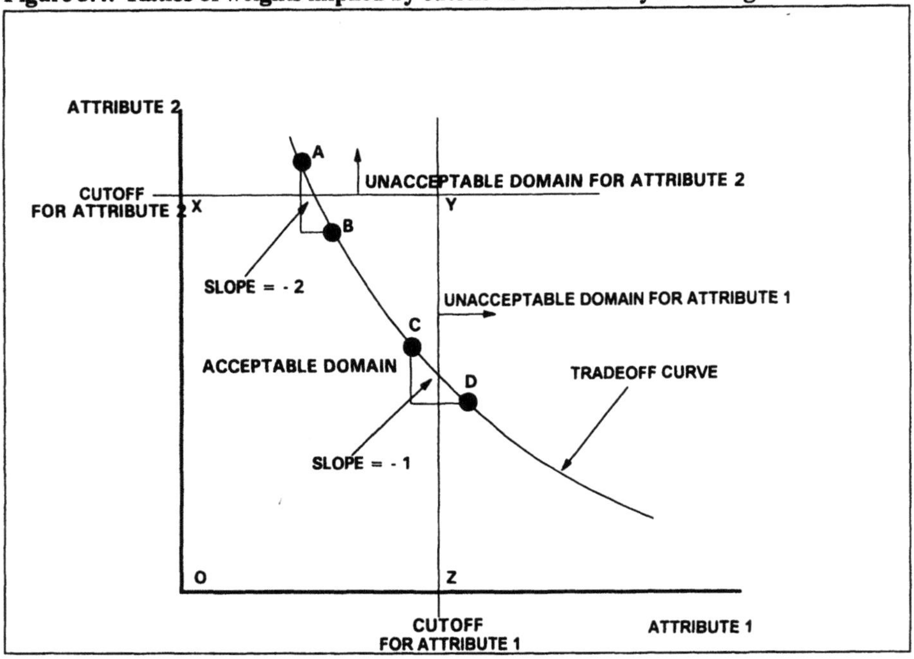

First, the cutoff for attribute 1 implies that C is preferred to D. The slope of the tradeoff curve at that point is -1. By noting the slope of the line segments connecting acceptable and unacceptable alternatives, we can infer that the decision-maker is (implicitly) willing to suffer an increase of more than one unit of attribute 2 to gain a unit decrease in attribute 1; that is, the weight of attribute 2 (w_2) is implied to be less than that for attribute 1 (w_1).[4]

Second, the cutoff for attribute 2 implies that B is preferred to A. As the slope of the tradeoff curve in the vicinity of A and B is -2, this means that (implicitly) w_1 must also be less than two times w_2. But if the decision-maker's true weights were outside of those ranges (i.e., either $w_1 < w_2$ or $w_1 > 2\, w_2$), either A or D should actually be preferred to C and B.

Hence, contrary to the assertions of some researchers, attributes are not necessarily weighted equally in exclusionary screening. A change in the cutoffs can change the implied weights. Consequently, if exclusionary screening is to be used, the exclusionary cutoffs should be chosen carefully and conservatively. One should avoid using discretionary attributes to exclude alternatives, because the result may be to prematurely drop options that might actually be preferred by some interests. If discretionary attributes must be used in exclusionary screening, then a sensitivity analysis should be undertaken to ascertain the effect of different cutoffs.

3.2 ANALYZING TRADEOFF PLOTS (STEP 5)

Tradeoff plots have long been used to aid understanding of the environmental dimensions of energy choices: the first example we encountered in the literature is Ferrell (1978), who examined the tradeoff between cost and SO_2 emissions.[5] Figure 3.5 shows a tradeoff plot of system cost versus GHG emissions for the Indian State of Bihar, taken from a recent study of environmental issues in the Indian power sector.[6]

Figure 3.5: Tradeoff plot for Bihar, India: system cost versus GHG emissions

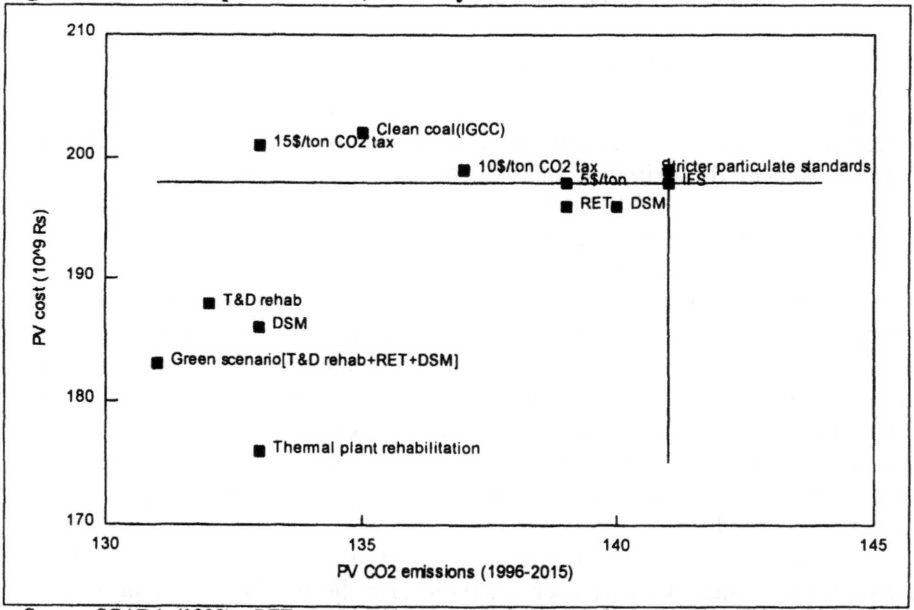

Source: SCADA (1998). RET = renewable energy technologies (bagasse cogeneration+mini-hydro). Imposition of CO_2 taxes results in the substitution of coal by imported LNG. IFS (inter-fuel substitution) scenario which defines the quadrants is the least utility-cost option taken as the reference case.

Unfortunately, the typical MCDM study has more than two attributes: between five and 15 attributes are typical for energy sector applications. Looking at *all* possible pair-wise comparisons in graphical form is essentially impossible when there are more than three or four attributes. For example, with 12 attributes, there are 66 possible pairs of attributes.[7] However, in most cases it will suffice to look at system cost versus each of the other attributes, plus selected tradeoffs among the other attributes (e.g., such as GHG emissions versus land use).

Moreover, just because an option is "lose-lose" (or "win-win") in one set of pair-wise comparisons, it does not necessarily follow that the option is "lose-lose" (or "win-win") in all of the other comparisons. A good example is demand side management (DSM) in resource planning studies: in many cases higher levels of DSM tend to be "win-win" for system cost and all environmental attributes, but require a tradeoff for rates (i.e., costs and all environmental attributes decrease, but short-term rates increase).[8]

However, for particular resource groups, attributes are often highly correlated, so that with some judgment and experience, looking at all comparisons is not necessary for the important issues and tradeoffs to be identified and discussed. For example, options that have low GHG emissions often perform well in other attributes related to air emissions. What one looks for in an MCDM study are options that fall on a majority of tradeoff curves. Finally, by using an appropriate *amalgamation* technique, performance across many different attributes can be combined.

3.3 DOMINANCE ANALYSIS (STEP 6)

In dominance analysis, an alternative is screened out if it is dominated by another option. An alternative is dominated if there exists another plan that is just as good in all attributes, and strictly better in at least one (e.g., C versus B in Figures 3.1 and 3.2).[9] Annex 3.3 presents an example of the use of dominance analysis for screening out dominated options at the early stages of a utility resource planning exercise.

Dropping dominated alternatives is logical because a valid multicriteria method will never choose a dominated alternative. However, the analyst should recognize that if there are other attributes that decision-makers have in mind in addition to those in the model, it is possible that a dominated alternative may actually be preferred. Another consideration is that sensitivity analysis of parameters (such as interest rates, fuel prices, or hydrologic probabilities) may reveal that an apparently dominated alternative is advantageous under some conditions.

Another advantage of dominance analysis is that no value judgments are needed, other than whether each attribute is to be minimized or maximized. One disadvantage is that relatively few alternatives may be dominated. In that case, value judgments and tradeoffs will have to be made to make a recommendation, perhaps with the aid of one of the other methods discussed in later sections.

Because of the uncertainty that surrounds quantification, there is a need to distinguish between *strict* dominance and *significant* dominance. To illustrate the difference between these two definitions, let the universe of initial options be denoted P_1, P_2, etc. Suppose for the sake of clarity that there are only two attributes: cost, and an environmental attribute reflecting, for example, the population exposure to SO_2. Figure 3.6 depicts the solution space for this problem, in which we plot the values of the two attributes for each alternative.

The alternative P_1 is said to *strictly dominate* P_2 if P_1 is better than (or equal to) -- i.e. dominates -- P_2 in every attribute, and is strictly better in at least one attribute. Thus, as illustrated in Figure 3.6, P_1 is better than P_2 in both cost and SO_2 exposure. In fact, P_1 strictly dominates all of the plans northeast of the boundary AP_1B. By repeating such comparisons for all pairs of plans and discarding all dominated plans, the remaining plans constitute the set of the so-called *non-dominated solutions* -- in our case the set of plans $\{P_6, P_4, P_1, P_{14}, P_{13}, P_{15}\}$. These points in turn define a tradeoff or non-dominated set (or "curve"), as indicated in Figure 3.6.

Figure 3.6: The two-attribute problem

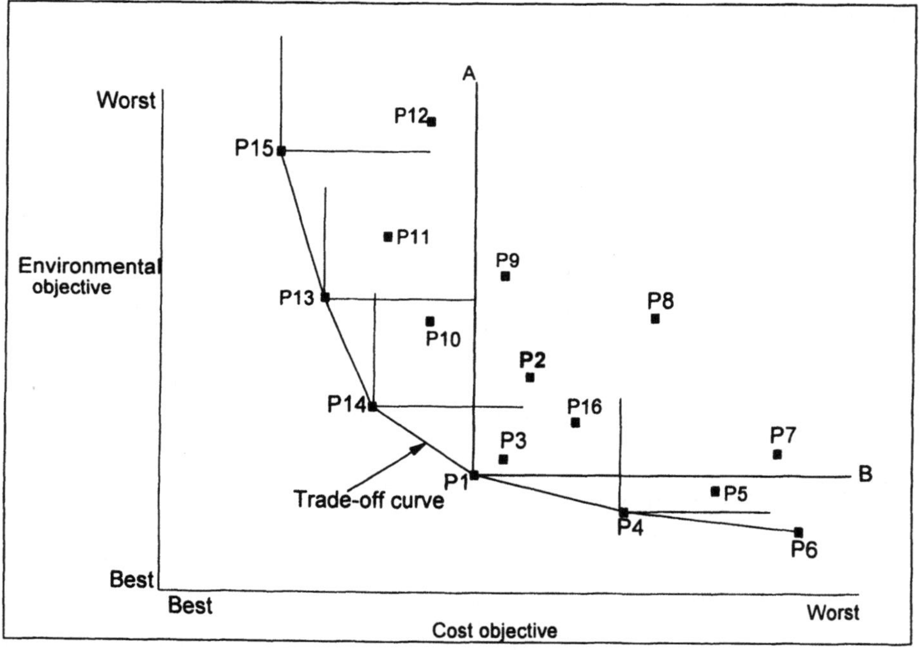

This procedure will in general provide a means for reducing a very large number of possible plans to some smaller number of plans -- a "short-list" or "candidate list" that is to be presented to decision-makers. In the example shown, using the concept of strict dominance, plan P_3 would not appear on the resulting short list. Yet one might argue that while P_3 is somewhat worse in both attributes than P_1, it is not *significantly* worse, (as opposed to, say, P_2 and P_8, which *are* significantly worse in both attributes). In particular, because of uncertainties associated with the calculations, one may be reluctant to discard a plan that is not significantly worse than another from a final short-list (especially if such an option is a portfolio of resources that involves a particular generation or demand-side alternative not included in the other plan).

This idea is captured by the concept of *significant dominance.*[10] P_1 is said to significantly dominate P_2 if $A_i(P_1) + m_i \leq A_i(P_2)$ for at least one i, and if $A_i(P_1) - e_i \leq A_i(P_2)$ for all i, where m and e are positive significance parameters or tolerances. m_i is the smallest difference in values for attribute A_i such that one plan is considered to be much worse than the other, and e_i is the largest difference between values such that one plan is considered essentially equivalent to another.[11]

Figure 3.7: Significant dominance

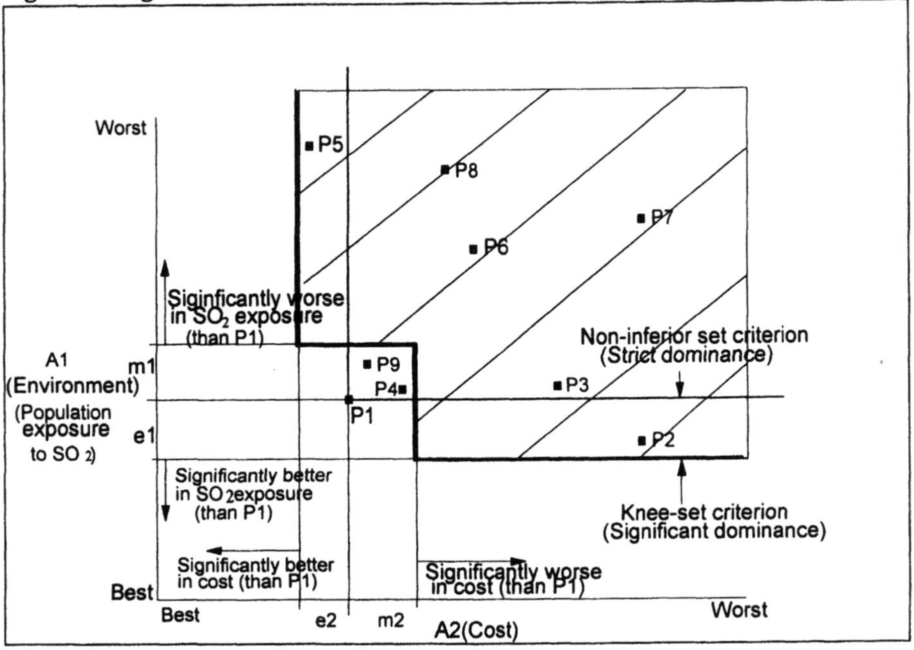

These concepts are illustrated in Figure 3.7. P_1 significantly dominates all of the plans in the shaded region. Note that P_2, which is not *strictly* dominated by P_1, is *significantly* dominated. That is, significant dominance also rejects solutions where the improvement in one attribute (in this case in the environmental objective) is bought at great deterioration in the other (in this case in cost objective). In other words, from the environmental perspective, for a plan to be significantly better than P_1, it must have a value less than $P_1 - e_i$.

The *knee set* is the set of plans that are not significantly dominated by any other alternative. In the example of Figure 3.7, redrawn in Figure 3.8, the knee set is $\{P_1, P_4, P_9\}$. Note that this set differs from the non-inferior set, defined by the tradeoff curve, which is the set $\{P_2, P_1, P_5\}$. From the practical standpoint of decision-making, the knee set is clearly the more useful, since the non-inferior set also includes solutions where slight improvements in one objective are bought only at great cost in the other -- such as P_5 and P_2. Another way of expressing this is that the knee set consists of that set of plans for which decision-makers representing different objectives are most likely to agree upon. In the above instance, for example, both P_5 and P_2 are unlikely candidates for agreement, even though both lie on the tradeoff curve.

Figure 3.8: Knee set definition

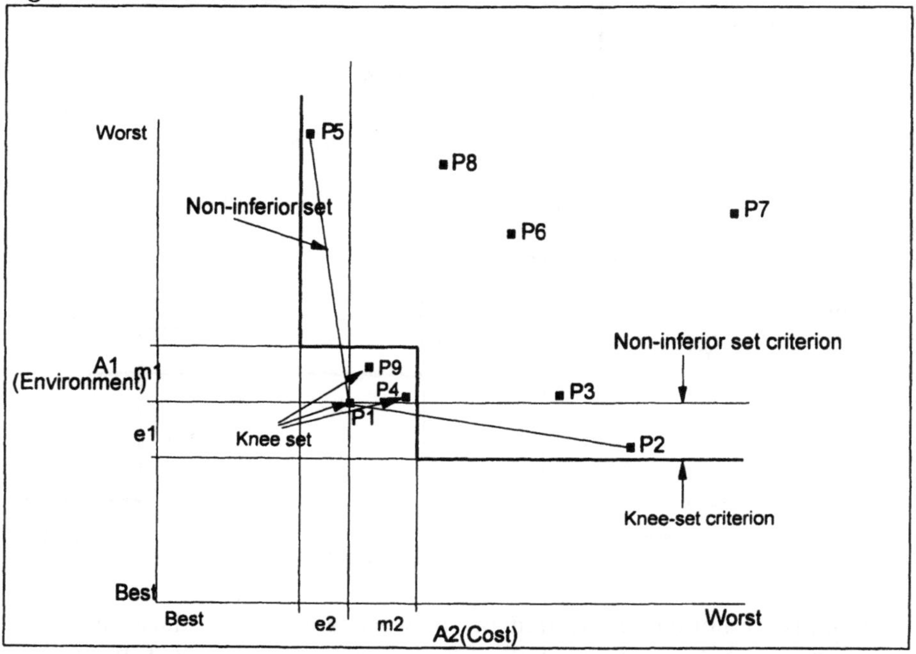

Obviously the knee set will be a function both of the values of the significance parameters used, and of the shape of the tradeoff curve. In the situation of Figure 3.8 the knee set is relatively compact, because the tradeoff curve exhibits a sharp rate of change of curvature at P_1. However, in the situation shown in Figure 3.9, there are no sharp changes in curvature of the tradeoff curve: here the knee set is the same as the non-inferior set.

All of the mathematical definitions of knee sets and non-inferior sets continue to be valid when there are more than two dimensions. In this case, one could properly refer to the tradeoff *set*, say $S = S(C, A_1, A_2, \ldots A_n)$, where C is the cost attribute, and A_i (i=1,..n) are n environmental attributes.

Figure 3.9: Knee set equivalent to the non-inferior set

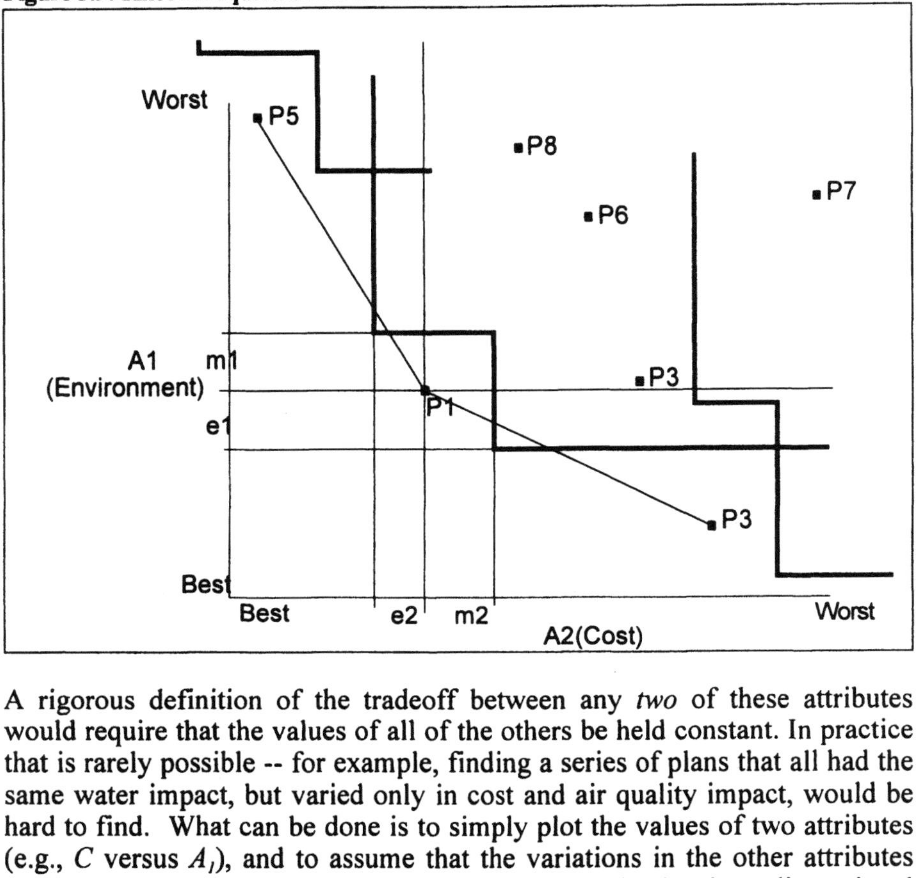

A rigorous definition of the tradeoff between any *two* of these attributes would require that the values of all of the others be held constant. In practice that is rarely possible -- for example, finding a series of plans that all had the same water impact, but varied only in cost and air quality impact, would be hard to find. What can be done is to simply plot the values of two attributes (e.g., C versus A_1), and to assume that the variations in the other attributes may be temporarily disregarded. In other words, in the three-dimensional case, the third dimension is in effect projected onto the two-dimensional plane of the two attributes being displayed, ignoring variations in the third dimension. An example is given in Annex 3.4.

* * * * *

In Steps 4 - 6, the user has displayed how the alternatives perform on the attributes, and eliminated the less attractive alternatives using dominance and other screening criteria. This screening has left a smaller number of alternatives, each of which has some advantages. To winnow the options further, value judgments will have to be made. The next chapter describes approaches for careful quantification and application of values.

ANNEX 3.1

Generating tradeoff curves

Although a review of the literature reveals a great many different approaches to the problem of identifying the non-inferior set, three methods account for the bulk of practical applications. The first two, the so-called constraint and weighting methods, are derived from the theory and practice of multi-objective mathematical programming (Cohon, 1978). These approaches are useful if a mathematical programming model is used for planning, such as Seattle City Light's ANN model, used in the case study of Chapter 10. Cohon (1978), Steuer (1986), and Zeleny (1982) summarize other methods for generating non-dominated solutions from mathematical programs.

The third approach is simply to analyze pre-defined alternatives using a simulation model, and determine the tradeoff curve and knee set by inspection -- as described earlier in this chapter. The latter approach is useful when a limited number of discrete alternatives are to be considered, as in the BC Hydro case study of Chapter 8.

The constraint and weighting methods are described in this Annex. Let *COST(X)* and $CO_2(X)$ be the two attributes under consideration in some planning problem. Their values depend on the values of the decision variables X, which, in general, include variables describing the design and operation of the system.[12] In the constraint method, one solves several optimization problems of the type

$$\begin{aligned}&\mathit{Min}\ COST(X)\\&\mathit{subject\ to}\ CO_2(X) \le CO_2^*\end{aligned}$$

one for each of a series of different values of the fixed right-hand side CO_2^*. The principle can be illustrated as follows (Figure 3.10).

Figure 3.10: Generating the non-inferior set

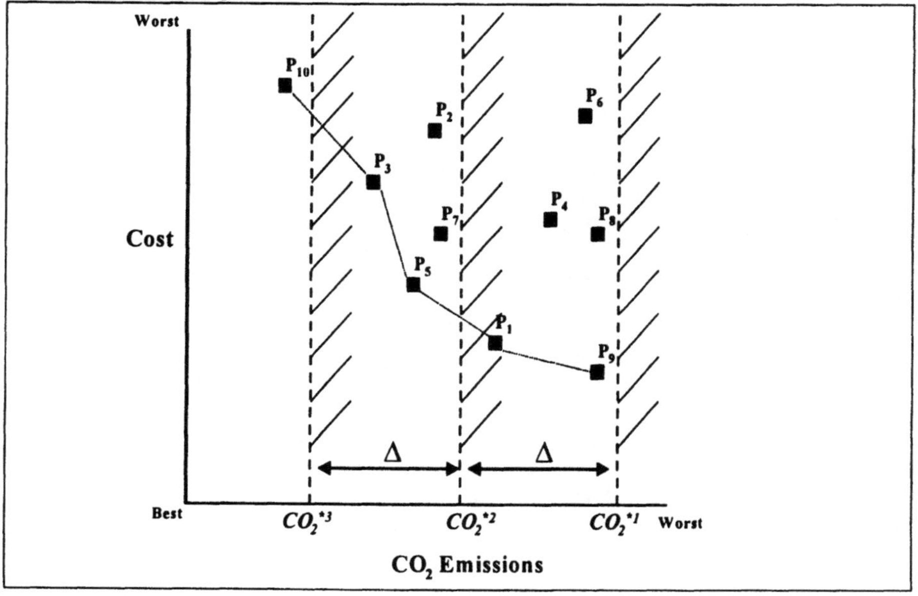

In iteration 1, the model minimizes cost subject to CO_2 emissions being less than or equal to CO_2^{*1}, for which the minimum cost plan is clearly P_9. In iteration 2, one tightens the CO_2

constraint by Δ to $CO_2^{*2} = CO_2^{*1} - \Delta$, for which the least cost solution is P_5. In the third iteration, based upon $CO_2^{*3} = CO_2^{*1} - 2\Delta$ as the upper limit, the least cost solution is P_{10}. In this way, by successive tightening of the emissions constraint, a non-inferior set $\{P_9, P_5, P_{10}\}$ has been generated.

In practice, this method has its hazards: as one can see from the figure, the non-dominated points P_1 and P_3 have been missed, given the way in which the successive limits CO_2^* have been defined. There are two ways of getting around this problem. The first is to make the increment Δ between successive values of CO_2^* very small, so that the probability of missing a point such as P_1 is minimized. For example, with Δ set at half the previous value, P_1 would be found to be non-dominated at the second iteration, and P_3 at the fourth iteration.

A second way of confirming the completeness of the non-inferior set is a two-phase procedure. In the first phase, one solves for some members of the non-inferior set, as described above. In the second phase, one reruns a series of optimizations with slightly modified CO_2 values of the initially defined non-inferior set step as the constraints. Thus, in phase one, the first optimization is subject to the constraint:

$$CO_2(X) < CO_2^{*1},$$

which yields P_9. Then in the second phase, the optimization is re-run subject to:

$$CO_2(X) < CO_2(P_9) - \delta,$$

where $CO_2(P_9)$ is the value of CO_2 resulting from plan P_9, and δ is some small increment ($\delta \ll \Delta$). This constraint eliminates P_9 and the optimization now yields P_1, the previously unrevealed member of the non-inferior set. P_1 is added to the non-inferior set, and process continues with the next optimization subject to:

$$CO_2(X) < CO_2(P_1) - \delta,$$

which reveals P_5 to be a member of the non-inferior set. This procedure will ultimately reveal the full non-inferior set $\{P_9, P_1, P_5, P_3, P_{10}\}$.

The weighting method solves a mathematical programming problem whose objective is:

$$MIN\ w_{COST}\, COST(X) + w_{CO_2}\, CO_2(X)$$

over a range of values of the weights w_{COST} and w_{CO_2}. Usually, the sum of the weights is constrained to equal 1. When $w_{CO_2} = 0$ and $w_{COST} = 1$, the result is the least cost solution (P_9 in Figure 3.10). If instead all of the weight is placed on the CO_2 objective, the model yields the feasible solution that has the lowest level of CO_2 emissions (which in practice might be an all hydro generation plant) -- P_{10} in Figure 3.10. More balanced sets of weights yield intermediate points of the tradeoff curve.

Figure 3.11 sketches a set of example "indifference" curves that result from one set of weights. An indifference curve defines a set of points that achieves the same value of the weighted objective function. The slope of the indifference curve is determined by the weights: in Figure 3.11, it is $-w_{CO_2}/w_{COST}$. The higher the weight on CO_2, the steeper the indifference curves are, which pushes the solution towards points with better CO_2 levels. The weighting method works by choosing the alternative that lies on the most desirable (here, the most south-westerly) indifference curve; in Figure 3.11, that alternative is P_5. Section 4.3 of Chapter 4 discusses the concept of indifference curves in more detail.

Figure 3.11: The weighting method with three indifference curves

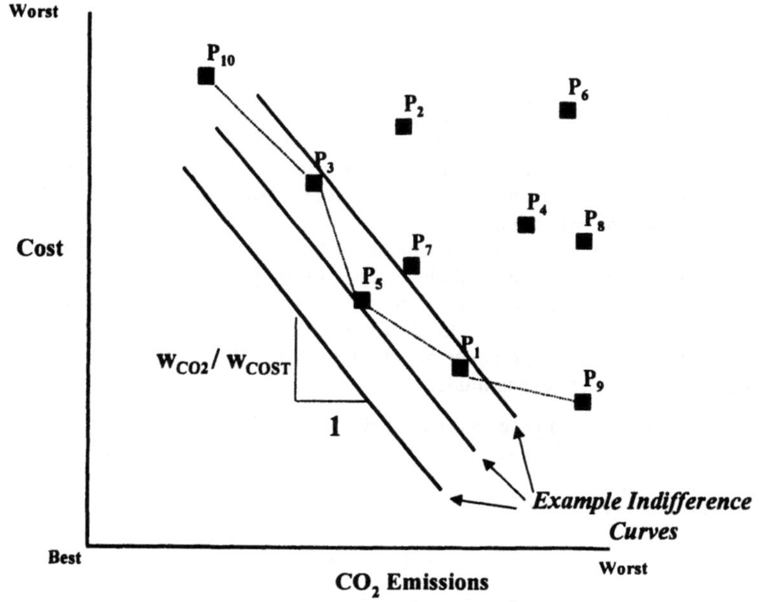

The weighting method was used in the Seattle case study of Chapter 10, in which a large scale linear programming model, the Annual Optimization Model (ANN), was applied. Tradeoffs between cost and CO_2 were obtained exactly in the manner described above. Since ANN is an optimization model of a type called "linear programming," the resulting tradeoff curve is actually a continuous piece-wise linear function that bends away from the origin.

A disadvantage of the weighting method is that it will miss non-inferior solutions that are so-called "gap points." Such a point is nestled between two or more non-inferior points in such a way that it is not best under any set of weights. In Figure 3.11, for instance, there is no positive set of weights w_{COST} and w_{CO_2} that can result in P_3 being chosen -- even though it is not dominated by any other alternative. Option P_4 in Table 3.2 of Annex 3.4 is another example. There are no gap points in linear programs such as ANN, but other resource planning models that are based on nonlinear or integer programming can have them. If there is reason to suspect that gap points exist, the constraint method can be used to uncover them.

ANNEX 3.2

Other tradeoff display methods

The Cartesian plots of this chapter are a popular way of portraying tradeoffs, especially when the number of attributes is small. However, there are many other techniques for displaying multidimensional information (Vetschera, 1994), several of which can be more useful than Cartesian plots when there are many attributes. In general, we advise use of more than one approach, as different people will react to them differently.

Some people will prefer tabular information, either numerical (Tables 9.3 and 10.3 in Chapters 9 and 10, respectively) or symbolic. As an example of the latter, Table 3.1 uses "+" symbols to indicate how four plans compare on eight criteria. More plusses indicate more desirable values of the attribute. The table shows, for instance, that alternative Z is dominated by Y. Consumers Reports™-type tables are similar devices that instead use red- and black-filled circles to indicate relative magnitudes.

Table 3.1: Four alternatives compared on eight attributes

Plan	Investment cost	O&M cost	Rates	CO_2	SO_2	Land use	Fish & wildlife	Water
W	+++++	++++	++++	+	+	+++	+++	+++
X	+++	++	++	+++++	+++++	++++	+++++	+++++
Y	++++	+++	+++	++++	+++	++	++	+++
Z	++	+++	+++	+++	+++	++	+	+

Another graphical approach is the "value path" (Bishop, 1974) and its close relative, the "spider web diagram." In the value path, vertical bars represent distinct attributes, which are rescaled so that the least desirable value of each attribute is at the bottom of the bar and the most desirable is at the top. Then individual alternatives are displayed as a line, or "path," that traces that alternative's relative performance on each of the attributes. Consider, for instance, Plan Y in Figure 3.12. Its value of every attribute (from Investment Cost to Water) is strictly better than Plan Z's value. On the other hand, a glance at Figure 3.12 reveals that there are important tradeoffs between Plans X and Y; the latter is better in Investment cost, O&M costs, and Rates, while the former is preferred in all of the environmental attributes. Such value paths are useful because many people will find these differences more apparent than in a table such as Table 3.1, and finer degrees of difference can be shown.

Figure 3.12: Value path display of tradeoffs among four alternatives

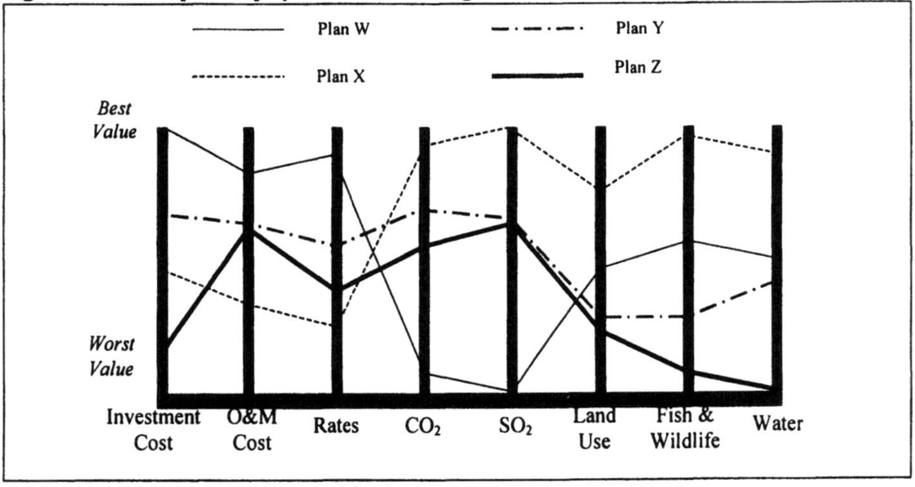

ANNEX 3.3

Screening resources at BC Hydro by dominance analysis

At an early stage in the BC Hydro case study (see Chapter 8), we faced the problem that some resources had been dropped by BC Hydro from the supply-side portfolios of previous studies, yet those resources remained of interest to some (but not all) members of the Consultative Committee. Examples included some large hydroelectric projects (such as Peace Site C), and the East Kootenay Coal project. There were also some technologies that BC Hydro considered quite impractical -- such as ocean wave energy -- but which were again of interest to some Committee members.

Although inclusion of these resources in any final portfolio was deemed unlikely by the Committee as a whole, it was considered important that the analysis spoke for itself, and that screening decisions were made by the stakeholder representatives, rather than by BC Hydro staff.

In the case of East Kootenay Coal and ocean wave, we were also interested in highlighting the impact of technological change. What would be the impact of cost reductions that might be achieved through widespread commercialization elsewhere? And in the case of coal plants, what would be the impact of successful commercialization of the new clean-coal technologies?

For ocean wave plants, we considered two cases. The first was a plant built in 2000, with an estimated capital cost of $4,000/kW; the second was based on a reduced capital cost of $2,000/kW, hypothesized to be achievable by 2007.

Similarly, for East Kootenay Coal, the conventional pulverized coal (PC) technology was assumed at a capital cost of $2,745/kW, and with heat rates and emissions characteristic of conventional technology. As a second option, we assumed commercialization of pressurized fluidized bed combustion/combined cycle (PFBC-CC) technology, at a capital cost of $1,500/kW. This technology has substantially higher efficiency, and significantly better emissions performance.

Figure 3.13 shows the tradeoff curve for cost versus GHG emissions. As expected, both ocean wave and coal perform poorly relative to the other options -- firmly in the lose-lose quadrant -- even when very favorable assumptions on cost and efficiency are made. We also show on this graph the Peace Site C, Brilliant, and Keenleyside hydro projects. While the latter two projects appeared in the 1994 Electricity Plan, with the signing of the Columbia Basin Accord in early 1995 these latter two options, along with the Waneta Expansion project, were no longer in BC Hydro's control. They were included in this analysis at the request of the Consultative Committee in the event that these resources were bid into a future BC Hydro RFP. With other options in the win-win quadrant, these hydro options were in any event dropped.

This supply-side screening exercise illustrates several important points about how an MCDM study should be conducted. First is that whenever possible, the analysis should speak for itself. Once the above figure had been presented to the Committee, the question of ocean wave technology, about which there had been heated discussion prior to the start of the formal MCDM analysis (and largely because BC Hydro took the view that the technology didn't merit *any* analysis), vanished. Second, the Committee member representing the interests of the Kootenay region, where unemployment is much higher than the Provincial average, needed to be satisfied that the gains to local employment of a coal project would be far outweighed by the high costs: presentations of the type illustrated here were persuasive. The willingness of the BC Hydro team to examine all options of interest to the stakeholders played a significant role in establishing a constructive working relationship among the participants -- who, as we shall see in Chapter 8, while not achieving a full consensus, were largely satisfied that their interests had been heard and considered.

Figure 3.13: Supply-side resource screening

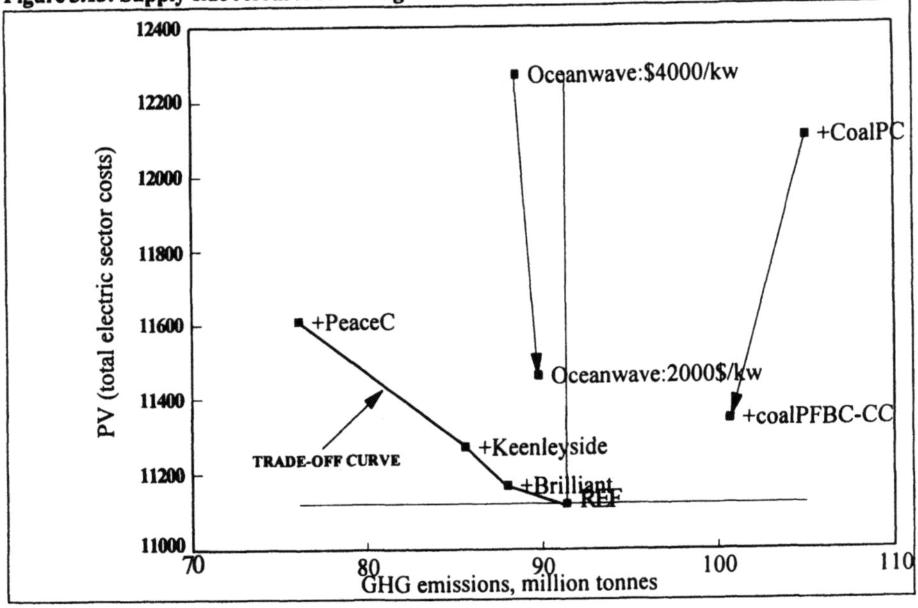

ANNEX 3.4

Tradeoff analysis for the three-dimensional case

As an illustration of the caution that needs to be applied when looking only at the two-dimensional representations of the tradeoff surface, consider a three-attribute example with attribute values for five plans as shown in Table 3.2.[13] The two-dimensional plots are shown in Figure 3.14. We assume, as elsewhere in this chapter, that we want to minimize each attribute.

Table 3.2: A three-dimensional example

	cost	air	water
P1	0.5	0.1	0.95
P2	1	0.5	0.1
P3	0.1	0.9	0.5
P4	0.6	0.6	0.6
P5	0.7	0.35	0.95

From Figure 3.14, one might infer that the non-inferior set is {*P1, P2, P3*}, since in two dimensions the members of the tradeoff curve are {*P1, P2*} for water versus air; {*P3, P2*} for water versus cost, and {*P1, P3*} for air versus cost. Both *P4* and *P5* appear to be strictly dominated.

But as inspection of Table 3.2 shows, in fact when one uses the mathematical definition of strict dominance in a simultaneous comparison of all three attributes, only *P5* is seen to be strictly dominated (by *P1*). However, Option *P4* is not dominated. *P4* is preferred over *P2* in cost, preferred over *P3* in air, and over *P1* in water. Indeed, it may in fact be an attractive compromise to decision-makers representing different interests because it avoids very bad outcomes.

For example, from the environmental perspective one may accept a second-best outcome for, say, water, if one can avoid the worst-case outcome for air quality. Such preferences on the part of decision-makers can be captured by an asymmetrical definition of the tolerance parameters in the knee-set definition: if $e >> m$ (see discussion above in Section 3.3), then it quickly follows by inspection from the above table that *P4* is the only member of the knee set! For example, if $e_i = 0.5$, and $m_i = 0.1$, then because *P1* is no longer *significantly better* than *P4* for the air quality criterion, but is significantly *worse* than *P4* for water, *P4* significantly dominates *P1*, and so on.

This does not imply that the two-dimensional plots of the tradeoff surface are useless: they still provide a helpful view of those plans that are closest to the origin (i.e., best) for the two attributes displayed. However, it must simply be recognized that both knee-set and non-inferior sets must be defined on the basis of the simultaneous, multi-attribute inspection of attribute values: looking *only* at the two-dimensional displays may not uncover all of the solutions of actual interest to decision-makers.[14]

Figure 3.14: A three-dimensional example

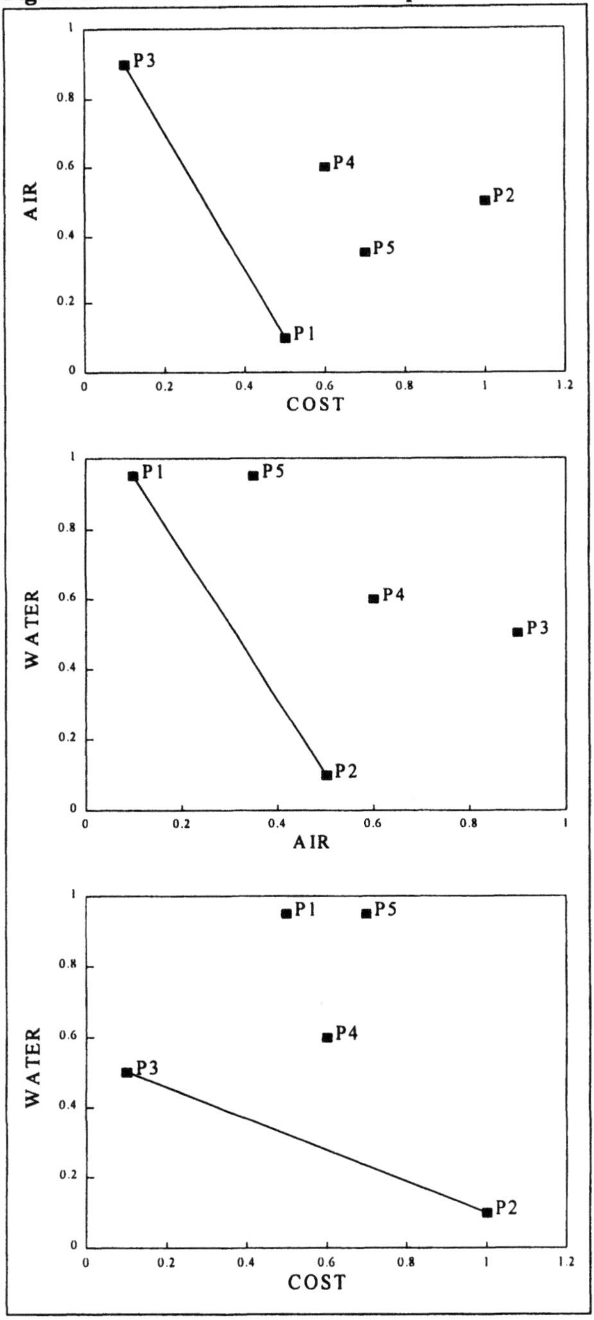

NOTES

[1] For example, in the Bihar case study (SCADA, 1998), rehabilitation of old thermal plants reduced costs, but increased air emissions. This occurred because the improved heat rates of the rehabilitated plants caused them to be dispatched more (but with emissions still higher than those from the new plants that would be required if the old plants were simply retired).

[2] For a full discussion of the concepts of dominance, and the technical aspects of tradeoff analysis, see Cohon (1978) or Steuer (1986); explanations in an energy-environment context are presented in Meier and Munasinghe (1994).

[3] The terms "non-inferior" and "efficient" are also used to describe the set of points on the tradeoff curve.

[4] As we discuss in the next chapter, weights should represent the willingness of decision-makers to trade off attributes.

[5] Other earlier examples include Meier and Ruff (1979) and Merrill, Schweppe and White (1982).

[6] World Bank (1998a). T&D loss rates in Bihar (as elsewhere in India) are close to 40%: rehabilitation of the T&D system brings losses down to 12-15%.

[7] The number of possible pairs among M attributes is $(M^2-M)/2$. See Annex 3.2 for alternative graphical representations that show all tradeoffs simultaneously.

[8] Further, it must be remembered that an option that appears to be dominated in all two-way tradeoff graphs may nonetheless be non-inferior in multi-dimensional space; and it is entirely possible that an option that lies on none of the two-dimensional tradeoff curves may still be one that a highly diverse group of interests can agree upon as a "second" best (see Annex 3.4). Indeed, in Chapter 8 we show an example of one set of portfolio assessments where the Reference Portfolio emerges as the least disliked, despite being nobody's first choice.

[9] Mathematically: assume that the attributes A_i are to be minimized. Then alternative j dominates alternative k if $A_{ij} \leq A_{ik}$ for all i, and the inequality is strict for at least one i.

[10] This discussion uses the definitions of Crousillat and Merrill (1992).

[11] Given that preferences for attribute levels might be non-linear, it could be argued that m_i and e_i depend upon the level of $A_i(P_j)$, rather than being independent of level. For simplicity, we disregard that complication here.

[12] For a detailed example of such a cost CO_2-tradeoff model for utility planning, see Hobbs (1997).

[13] This point was first made by Loucks and Somlyody (1988).

[14] It is of course also true that we have used a contrived example here to make our point.

4 SCALING, WEIGHTING AND AMALGAMATION

Though this be madness, yet there is method in it
SHAKESPEARE, Hamlet: II(ii) 21

A false balance is abomination to the Lord: but a just weight is his delight
OLD TESTAMENT, Proverbs 11:1

Steps 1 - 6 of our general MCDM procedure emphasize the generation and display of tradeoff information, a major purpose of MCDM methods. A second major purpose is the quantification of value judgments that can lead to a choice from among the alternatives. This purpose is the subject of Steps 7 - 9, which we discuss in this chapter.

4.1 VALUE SCALING (STEP 7)

Most methods for amalgamation require that the decision-maker *scale* attributes. Value scaling can be viewed as the creation of a value function $V_i(A_{ij})$ which translates A_i, which is a social, economic, or environmental attribute, into a measure of worth or desirability. If A_{ij} is the level of attribute i for alternative j, then $V_i(A_{ij})$ is the worth of that alternative for that attribute. $V_i(A_{ij})$ is usually defined so that the best outcome is assigned a value of one, the worst a value of zero.

For example, the widely used additive value function method of amalgamation -- sometimes described as "weighting summation" -- and discussed further below in Section 4.3.1, ranks options by the score:

$$TV_j = \sum_{i=1}^{I} w_i V_i(A_{ij}) \qquad \text{Eq.[4.1]}$$

where TV_j is the total value of option j, and w_i is the weight of attribute i.

Both the value functions and weights can be obtained by several means, as described below. These methods for weighting and scaling apply also to other amalgamation rules (such as goal programming, discussed in Section 4.3.2). A value function usually describes a person's preferences regarding different levels of an attribute under certainty; issues involved in assessing them are the subject of Section 4.1.1. In contrast, *utility* functions (a specific type of value function) also reflect attitudes toward risk (Section 4.1.2).

In the example of Figure 4.1, we show alternative representations of a value function for damages resulting from the discharge of thermal effluents into an estuary. The simplest representation (attribute value function 1) is a simple linear scale that has the implication, for example, that a 1 degree rise in temperature from 26 to 27 °C produces the same damage as a 1 degree rise from 29 to 30 °C. Attribute value function 2, on the other hand, implies a non-linear relationship, and in particular a threshold at T^*. Temperature increases to below this value are assumed to cause little or no damage, while any temperature greater than T^*, all the way to the maximum observed, is perceived as equally bad.

Figure 4.1: Attribute scaling

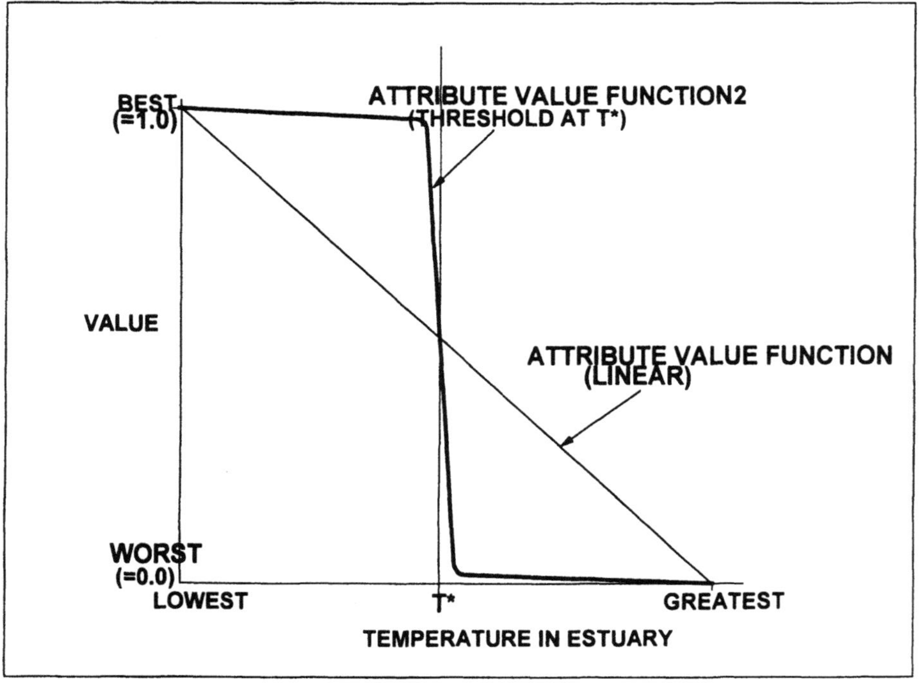

4.1.1 Assessing value functions

Most amalgamation methods presume that attribute value functions are on an interval level of measurement with respect to value. In a valid interval scale, differences between numbers are meaningful, but the zero point is arbitrary. For example, on the interval Fahrenheit temperature scale, it makes sense to say that 10° is halfway between 0° and 20° -- but it is incorrect to state that 20° is twice as warm as 10°. Likewise, in an appropriately scaled attribute value function, the difference between a value of 0.0 and one of 0.5 should be the same as the difference between 0.5 and 1.0. Only such interval scales can be weighted and otherwise manipulated algebraically.[1]

A number of methods can be used to create interval-scaled value functions (Fishburn, 1967; Hobbs 1979). For example, the decision-maker might be asked the following question:

> "What level of net benefits is halfway between \$0/yr (having a value of 0) and \$100,000,000/yr (having a value of 1) in value or desirability?"

Note that the answer does not have to be \$50,000,000; non-linear value functions are possible and, indeed, likely. If linear, then the answer to that question would be a value of 0.5. Scaling techniques that fail to assure an interval level of measurement are widely applied, however, and are in violation of measurement theory. An example would be to simply ask the decision-maker to sketch a value function for, say, reliability or solid waste impacts.

However, there has been no reported research that shows that interval scaled value functions are significantly different from value functions created using methods which do not guarantee interval scales (Hobbs, 1979). By "significantly", what is meant is that choice of scaling method makes an important difference in the final ranking of alternatives. Likewise, there has been no significant difference shown between the results of utility functions (which include a decision-maker's attitudes towards risk, Section 4.1.2) and deterministic value functions, such as those derived from the methods just discussed (which do not include risk attitudes), although some studies have shown that the shapes of the functions have differed (Barron *et al.*, 1984; Krzysztofowicz and Duckstein, 1984).

Therefore, it is suggested that scaling be done using the method with which the decision-makers feel most comfortable. Where scaling involves significant technical expertise, it might be best to allow experts in the relevant fields to scale the attributes with which they are familiar (Dyer and Sarin, 1979). For example, a biologist might choose the value function for wetland impacts, while an economist would scale net monetary benefits. However, the functions should be reviewed and, if appropriate, modified by managers, stakeholders or others responsible for the final recommendations, since the scales may embody important value judgments (such as thresholds for what constitutes an acceptable impact).

Although non-linear scales are unlikely to affect the final option rankings, we recommend that the analyst not automatically assume that value functions are linear; there may be thresholds or other non-linearities. Indeed, in both the Centerior and SCL case studies, some of the functions were strongly non-linear. In both cases the participants were presented with postulated linear value functions, which displayed the individual options and their values on the postulated linear scale, and were then asked about whether linearity reflected their actual assessment of damages and values. Figure 4.2 shows such postulated linear scales as used in the questionnaire of the Seattle case study: the abbreviations used to identify the options in this figure are described in Table 10.2 of Chapter 10.

Figure 4.2: Scaling questionnaires in the Seattle case study

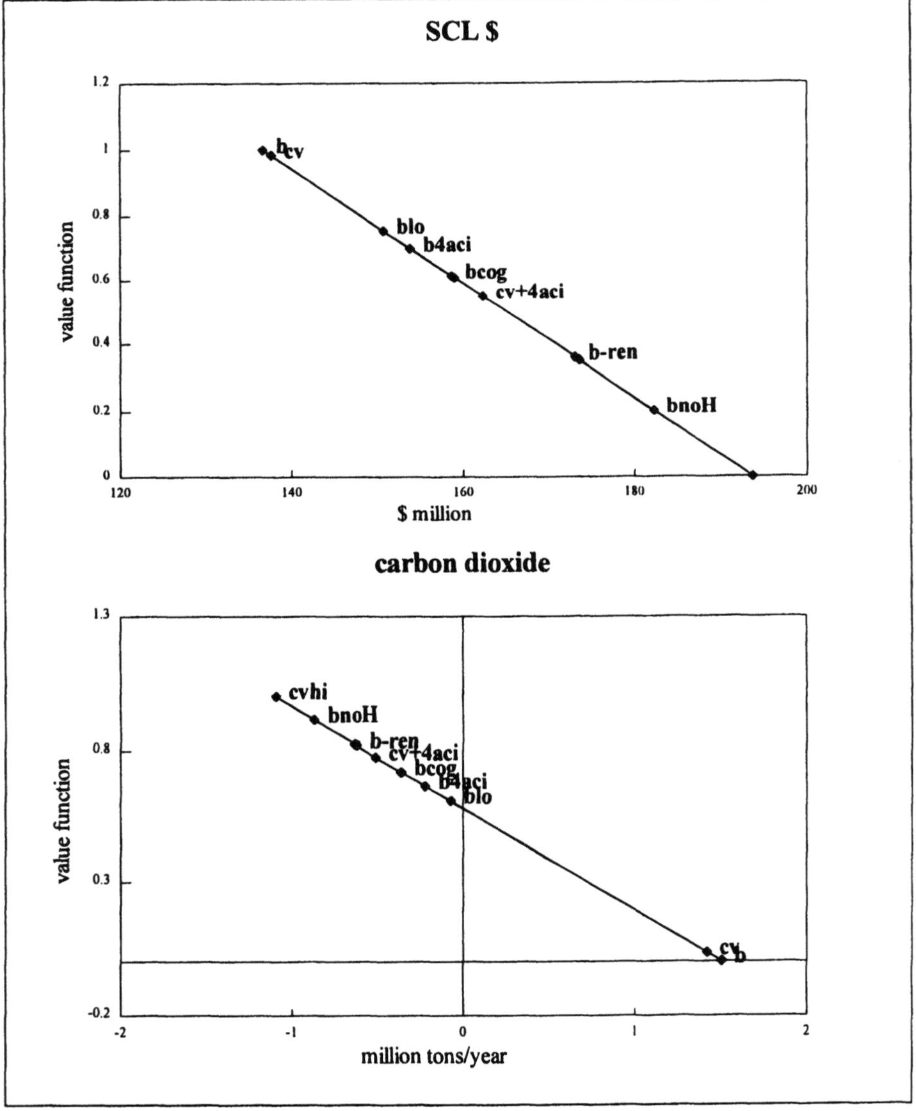

In Figure 4.3 we show some results of the group discussions on scaling in our case studies.[2] Figure 4.3(a) represents the scaling curve for short-term price impacts in the Centerior study. The sharp threshold at zero reflected the low tolerance in the group for any short-term rate increase: whether the rate increase was very small or more significant, in both cases a rate submission would have to be made, which was perceived as having a high transaction cost.

Figure 4.3: Electricity rate and hydro single-attribute value functions

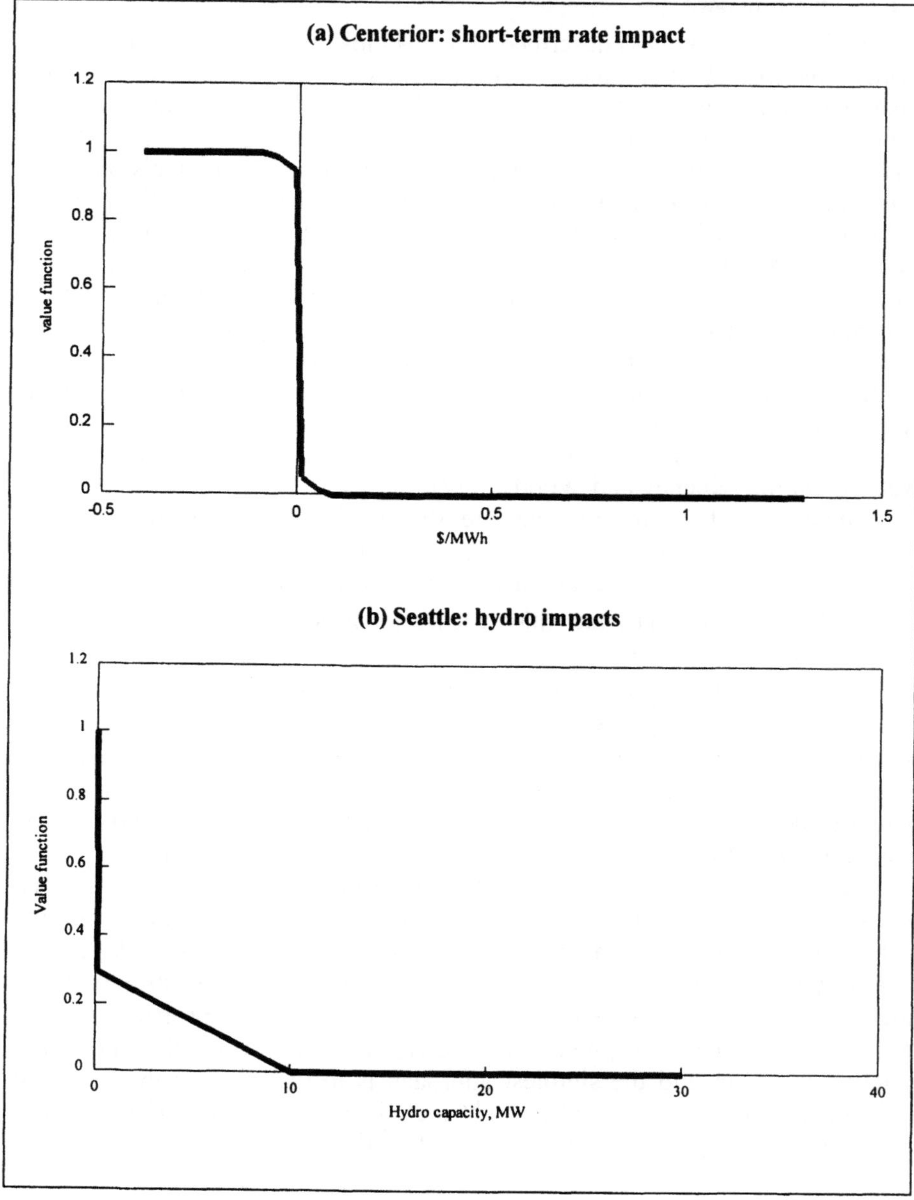

Figure 4.3(b) represents the scaling curve for the attribute *hydro MW* in the Seattle study. The shape of this curve -- strongly convex, indicative of the judgment that any positive amount is bad -- reflected the group consensus that the critical hydro impact was limitation of white-water recreation opportunities. Thus the extent of river affected was unlikely to be a function of the installed capacity (in other words, a 15 MW hydro plant would have the same undesirable effect in terms of miles of river affected as a 30 MW plant).

4.1.2 Utility functions

The value scaling process described above is closely related to the concept of *utility functions.* Utility functions, however, are more general because they also capture the user's attitude towards taking risks. The purpose of utility functions is to find some function $U_i(A_i)$ for each attribute A_i, such that if $E\{U(A_i)\}$ for option X is greater than for option Y, then the decision-maker prefers option X, and vice versa.[3] Such a function can reflect both the innate value of different levels of the attribute, as well as the decision-maker's attitudes towards risk (risk seeking, risk neutral, or risk averse, defined below). There is a well-developed body of theory which describes under which circumstances utility functions accurately characterize decision-maker preferences (Keeney and Raiffa, 1976). This approach is widely used: besides applications to energy planning and policy (see particularly the various studies by Keeney and his colleagues in the References), it has been recommended by many risk analysts (e.g., Sage and White, 1980), and for use in water planning (Keeney and Wood, 1977).

Single-attribute utility functions are often applied in three steps:

1. Characterize the probability distribution $P(A_{ij})$ of A_i for each option j. (This can be viewed as an extension of Step 3, Section 2.3.)
2. Create a single-attribute utility function for the attribute by the procedures outlined below.
3. Calculate the expected utility $E\{U_i(A_{ij})\}$ of attribute i for each j as:

$$\int P_{ij}(A_{ij})\, U_i(A_{ij})\, dA_{ij},$$

where $P_{ij}(A_{ij})$ is the probability density function of attribute i for alternative j. $E\{U_i(A_{ij})\}$ is then inserted in a multi-attribute utility function (Section 4.3.4), which trades off attribute i against the other attributes. This yields an expected overall utility for option j. The options can then be ranked.[4]

If a decision-maker is risk neutral, then he or she will prefer the alternative with the best expected values of the attributes $E(A_i)$, no matter how little or how much risk there is associated with each option. For example, if A_i is an attribute for flood risks from a hydro plant, a single rare catastrophic flood might implicitly be weighted far less (by virtue of its low probability) than more frequently occurring medium-sized floods. This means that the utility function is linear, and the simplest approach is to directly use the expected value of the attribute. Examples might include expected generation cost, expected flood damages, and expected loss of wildlife habitat.

But decision-makers are not generally risk neutral. In that case, utility functions are non-linear, and the alternative with the best $E(A_i)$ might not be preferred because it might be riskier than other options (for example, because it has a higher standard deviation). An example is the exponential utility function:

$$U_i(A_i) = a + be^{cA_i} \qquad \text{Eq. [4.2]}$$

in which a, b, and c are parameters, and e is 2.7128... .

A single-attribute utility function such as Eq. [4.2] can be calibrated by asking a decision-maker to compare a "sure thing" (an option whose attribute level is known) and a "gamble" (an option whose attribute value is random). The value of the sure thing or the probabilities in the gamble are then varied until the user is indifferent to the outcome between the two options. As a result of this process, both preferences for levels of an attribute and risk attitudes can be captured. For instance, the most common version of this process phrases the sure thing and gamble as follows:

"Sure thing": achieves $A_i^{c.e.}$ for sure

"Gamble": 0.5 chance of A_{i*} and 0.5 chance of A_i^*

where $A_i^{c.e.}$ is the "certainty equivalent" of the gamble. A_{i*} is the worst possible value of A_i, and is assigned utility $U_i(A_{i*}) = 0$; A_i^* is the best possible value, and is defined as having a utility of $U_i(A_i^*) = 1$. If the value of $A_i^{c.e.}$ is adjusted up or down until the two options are equally attractive, then the utility of the "sure thing's" utility must be equal to the expected utility of the gamble, which is 0.5*0+0.5*1, or 0.5. This gives three points on the utility function. Additional points can be obtained in an analogous manner. This procedure is illustrated in Section 6.3. Other approaches to calibrating utility functions are presented by Keeney and Raiffa (1976).

Table 4.1 gives an example of the application of such a utility function. It shows the calculation of expected flood damages (in $) and the expected utility (in "utils") of flood damages associated with two different hydropower facility designs. Option X has lower expected damages than Option Y ($1,854,000/year vs. $1,961,000/year). Thus a risk neutral decision-maker would also prefer X. But X could be considered to be more risky, since it has a higher probability of floods causing $15,000,000 in damage or more (0.06/year, vs. 0.04/year for Y). Indeed, under the risk-averse utility function assumed in the table, X's higher risk of severe damages causes it to have a worse expected utility than Y. Table 4.1 shows that X's expected utility is 0.940, which is lower and, hence, less desirable than Y's expected utility of 0.946. The more risk averse the utility function, the lower X's expected utility would be relative to the less risky option Y.[5]

It must be noted, however, that non-linearities in utility functions arise for two reasons (Dyer and Sarin, 1979; Chankong and Haimes, 1983):

Table 4.1: Expected utility calculation for flood damages under two hydropower options

		Option X			Option Y		
Flood damages *FD* ($1000/yr)	Utility of damages	Probability	Probability times damages ($1000/yr)	Probability times utility	Probability	Probability times damages ($1000/yr)	Probability times utility
$0	1.0000	0.7	0	0.70000	0.65	0	0.65
$100	0.9988	0.03	3	0.02996	0.02	2	0.01998
$200	0.9976	0.03	6	0.02993	0.02	4	0.01995
$500	0.9940	0.03	15	0.02982	0.03	15	0.02982
$1,000	0.9876	0.03	30	0.02963	0.04	40	0.03950
$2,000	0.9736	0.03	60	0.02921	0.05	100	0.04868
$5,000	0.9197	0.03	150	0.02759	0.06	300	0.05518
$8,000	0.8419	0.03	240	0.02526	0.05	400	0.04210
$10,000	0.7719	0.03	300	0.02316	0.04	400	0.03088
$15.000	0.5000	0.03	450	0.01500	0.02	300	0.01000
$20,000	0.0000	0.03	600	0	0.02	400	0
	sum	1	1,854 (=expected damages)	0.93955 (=expected utility)	1	1,961 (= expected damages)	0.94608 (= expected utility)

Note: Assumed utility function for *FD*: $U_{FD}(FD) = 1.0957 - 0.0957e^{(0.0001219\,FD)}$

1. The decision-maker's preferences for levels of an attribute (as represented by a deterministic value function) are non-linear. As an example, preferences for levels of dissolved oxygen in a stream are non-linear because of threshold effects.
2. The decision-maker is either relatively risk averse or risk seeking in general. An example involving bets with money might clarify the latter type of preference. Assume that a person's deterministic value function for money is linear. If the person is indifferent to the outcome between $50 and a 50-50 chance at $100 (expected value = $50), he is considered to be relatively risk neutral. If he would rather have the bet, he is relatively risk seeking. However, he would be risk averse if the certainty of $50 were preferred.

Formally, a person is said to be relatively risk averse if the expected value of their deterministic value function is greater than the expected utility. They are relatively risk preferring if the opposite is true. Utility functions are unnecessary, and easier to assess deterministic value functions can be used instead under the following conditions:

- the decision-maker is relatively risk neutral, *or*
- the level of each attribute for each alternative is specified as a single number rather than as a probability distribution--which is (perhaps unfortunately) the case for most energy decision analyses.

Using the utility function method presents several challenges. Decision-makers may not be able to respond to lottery questions or their answers may be inconsistent, leading to invalid functions. Krzysztofowicz and Duckstein (1984), for example, show that a person's utility function can depend on purely arbitrary aspects of how the questions needed to calibrate that function are posed. And even if decision-makers can choose meaningful functions, whose values and risk attitudes should be used? The utility functions of different people can differ drastically. Of course, this problem is not unique to utility functions. Different people's value functions can also differ, as can their weights. In Step 10 (Chapter 5), we address the question of how to resolve such conflicts.

Even if deterministic value functions are adopted instead of utility functions, we believe that it is prudent to assess the effects of different risk attitudes upon the results if attribute values are uncertain. We recognize that there is little empirical evidence that choice of scaling method makes a significant difference in decisions, but in theory differences could result. Such a sensitivity analysis could be done as follows (Dyer and Sarin, 1979). Let $V_i(A_i)$ be a deterministic value function, scaled from 0 to 1, for attribute A_i. The following exponential utility function can be used to model different risk attitudes:

$$U_i(A_i) = a + be^{cV_i(A_i)} \qquad \text{Eq.[4.3]}$$

where the parameters a, b, and c are chosen so that $U_i(A_i)$ is also on a 0 to 1 scale and that the desired amount of risk aversion results. If c is negative, then the greater c is in absolute value, the more risk averse (concave) the utility function is. Different degrees of risk aversion can be tested to see if they alter the ranks of alternatives.

4.2 WEIGHTING METHODS (STEP 8)

Almost all ranking methods (such as additive value functions) require weights. Because the relative weights can make a large difference in the resulting ranks of alternatives, and because the literature amply documents how commonly used weighting methods can bias or distort preferences (e.g., Weber and Borcherding, 1993; Poyhonen, 1998), this task should be approached carefully.

Many weighting methods have been proposed, and new ones continue to be introduced (e.g., Salo and Hamalainen, 1997a). After introducing several of the more popular approaches below (Sections 4.2.1 - 4.2.6), what constitutes a "valid" weight is carefully defined, and the validity and ease of use of different techniques for picking weights are then contrasted (Section 4.2.8). The methods are discussed as if they were used in additive value functions; however, the discussion applies equally to other amalgamation methods that require weights. Recommendations concerning which weighting methods should be used conclude the section (Section 4.2.9). The calculations involved in some of the methods are illustrated by a simple example in Chapter 6. Detailed descriptions of how several methods have been used in energy resource evaluation are given in the case studies of Chapters 8 through 10.[6]

4.2.1 Equal weights

The simplest approach to weighting is to use equal weights. However, to say that when two attributes are given equal weight, this makes them "equally important" is incorrect. For example, say that $w_1 = w_2$. Then if the difference between $V_i(A_i)=0$ and $V_i(A_i)=1$ represents only a trivial change in A_1 but a significant change in A_2, then, implicitly, a *higher* importance is assigned to A_1. This is because a large improvement in A_2 is required to compensate for a small worsening in A_1. The relative importance of two attributes depends not just on their weights, but also their ranges, as reflected in the $V_i(A_i)$. We return to this point later in this section.

4.2.2 Observer derived weights

This method estimates weights using multiple regression or linear programming to analyze unaided subjective evaluations of alternatives in order to determine which weights best predict those evaluations. For example, someone might be asked to rate each of 20 options on a scale of 0 to 100. Let the total stated value of option *j* be TV_j, then this method solves for the w_j by regressing the $V_i(A_{ij})$ against the TV_j; that is,

$$MIN \qquad \Sigma \varepsilon_j^2$$

$$\text{subject to } TV_j = \sum w_i V_i(A_{ij}) + \varepsilon_j \quad \text{for } j = 1,...20, \qquad \text{Eq. [4.4]}$$

where A_{ij} is the value of attribute *j* for option *i*, and ε_j is the predictive "error." Standard least-squares methods can be used in this case to estimate the weights.

This method is sometimes called "policy capturing," as it calculates the weights implicit in a person's unaided judgments (Hammond *et al.*, 1977). Psychologists and pollsters like this method because it yields the weights that best *predict* unaided opinions. We believe, though, that the purpose of multi-criteria analysis should be to *improve*, not simulate, such holistic judgments. Shepard (1964) and other researchers since have noted that people tend to ignore all but a handful of attributes when ranking multi-dimensional objects, treating other, yet still important criteria inconsistently.

Most of the methods below, in contrast, use a "divide and conquer" strategy, asking users to consider the attributes, but only one or two at a time. The result can be more balanced, consistent, and defensible rankings of options. These ranks might differ from unaided ranks -- but that is to be expected, even encouraged. Such contradictions give people cause to pause and ponder the reasons for these differences; such reflection in turn can build insight and confidence in the decision. In the case studies of Chapters 9 and 10 we gave stakeholders this opportunity. They felt that this resolving of conflicts in ranks was perhaps the most valuable part of the MCDM process.

4.2.3 Direct weighting

Direct weighting methods, in contrast, ask decision-makers to specify numerical values for weights directly. Examples of direct methods include point allocation (allocate 100 points among attributes in proportion to their "importance"), categorization (assign attributes to different categories of importance, each carrying a different weight), ranking (the least important attribute receives a weight of 1, the next lowest a weight of 2, etc.), ratio questioning (which asks for ratios of the importance of attributes, two at a time), and rating (each attribute's importance is rated on a scale of, say, 0 to 10).[7] These methods tend to yield similar weights, or at least the same rank order for the importance of the attributes.[8] Unfortunately, direct methods often fail to yield weights that correspond to tradeoffs people are willing to make: the reason is that subjective notions of attribute "importance" may have little to do with the rate at which a person is willing to trade off the $V_i(A_i)$ for two attributes.

Direct methods can assign weights to all criteria at once. But if there are many criteria, such as the 12 in our Seattle case study, then a hierarchical approach is often easier (see Figure 10.1, Chapter 10). A hierarchical approach can also lessen the danger of double counting. In the hierarchical procedure, the user groups criteria into general categories. In the top-down version, the user first assigns a weight to each general category. Then, the user allocates each category's weight among its criteria. The other, bottom-up procedure first assigns weights within categories and then weights the categories themselves. Hierarchies can also be defined with several levels, as in Figure 10.1.

In Figure 4.4 we show the "peakiness" of weights elicited by a variety of different methods in our Centerior case study. "Peakiness" $p(w)$ is defined as the difference in weight between the highest weighted attribute, and the lowest weighted attribute: if all were given equal weight this measure would be zero. This measure was defined for each individual, and then averaged across individuals for each method.[9]

The figure reveals that the peakiness of hierarchical rating weighting ($p(w)$ = 0.29 and 0.37) is (statistically) significantly higher than for non-hierarchical rating weights ($p(w)$ = 0.19 and 0.25). Many other studies (e.g. Stillwell *et al.*, 1987) have also shown that hierarchical weights are less uniform. This is at least in part because individuals tend to avoid giving extremely low or extremely high values to categories; as a result, a category with fewer attributes will have to split its weight less, resulting in higher per-attribute weight than categories with few attributes. Indeed, the statistical tests described in the Seattle case study confirm this trend (see Chapter 10, where we discuss the issue further). The fact that a seemingly arbitrary decision (such as the way a hierarchy is structured) affects rating weights suggests that the method is not a fully valid way of gauging priorities. Figure 4.5 also shows results for some other methods we discuss below: AHP weights (see discussion below of this method) are very peaked.[10]

Figure 4.4: Peakiness of weights in the Centerior study

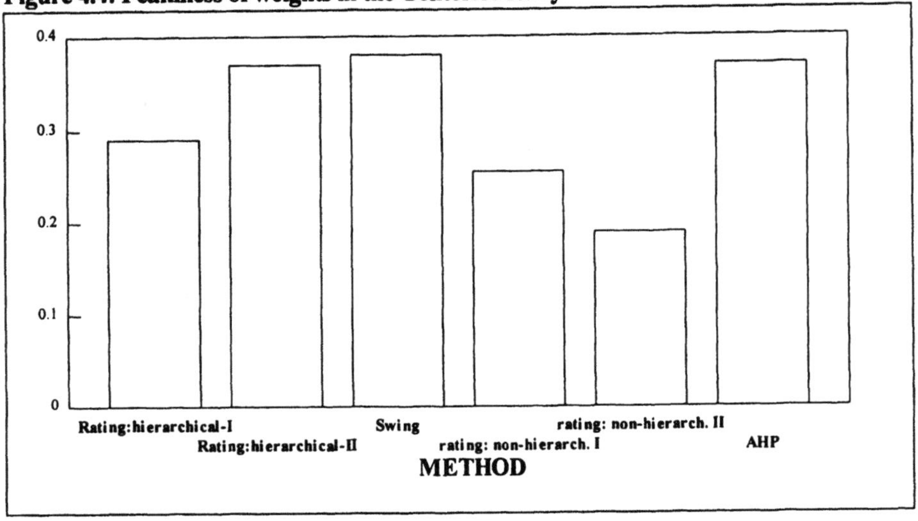

4.2.4 The Analytic Hierarchy Process

A popular version of ratio questioning is the Analytic Hierarchy Process (AHP; Saaty, 1980). For example, Rahman and Shrestha (1992) and Pan and Rahman (1998) use AHP to evaluate the environmental impacts of power planning options, while Ramanathan (1999) proposes its use for selecting greenhouse gas mitigation options.[11] AHP asks decision-makers to compare every possible pair of criteria, stating the ratio of their importance. For instance: what is the ratio of importance of air quality impacts and cost?

In assessing relative importance ratios, it is often difficult for the assessor to give precise numbers. The scale in Table 4.2 has been suggested to be a useful guideline for making pair-wise comparison judgments. Obviously, if attribute II is actually more important than I, an appropriate reciprocal of the number in the table should be used.[12] The AHP was used in the Centerior case study: Figure 4.5 shows the questionnaire that was used.

Table 4.2: Suggested scale for AHP ratio assessments

The ratio for attribute I over attribute II should be:	
1	if the two attributes are judged to be equally important
3	if attribute I is judged to be slightly more important than Attribute II
5	if attribute I is judged to be moderately more important than Attribute II
7	if attribute I is judged to be strongly more important than Attribute II
9	if attribute I is judged to be extremely more important than Attribute II
2,4,6,8	if intermediate values between two adjacent judgments are needed

Figure 4.5: The AHP weighting questionnaire in the Centerior case study

In this questionnaire we ask you to compare the relative importance of the criteria two at a time. The criteria measure how a plan performs relative to the "Reference" case. A weight set consistent with the comparisons will then be derived using the "Analytic Hierarchy Process." These weights will then be used in an additive value function. Criterion "importance" can generally be defined thus: *A criterion X is more important than a criterion Y if option A is preferred to option B*, where:

A: X is at its best value, Y is at its worst value

B: X is at its worst value, Y is at its best value

The ratios of importance you state should represent your willingness to trade off the criteria. The following scale is suggested for ratio judgments:

The ratio of Criterion X over Criterion Y should be:	*IF:*
1/9	criterion Y is extremely more important than criterion X
1/7	criterion Y is strongly more important than criterion X
1/5	criterion Y is moderately more important than criterion X
1/3	criterion Y is slightly more important than criterion X
1	the two criteria are equally important
3	criterion X is slightly more important than criterion Y
5	criterion X is moderately more important than criterion Y
7	criterion X is strongly more important than criterion Y
9	criterion X is extremely more important than criterion Y
1/8, 1/6, 1/4, 1/2, 2, 4, 6, 8	intermediate values between two adjacent judgments

Please fill in the below blanks: What is the ratio of importance of criterion X to criterion Y?

Criterion X (range)	Criterion Y (range)	Answer (1/9 - 9)
economic criteria:		
Δrevenue requirements (-41 to 339 $M/yr)	Δcapital expend. (-86 to 220 million$/yr)	____
Δrev. requirements (-41 to 339 $M/yr)	Δrates (1993-99) (-3.3 to 10.8 $/MWh)	____
Δrev. requirements (-41 to 339 $M/yr)	Δrates (1993-2013) (-3.2 to 24.3 $/MWh)	____
Δcapital expenditure (-86-220 $M/yr)	Δrates (1993-99) (-3.3 to 10.8 $/MWh)	____
Δcapital expenditure (-86-220 $M/yr)	Δrates (1993-2013) (-3.2 to 24.3 $/MWh)	____
Δrates (1993-99) (-3.3 to 10.8 $/MWh)	Δrates (1993-2013) (-3.3 to 24.3 $/MWh)	____
air criteria:		
ΔSO_x (-28,564 to 17,876 tons/yr)	ΔCO_2 (-4.44 to 3.04 Mtons/yr)	____
ΔSO_x (-28,564 to 17,876 tons/yr)	ΔNO_x (-10,000 to 7,000 tons/yr)	____
ΔCO_2 (-4.44 to 3.04 Mtons/yr)	ΔNO_x (-10,000 to 7,000 tons/yr)	____

A matrix R of such ratios can then be set up as shown in Table 4.3 for a four-criteria example.

Table 4.3: Matrix *R* of importance ratios

Criteria	A	B	C	D
A	1	1/3	3	5
B	3	1	7	9
C	1/3	1/7	1	3
D	1/5	1/9	1/3	1

This table shows, for example, that criterion A is three times more important than C (and, thus, C is 1/3 as important as A). These ratios may not be completely consistent -- for example, A is 3 times as important as C, C is three times as important as D, but A is not 9 (=3 x 3) times as important as D.

Such inconsistencies are to be expected, since people's values, or their quantitative expression, are often inconsistent, contradictory or not carefully thought through. AHP solves for the set of weights that are, in some sense, most consistent with these ratios by a mathematical procedure called eigenvector analysis. With R being the above matrix of ratios, then W is the vector of weights that solves the equation

$$R\,W = \lambda\,W \qquad \text{Eq. [4.5]}$$

while simultaneously yielding the largest possible value of λ (which is called the eigenvalue of the matrix R).[13] In the above case, $W = \{0.24, 0.61, 0.10, 0.05\}$, and $\lambda = 4.09$. The closer λ is to the number of attributes (here four), the more consistent is the assessor. Additional numerical examples are given in Section 6.4.

For a problem with n attributes, $n(n-1)/2$ pairwise comparisons must be performed to determine weights of all n attributes. Thus, for a problem with 10 attributes, 45 pairwise comparisons are needed. Further, with so many pair-wise comparisons, there is a potential for large inconsistencies, which in turn means that the pairwise comparisons should be revisited. This could be taxing to the assessor. The use of a hierarchy helps reduce this burden. For example, if 10 attributes were broken up into one group of four and two others of three apiece, then six paired comparisons would be required in the first group of attributes, three each in the second and third groups of attributes, and three more to compare the three aggregate groups. This brings the total comparisons to 15, as compared to 45 for the ungrouped case. In such hierarchical assessments, the desired weight of an attribute is obtained by multiplying its conditional weight assessed within its own group by the weight of the group itself. An example of this process is presented in Chapter 9.[14]

Because only two attributes are compared at a time, the AHP procedure may result in a more diverse set of weights than non-hierarchical rating -- which is confirmed by Figure 4.4. Yet AHP suffers the same problem as rating and most other direct methods; with no explicit reference to the ranges of the attributes, assessors may apply some notion of importance other than their willingness to trade off attributes. To combat this, we combined AHP ratio questions with swing weighting in the BC Gas analysis of Chapter 9; the latter method, as we explain next, attempts to ensure that users have ranges of attributes in mind when they choose weights.

4.2.5 Swing weights

A final variant of direct weighting is swing weighting (von Winterfeldt and Edwards, 1986). In this method, the user considers a hypothetical alternative where attributes are all at their worst value. The user is then asked "which of the attributes would you most prefer to 'swing' from its worst value up to its best value?" Then the user is asked which attribute he or she would swing up second, and so on. After ranking the attributes in this manner, the user is asked to give the most important attribute a weight of 100, and then assign weights to the other attributes in proportion to the importance of their ranges, while being sure that the most highly ranked attributes receive the highest weight. Many decision-analysts favor this method over other direct techniques because the user is asked to directly consider each attributes' ranges when ranking them, which is not overly difficult.

4.2.6 Indifference tradeoff weights

As we will explain in Section 4.2.8, weights should be consistent with tradeoffs people are willing to make among attributes. One way to ensure this is to ask assessors to make tradeoffs and then derive the implied weights. Eq. [4.6], below, is used to make this calculation: for an example, see Section 6.4. Thus the tradeoff method is an indirect weighting technique. Many decision-analysts (e.g., Keeney and Raiffa, 1976) prefer it because weights will not be contaminated by incorrect notions of attribute importance.

However, the method is difficult compared to direct weighting. In our experience, physicists and economists like it very much, but others struggle with it. People are used to rating items on a scale of 0 to 10, but stating how much of one valued item they are willing to sacrifice for a given amount of another is very difficult unless they have very definite preferences (or have often made such choices before). Consequently, most participants in our Seattle and BC Gas case studies preferred direct methods, although a majority also felt that tradeoff questions provided valuable consistency checks.

4.2.7 Gamble method

This weighting method is most commonly used in additive or multiplicative utility functions (Section 4.3.4). It chooses a weight for one attribute at a time. This is done in a manner similar in philosophy to the single attribute utility function approach of Section 4.1.2, in that the assessor is asked to compare a "sure thing" and "gamble." In this case, the assessor adjusts the probability in the gamble until he or she finds the two options to be equally attractive (unattractive). If the weight is to be determined for attribute A_i, then the options are defined as follows:

"Sure Thing": $A_i = A_i^*$; $A_h = A_{h*}$ for all h not equal to i.

(attribute i is at its best level, and the other attributes are at their least desirable levels.)

"Gamble": Chance p of all attributes $A_h = A_h^*$;
Chance $(1-p)$ of all attributes $A_h = A_{h*}$

(i.e., a gamble in which one outcome is the best possible alternative with all attributes at their most desirable levels, and the other is instead an alternative with all attributes at their worst values)

If $U_h(A_{*h}) = 0$ and $U_h(A_h^*) = 1$ for every attribute h, and if either additive or multiplicative utility functions are used (Section 4.3.4), then attribute i's weight w_i will be precisely p. A demonstration of this is provided in the illustrative application in Section 6.7.

4.2.8 Desirable properties of weights

No matter which method is used to choose weights, the resulting values must be ratio scaled. If attribute A_1 is twice as "important" as A_2, w_1 should be twice as large as w_2. Furthermore, weights should represent the relative importance of unit changes in their attribute value functions. If a decision-maker is indifferent between a change in $V_1(A_1)$ of 1 and a change in $V_2(A_2)$ of 1/3, then w_1 should equal 0.333 w_2. Only this definition of importance should be used.

Another way to put this is that weights should represent the rate at which the user is willing to trade off one criterion for another. This can be seen by imagining two alternatives *X* and *Y* that differ in just two attributes, say *h* and *i*. If a decision-maker finds *X* and *Y* equally preferable, then, if we are assuming an additive value function, the value of Eq.[4.1] for *X* should be the same as the value of Eq.[4.1] for *Y*. After some algebra, one may derive

$$w_h/w_i = -[V_i(A_{iX})-V_i(A_{iY})]/[V_h(A_{hX})-V_h(A_{hY})] \qquad \text{Eq.[4.6]}$$

That is, the ratio of the weights of two criteria should be inversely proportional to the rate at which the decision-maker is willing to trade them off. For instance, if NO_x gets three times the weight of CO_2, then an improvement of amount Δ in $V_{NOx}(NO_x)$ should be valued just as much as an increase of 3Δ in $V_{CO2}(CO_2)$.

The above example underscores the point that an attribute's weight is inextricably linked to its value function. One cannot meaningfully choose a weight unless the value function (or at least its upper and lower bounds) has been defined; otherwise, tradeoffs may be made which do not reflect the "true" preferences of the decision-maker. For example, one might assert, *a priori*, that "environment" is just as important as "economics." However, depending on the actual ranges of the environmental and economic attributes and how their value functions are defined, equal weights could imply that each ton of NO_x emissions from a fossil fuel power plant is worth \$200 -- or \$2,000,000![15] Many applications have made the mistake of setting weights without defining or presenting attributes and their value functions (see Hobbs and Rowe, 1979, p. 122 for a number of citations in the context of power plant siting). The resulting additive value functions are, at very best, misleading representations of the decision-makers' preferences, as there is no way to determine which tradeoffs they are actually willing to make.

Some methods for choosing weights are specifically constructed to produce valid weights -- that is, weights on a ratio scale representing the correct type of importance. Others are not. The former group of methods generally ask a decision-makers to trade off attributes (e.g., "how many dollars are you willing to sacrifice in system cost in order to decrease CO_2 emissions by 1 million tons?"), and then calculates weights that are consistent with the tradeoffs he is willing to make. An example is the indifference tradeoff method. In contrast, methods that do not assure valid weights usually directly ask for judgmental assessments of an attribute's importance. However "importance" is ambiguously defined and may have nothing to do with the decision-maker's willingness to trade it off against other attributes (Schoemaker, 1981; Goldstein and Mitzel, 1992; Belton and Gear, 1997). Direct weighting methods, such as rating, are methods of this type. Such methods make no effort to determine if weights are consistent with the tradeoffs decision-makers are willing to make.

However, differences in the theoretical validity of weighting methods do not matter if, in practice, different weighting methods yield the same weights and rankings of alternatives. If that were true, the most easily applied method could be used. Unhappily, a number of experiments have shown that different methods can choose different weights and, ultimately, yield different decisions (Hobbs, 1986; Weber and Borcherding, 1993). In general, observer-derived (multiple regression) weights differ strongly from those chosen by other methods. This is perhaps because, as mentioned above, decision-makers tend to focus erratically on just two or three of the attributes when making the holistic evaluations of alternatives that are required by regression (Shepard, 1964). For this reason, we cannot recommend the use of regression methods.[16]

In contrast, different direct methods, such as rating and point allocation, generally yield similar weights and evaluations of alternatives. But several experiments (e.g., Hobbs, 1986) have shown that weights and evaluations resulting from indifference tradeoff and direct assessment can be significantly different, not even agreeing on the rank order of the attributes' importance. This result is in part confirmed in the Centerior, Seattle and BC Gas case studies as well. By "significantly," we mean that choice of weighting method can make as much, or more, a difference in the results as which person chooses the weights. In Chapters 9 and 10 we explore these issues of validity and differences in decisions more thoroughly by statistically analyzing the results of several methods applied by a number of energy planners and stakeholders.

4.2.9 Weighting conclusions

Thus our recommendations on weighting methods are as follows. If decision-makers feel comfortable in answering tradeoff questions, and just one method is to be used, then indifference tradeoff weights are to be preferred. Tradeoff questions force decision- makers to think about what they value in very concrete terms; as a result, learning takes place and they feel that they know more about the problem.

However, trading off attributes is an unfamiliar task, and often involves conflicting, strongly held values. For instance, if users strongly wish to minimize both costs and emissions, they may find it hard to say how much they are willing to worsen one to gain an improvement in the other. As a result, the indifference tradeoff weights they choose may be unstable (Fischoff *et al.*, 1980). It is therefore important to check for consistency by asking more than the minimum number of questions needed to determine weights. Consistency checks also aid the learning process by encouraging decision-makers to reflect on their values and the problem.

On the other hand, our experiments with energy planners and stakeholders show that many (and perhaps even a majority) of decision-makers are distinctly uncomfortable with tradeoff questions. As a result of this discomfort, the answers may not be meaningful and the decision-makers have little confidence in the results. In that instance, weights might best be chosen by a direct method.

Despite the misgivings we have voiced above about direct weighting, we agree with Poyhonen (1998) that "there is hope in direct weighting." She argues that with the help of skilled facilitators, decision-makers can directly chose meaningful weights that are relatively free of the biases we explore in Chapters 8 - 10. We think that this is more likely if precautions are taken and direct methods that are less prone to bias are used.

Among the direct methods, hierarchical rating is certainly to be preferred over non-hierarchical rating, as our participants found the former easier and more logical than the latter. Further, we prefer swing weighting to other direct methods, because it at least confronts users with the ranges of the attributes, even if explicit tradeoffs are not made. A hybrid method, such as the swing-AHP technique presented in detail in Chapter 9, can combine the best features of different methods.

There are several precautions that we recommend be taken if direct methods are used. First, those who are choosing the weights should be informed about the necessary properties of weights (ratio scales and the correct type of importance). Second, when picking weights, decision-makers should firmly keep in mind the ranges of the attributes. For example, if system cost has a range of $0 to $10,000,000 and CO_2 emissions have a range of 1 to 20 million tons, benefits should receive roughly twice the weight as would be the case if the range of system cost was only $0 to $5,000,000. (We assume here that the value functions $V_i(A_i)$ are always on a 0 to 1 scale, with 0 being the value of the least-desired outcome and 1 the value of the most desired). Third, some tradeoff questions should be asked as checks upon the weights. For instance:

> "You stated that the weight for system cost is 75 and that the weight of CO_2 emissions should be 50. Based on the ranges of the attributes, this implies that you would prefer an alternative with a cost of $10,000,000 and 1 million tons of CO_2 emissions to one costing $20,000,000 and 50,000 tons of CO_2 emissions. Is this true?"

If time allows, we strongly recommend the use of two or more methods. Our experience in the case studies of Chapters 8 - 10 is that people like to view the problem from several points of view, and that the resulting contradictions in weights or option ranks are valued opportunities for learning.[17] Our BC Gas participants, for example, said that the resolution of differences was essential to the success of that study, because it made people feel more in control of the process and confident that the outcomes truly represented their values. We discuss this further in Chapters 5 and 9.

4.3 AMALGAMATION METHODS (STEP 9)

Amalgamation methods can be viewed as methods for approximating decision-makers' "indifference curves," which show which alternatives are equally preferred. For example, Figure 4.6 shows a set of hypothetical indifference curves for one decision-maker. Alternatives P1, P13 and P14 are equally preferred and therefore lie on the same indifference curve. So are P12, P9 and P8. Curves below and to the left represent more-preferred combinations of the two attributes (assuming that less of each is better). From this perspective, the problem faced in amalgamation is to measure a person's values (as crystallized by the indifference curves) and then to find the most-preferred option (the one that lies on the lowest curve) -- i.e., in the example if Figure 4.6, the options P1, P14 and P13, all of which happen to lie on the same indifference curve.[18]

Figure 4.6: Indifference curves

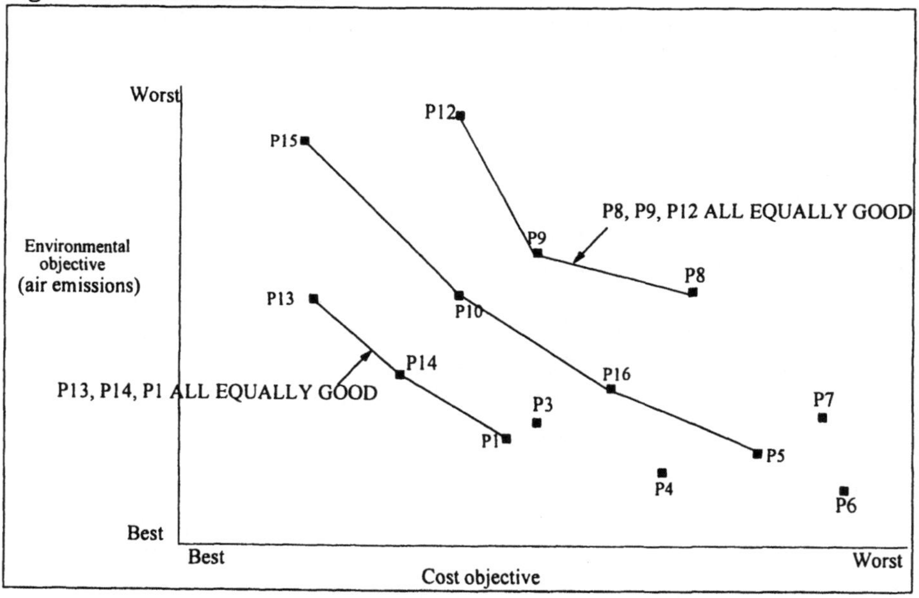

Six amalgamation methods are presented in this section. The first four (additive value functions, goal programming, the Power Law, and utility functions) yield a single index of alternative worth, which allows the options to be ranked. As explained in Section 4.3.5, these methods can also be used as a screening device, in which just the worst-ranking options are eliminated.

The fifth amalgamation approach does not calculate a single index, but constructs "outranking" relationships among options by comparing them pairwise. This can be a useful screening approach, because outranking can produce a subset of attractive but diverse options for further study.

The last method, successive elimination, is actually a refinement of other amalgamation methods that allows users to be imprecise about attribute values and weights. For example, one might state that "air quality is definitely more important than water quality, but I'm not sure how much so." Successive elimination uses such judgments to delete alternatives that are not better than other options under any weights or attribute values within ranges specified by the user.

Thus most amalgamation methods, in essence, estimate a value function which represents the decision-maker's priorities. The isoquants of such a value function are indifference curves, and the questions the method asks of decision-makers fix the slopes and shapes of those curves. Figure 4.7 shows the functional forms of indifference curves assumed by two amalgamation methods: additive value functions (in which the weighted sum of scaled attributes is maximized), and goal programming (in which the distance to an "ideal" alternative or set of goals is minimized).

4.3.1 Additive value function

We have already made a number of references to this popular amalgamation approach, several energy applications of which are described in Chapters 7 and 10. This ranks alternatives by the score shown in Eq. [4.1], which we repeat again here:

$$\text{Maximize } TV_j = \sum_{i=1}^{I} w_i V_i(A_{ij}) \qquad \text{Eq.[4.1]}$$

The resulting indifference curves are shown in Figure 4.7a.[19] This is unquestionably the amalgamation method in widest use, for two very good reasons: its simplicity, and the fact that its results tend to be robust with respect to mild non-linearities (Stewart, 1996).

Yet we have already pointed to the need to pay attention to how the weights and the value functions are derived in order for the results to be valid. Unhappily, as we shall see in our review of applications in Chapter 7, many users appear lacking in their understanding of the requirements: the problem appears to be that the concept and algebra of the method are so simple that the issues of validity seem esoteric.

But in fact they are not, as well documented in a number of studies. Two decades ago the US Nuclear Regulatory Commission examined the application of such site selection models to nuclear power plant siting (Rowe *et al.*, 1979): ten of thirteen applications of additive value functions violated theoretical requirements (such as avoiding use of ordinal scales for single attribute value functions). Further, eight of thirteen violated the requirement for preference independence of attributes, and *none* acknowledged the existence of *any* theoretical requirements for validity.

Figure 4.7: Indifference curves implied by two amalgamation methods

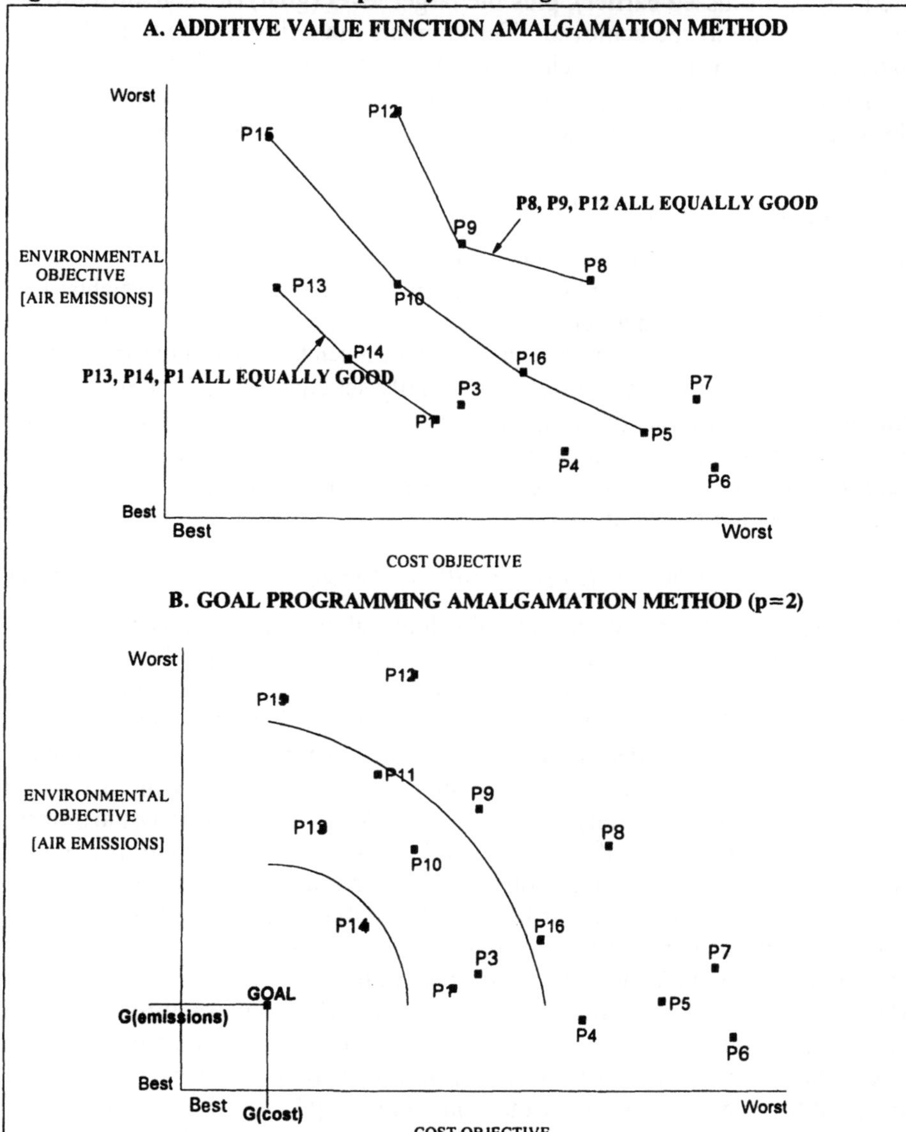

4.3.2 Goal Programming

Goal programming (GP) is another popular amalgamation method that yields a ranking of alternatives. One literature review cites 280 applications of the approach (White, 1990). It has been proposed for use in energy and resource planning in India (Ramanathan and Ganesh, 1995); for electric resource bidding systems (Stanton, 1990); and for dispatching energy systems (Chang and Fu, 1998; Hota *et al.*, 1999). The user provides weights that reflect the relative importance of each attribute (using an appropriate weight assessment method, as discussed earlier), plus the target level (goal) of each attribute the user would like to see achieved. These inputs make it possible to rank alternatives according to the weighted deviation from the goals: the smaller the deviation, the more preferred the alternative. The idea, therefore, is to choose the option j closest to the goals by minimizing a distance measure of the form:

$$\{\sum_{i=1}^{I} [w_i \ |G_i - V_i(A_{ij})|]^P \}^{(1/P)} \qquad \text{Eq.[4.7]}$$

where $|\ |$ = the absolute value operation

w_i = weight for attribute i (normalized so that the sum of the weights=1.0)

G_i = goal for attribute i, expressed here as the desired value of $V_i(A_i)$

P = a positive constant signifying how much large deviations from goals should be penalized. P is usually set equal to 1, 2, or ∞.

An alternative form has the weights w_i located outside the parentheses so that they are not raised to the power P. A third form, which participants in our Seattle study preferred (Chapter 10) because they objected to penalizing over-achievements of goals, is

$$\textit{Choose } j \textit{ to minimize } \{\sum_{i=1}^{I} [w_i \ Max(0, G_i - V_i(A_{ij}))]^P \}^{(1/P)} \qquad \text{Eq.[4.8]}$$

If P = 1, the result represents the absolute value of deviations form ("city block distance"), which penalizes each size of deviations equally. When P = 2, the squared deviations form ("Euclidean" distance) results. Here, a doubling of a deviation imposes four times the penalty. Figure 4.7b displays the implied difference curves in that case. Finally, when $P = \infty$, the "minimax" distance results, in which the option that minimizes the maximum (worst) weighted deviation from any goal is chosen. The value of P is not arbitrary, as it can affect the ranks of options. Guidelines for choosing appropriate values of P have been suggested (Hobbs, 1979; Karni and Werczberger, 1995).

The target level G_i of an attribute is often set at an ideal value or the best-possible value achievable by any alternative. For example, ideal goals for cost, hydropower MW, and SO_2 emissions might be set at, respectively, \$0, 0 MW, and 0 tons. Using these ideal goals, the only inputs required from the user are attribute weights, the value of P, and the ranges of the attributes (for determining the $V_i(A_{ij})$). As a result, goal programming can be viewed as a linear additive model with all value functions $-1*|G_i - V_i(A_{ij})|^P$. For example, when P = 1, goal programming using ideal goals is completely equivalent to an additive value function. Therefore, the former method is clearly subject to any assumptions, pros, and cons of the latter.

Likewise, if P = 2, Eq. [4.7] is equivalent to an additive model with $V_i(A_{ij})$ redefined as $-1*(G_i - V_i(A_{ij}))^2$ and weights w_i^2. The indifference curves will be a set of concentric ellipsoids centered at the goal point $(G_1,...,G_I)$, as shown in Figure 4.7b. Here, the further away an attribute is from its goal, the less of it one is willing to give up for an improvement in other attributes. Compared to the case P=1, such concave indifference curves favor options with balanced impacts over ones with extreme high and low impacts.

Finally, in the extreme case where $P = \infty$, Eq. [4.7] reduces to:

$$\textit{Choose } j \textit{ to minimize } [Max\{w_1(G_1 - V_1(A_{1j})), \ldots, w_n(G_n - V_n(A_{nj}))\}] \quad \text{Eq. [4.9]}$$

Here, the user is not willing to worsen the attribute furthest from its goal even for large increases in other attributes. Consequently, the solution will be very sensitive to the goals (see Section 6.6 for an example). With equal weights and a large number of options, each goal will usually be achieved to about the same degree. That is, for the best option j, $w_i(G_i - V_i(A_{ij})) \cong w_k(G_k - V_k(A_{kj}))$ for all attributes i and k.

In practice, the user selects targets G_i that are "acceptable" rather than "ideal". In this case, it is possible that the target levels will be chosen so that the option chosen is indeed a dominated alternative (Cohon, 1978). This is particularly likely if the goals are chosen in the absence of a complete knowledge of the set of choices and their impacts. The problem of dominated alternatives being chosen by goal programming is a well-known and serious drawback of the method.

4.3.3 Power Law

The Power Law is on occasion used to amalgamate criteria (e.g., Section 7.4.1), and is based on the following formula (Gum *et al.*, 1976):

$$\text{Total value of alternative } TV_j = \prod_{i=1}^{I} V_i(A_{ij})^{w_i} \quad \text{Eq.[4.10]}$$

The Power Law assumes logarithmic perceptions of relative importance of criteria; that is, people's expressed weights are in fact a logarithmic function of the actual importance. We discuss these assumptions more thoroughly in the context of the case study in Section 7.4.1.

There are relatively few applications of this method and we believe that Occam's Razor should be applied (Box 4.1): a method should be no more complicated than is necessary to accomplish its purpose. Because the Power Law is less familiar and more complex than additive value functions, and because little theoretical or empirical evidence exists that the Power Law is a better representation of people's values, we cannot argue for its use.[20]

Box 4.1: Occam's Razor

William of Occam (Ockham) was a medieval English philosopher (1290-1349) to whom is attributed the maxim (Occam's Razor) that "Entities are not to be multiplied without reason."

According to Betrand Russell's History of Western Philosophy (1945), this quote was never actually found in his published works. What he did in fact say is that "it is vain to do with more that which can be done with fewer," which has led to the modern decision science interpretation of this principle -- if two competing theories are equal in their ability to explain some phenomenon, then the simpler theory should be chosen as being more likely to be true.

Occam, a Franciscan friar, was excommunicated in 1328 not for his logic, but for his allegedly heretical views on transubstantiation -- though at the time the papacy was resident in Avignon, not Rome; and noted for its venality and corruption, not theological innovation.

4.3.4 Multiplicative and additive utility functions

As noted previously, single-attribute utility functions capture not only preferences regarding levels of the attributes, but also relative risk attitudes. Multi-attribute utility functions do the same for several attributes simultaneously, usually by weighting and amalgamating single-attribute utility functions. The expected value of the multi-attribute function is then used to rank the alternatives.

The multiplicative form is the most widely applied multi-attribute utility function (Keeney and Raiffa, 1976), and has found use in utility planning (Keeney and Nair, 1977; Keeney and Sicherman, 1983; Keeney and von Winterfeldt, 1987). It can be shown that a special case of the multiplicative form is the additive form, which happens when a user's preferences satisfy a particular type of risk neutrality. The remainder of this section describes in general how such a function is constructed. In Section 6.7 we present a simple illustrative example.

Two types of information are required from the decision-maker:

1. single-attribute utility functions $U_i(A_{ij})$ (defined in Section 4.1.2), and
2. a set of weights w_i chosen by a combination of the gamble and indifference tradeoff methods (Sections 4.2.6 and 4.2.7).

The utility functions $U_i(A_{ij})$ are scaled from 0 (worst) to 1 (best), and the weights each lie between 0 and 1. After they are chosen by one of the methods described above, the alternative that maximizes the expected value of one of the following multi-attribute utility functions is picked:

Additive form: $\textit{Maximize} \sum_{i=1}^{I} w_i \, U_i(A_{ij})$ Eq.[4.1]

Multiplicative form: $\textit{Maximize} \; \{[\prod_{i=1}^{I} (1+Kw_i \, U_i(A_{ij}))] - 1\}/K$ Eq.[4.11]

Which form is used depends on the sum of the weights w_i. If the sum is 1.0 (indicating multi-attribute risk neutrality), then the additive form applies; if not, then the multiplicative form is used (reflecting multi-attribute risk aversion or seeking). The scaling parameter K is set to the value that causes the value of the multiplicative form to lie between 0 (when $U_i(A_{ij})=0$ for all i) and 1 (when $U_i(A_{ij})=1$ for all i).

The assumptions made by both the additive and multiplicative utility functions are:

1. *Utility independence*: the answers to gamble questions involving just one attribute do not depend on the levels of the other attributes (i.e., each attribute's utility function $U_i(A_{ij})$ is not a function of the level of the other attributes); and
2. *Preference independence*: this holds if the tradeoffs one is willing to make among any two attributes do not depend on the levels of the other attributes (that is, the weights are constant -- see Section 2.1.1).

In general, numerical experiments show that it is rare that the multiplicative and additive forms differ in their rankings of alternatives (Hobbs, 1986). Usually, how the weights are chosen and, more importantly, who chooses the weights matters more. Therefore, in general, we suggest that the simpler additive form be used because of its transparency. Further, we recommend that the gamble methods of Sections 4.1.2 and 4.2.7 for including risk preferences be used only if 1) probability distributions are quantified for each alternative's performance on the attributes and 2) decision-makers are demonstrably risk averse. Otherwise, one may as well use simpler and more readily understood deterministic scaling and tradeoff weighting methods.

4.3.5 Using additive value functions, goal programming, Power Law, and utility functions for screening

As mentioned in Section 3.1, one use of amalgamation methods is to screen alternatives. Poorly performing options are rejected, yielding a subset to be subjected to more detailed study. But a problem with using a ranking method such as additive value functions to screen options, is that the top few alternatives may turn out to be very similar. This occurred when additive functions were applied by water planners in Hobbs *et al.* (1992). The top alternatives were either all dam-based or entirely excluded dams, depending on the weights chosen. Screening out the "worst" options resulted in a narrowing of the range of choice to a few options that were basically alike. This violates one of the criteria for an appropriate screening method, which is that it should not oversimplify the problem.

A means of preventing the range of choice from becoming too narrow is to systematically vary the attribute weights. For example, several weight sets could be chosen by different people that would emphasize different perspectives. Then, only the options that are poorly ranked by all weight sets would be screened out. Alternatives that are ranked highly by any one weight set would be retained. This procedure is more likely to result in a diverse set of alternatives that survive the screening. ELECTRE I is another approach which can accomplish this objective, as discussed in the next subsection.

Further, such a process may identify "robust" alternatives that perform well from a variety of perspectives. Identification of such options is desirable for two reasons. First, a robust alternative is more likely to be accepted by most or all interests because it is attractive under a variety of attribute weights. Second, when examining a robust alternative, affected interests may learn that it has advantages over other options that they initially favored.

4.3.6 Outranking Methods: ELECTRE I

Outranking methods involve comparisons of two alternatives at a time to determine if one is sufficiently more attractive than the other. There have been several energy applications of this approach. In this section, an early and simple example of an outranking method, ELECTRE I, is discussed. Other, more recent outranking methods include ELECTRE II, III, and IV (Pirlot, 1997, Hokkanen and Salminen, 1997); ORESTE (Pastijn and Leysen, 1989) and PROMETHEE (Brans and Mareschal, 1986; Briggs *et al.*, 1990).[21]

The purpose of ELECTRE I (henceforth simply called ELECTRE) is to establish a set of "outranking" relationships among alternatives (Duckstein and Opricovic, 1980). It can be used to screen options, and it has the advantage that the set of alternatives that passes the screen is likely to be diverse.

The basic procedure of ELECTRE is as follows. An option X is found to be superior to another option Y ("X outranks Y") only if the following two conditions hold:

1. The concordance index $C(X,Y)$ exceeds a threshold p. $C(X,Y)$ is defined as the sum of the weights in which X is better than Y. The concordance index is normalized to a range of 0 and 1 by having the sum of the weights equal one.
2. The discordance index $D(X,Y)$ is zero. $D(X,Y)$ is the number of attributes i in which Y is better than X by an amount greater than a tolerable threshold q_i.[22] This "pain" threshold can be different for different i.

The decision-maker chooses the p and q_i thresholds. An interpretation of the above conditions is that a solid majority of attributes must "vote" for X over Y, without too strong a "dissent" from any of the minority attributes that instead favor Y.

The output of ELECTRE is a set of "outranking" relationships showing which options "outrank" which. These relationships are used for screening by searching for alternatives that are not outranked by any other option. These options belong to the "kernel."[23] Often, there are several such options, none outranking the other (see the numerical example in Section 6.8). The reason is that each of these alternatives excels in different attributes. The philosophy of ELECTRE is that each such alternative has unique virtues (for example, provision of energy supply at low cost or low environmental impact) that makes it difficult to compare it with other alternatives. Such options should not be screened out because a diversity of choice should be preserved for presentation to the those responsible for the final decision. Alternatives that are outranked can be dropped since they have no important advantages compared to the alternatives in the kernel.

A disadvantage of ELECTRE is that the p and q_i thresholds are essentially arbitrary, although Rogers and Bruen (1998) suggest some guidelines for their selection. One way these parameters are used in practice is that the decision-makers manipulate them in order to screen out as many alternatives as possible. Hobbs *et al.* (1992) found that many US Army Corps of Engineers planners were confused by this process, while some others did not mind it. An advantage is that using it for screening may yield a subset of desirable yet diverse alternatives; this appealed to some participants in the Corps experiment.

ELECTRE II, which distills a complete rank order of alternatives from outranking relationships, has been used to rank energy supply plans under uncertainty (Samouilidis *et al.*, 1988). However, its use under multiple scenarios is complex, and we would not recommend its use in that case except in a sensitivity analysis mode (i.e., see if the outranking relationships change under different scenarios). Other, yet more advanced versions of ELECTRE have also been developed. But we entirely agree with Stewart's (1992) assessment that the advanced versions introduce "further *ad hoc* functional forms, which although intuitively appealing, are difficult to verify empirically as models of human preferences." We further agree with his statement that ELECTRE I "is particularly valuable as a descriptive device when the number of alternatives remaining under consideration is small . . . The outranking graphs help (users) focus attention on critical issues. It can also assist in understanding how and why other MCDM methods generate the answers they do."

4.3.7 Successive elimination

Recognizing that it is difficult for decision-makers to specify precise value judgments, some multi-objective techniques, called iterative methods, ask the user to make some tentative value judgments and then show the choice that would result. The user then considers the tentative decision, and if unsatisfied, modifies his or her judgments. The process is repeated until the user is satisfied (Cohon, 1978). Climaco *et al.* (1995) presents an application of this type of method to power generation planning.

Here we consider an iterative approach called "successive elimination" (Sarin 1977a; Sarin 1977b; Kirkwood and Sarin, 1985). It can be used to formally evaluate the robustness of decisions relative to attribute levels and weights. Some procedures of this type are also called "fuzzy programming." Goicoechea *et al.* (1992) evaluate one such method (ARIADNE, White *et al.*, 1984), in an experiment involving water planners similar to the comparison in Chapter 10. The planners found that it was the most well-liked of the methods tested. Hamalainen *et al.* (1999) recommend the use of a related approach called "preference programming" to identify and resolve value disagreements among stakeholders that affect the decision.

One version of successive elimination deals only with uncertain attribute levels (Sarin, 1977a). It is appropriate for alternatives for which data is uncertain and/or value functions have not been specified. Weights must be identified beforehand. The stages of the method are:

1. Characterize attribute uncertainty for each option j by stating the range of possible levels of each attribute i for that option; let the worst value in this range be called A_{ij^*}, and define A_{ij}^* as the best value in that range.
2. Calculate upper and lower bounds for the total value of each option using an amalgamation method (we recommend additive value functions) using the worst and best levels, respectively. For alternative j, these bounds are defined as TV_{j^*} (which results from the A_{ij^*}) and TV_j^* (resulting from A_{ij}^*), respectively.
3. Eliminate any option j whose upper bound TV_j^* is below the lower bound TV_{k^*} of the utility of any other option k. If too few alternatives are eliminated, return to Stage 1 and tighten the bounds, if possible, and repeat.

This approach is useful for analyzing the robustness of alternatives relative to uncertainty in the levels of the attributes.

A second version of successive elimination in addition considers robustness relative to weights. Its stages are as follows:

1. State upper and lower bounds on weights or, better yet, ratios of weights. For instance, attribute 1 may be between two and three times as important as attribute 2, implying that $2w_2 < w_1 < 3w_2$. Usually, the sum of the weights is constrained to equal 1.
2. State upper and lower bounds for each attribute for each alternative, as in the first version of successive elimination just discussed.
3. For each possible pair of options j and k, calculate the following index:

$$D_{jk} = \underset{\{w_i\}}{Maximize}\ (TV_j^* - TV_{k*})$$

subject to the stated bounds on weights, the constraint that the sum of weights must equal 1, and definitional constraints:

$$TV_j^* = \sum_i w_i V_i(A_{ij}^*)$$

$$TV_{k*} = \sum_i w_i V_i(A_{ik*}), \text{ for all options } k \text{ other than } i$$

The latter two constraints assume that the amalgamation method is an additive value function; other amalgamation forms can also be used. The symbol "Maximize over $\{w_i\}$" means that an optimization problem is solved that chooses the weights w_i in order to maximize $TV_j^* - TV_{k*}$. Note that TV_j^* uses *j's* best attribute values A_{ij}^*, while TV_{k*} uses *k's* worst, A_{ik*}. The optimization problem will be a linear program if an additive value function is used as the amalgamation method; a non-linear program results if a non-linear amalgamation method, such as the Power Law or a multiplicative utility function, is applied instead.[24]
4. Likewise, solve for D_{kj} in a similar manner.
5. Screen out any alternatives j for which D_{jk} is negative for any k. A negative D_{jk} indicates that under the best possible circumstances for j, it is still inferior to k; i.e., TV_j^* (the best possible value for j) is lower than TV_{k*} (the worst possible value for k) under all feasible weight sets. If too few options are eliminated by this method, the above stages could be repeated after tightening the bounds on the weights and attribute values.

The alternatives surviving this process would be robust in terms of both the weights and the uncertainties regarding each attribute. A simple numerical example is presented in Section 6.9.

Although successive elimination seemingly takes more time than a "one-shot" application of an amalgamation method, in many cases it will actually take less effort. Consider Table 4.4, which summarizes a multi-objective transmission routing analysis conducted for the US Bureau of Land Management (Sicherman, 1979). Clearly, alternatives II, IIIb, and IVb drop out, as they are dominated by I. Among the remaining corridors, I is best in attributes A_1 and A_3, while III and IV are better in A_2. For reasonable ranges of weights, successive elimination would drop all but I in the first iteration. The elaborate utility function-based analysis of these alternatives that was performed by Sicherman (1979) was unnecessary.

The advantages of successive elimination are obvious; users do not have to be more precise than they feel comfortable being, and the method can be used both for choosing one best option and for screening out the least desirable alternatives. Further, as will be discussed in Section 5.2, the approach can be used to attempt to identify consensus recommendations among different people. Two disadvantages are also evident; computational intensiveness and a level of sophistication that may be more than is digestible by some stakeholder groups. But if time and computational resources are available, the method is worth considering.

Table 4.4: US Bureau of Land Management transmission corridor alternatives

corridor	A_1 Bighorn sheep habitat crossed	A_2 Aesthetics (miles within 1/4 mi. of highway)	A_3 Differential cost
I	13.5	15.20	0
II	34.5	51.25	792
III	52.5	13.85	3,312
IIIb	27.8	18.27	3,702
IV	57.0	12.81	1,405
IVb	32.3	16.2	1,795

Source: Sicherman (1979)

4.3.8 Amalgamation conclusions

Choosing a multicriteria method is itself a multicriteria problem. Considerations involved include appropriateness, ease of use, whether different methods yield different results, and validity. Experimental evidence shows that method results can diverge.[25]

We tend to prefer simpler methods because of their clarity, unless more complex ones are demonstrably more valid. However, such demonstrations are rare. Even better, as recommended in the next Chapter (and by others, e.g., Salminen *et al.*, 1998), is to apply more than one amalgamation method, and then use any differences in alternative ranks as an opportunity for reflection and learning.

NOTES

[1] The only important exception is utility functions, which, when correctly assessed, are not interval scaled in terms of value: see Dyer and Sarin (1979). Utility functions are discussed below.

[2] What was particularly noteworthy in both of the case studies was the relative ease with which consensus was attained. As each attribute was discussed in turn, initially there was often wide disagreement about the shape of the curve, where thresholds might lie, etc. However, after several rounds of letting each participant express his or her views, with one or two exceptions where one or more individuals insisted on a dissent, a consensus was reached.

[3] *E{Z}* is defined as the expected value or mean of the variable *Z*.

[4] Strictly speaking, this procedure is only appropriate if either 1) the probability distributions of the different attributes for a given option are statistically independent and the multiplicative multi-attribute utility function is used, or 2) the additive multi-attribute utility function is applied. In practice, one or another of these assumptions is usually made. In the more general case (e.g., correlated attributes together with a multiplicative utility function), the expectation calculation should be made *after* amalgamation takes place, not with respect to individual attributes before amalgamation.

[5] If we applied portfolio analysis (Section 2.1.2) to X and Y, we would find that the standard deviation and, thus, the risk of the costlier option Y (which can be calculated as $1,961/yr) is considerably less than X's ($4,561/yr). Thus, there is a tradeoff between expected benefit and risk; therefore, it should not be surprising that a sufficiently risk-averse utility function (such as that in Table 4.1) would find Y to be more attractive.

[6] For other reviews of the weighting methods presented here, see Keeney and Raiffa (1976), Eckenrode (1965), and von Winterfeldt and Edwards (1986).

[7] We should note that there are many variations upon the direct weighting theme, all of which have avid proponents. For instance, Barron and Barrett (1996) describe no less than four different ways to calculate ranking weights. However, we believe that how direct weighting is done matters less than the difference between direct weighting and other general philosophies of weighting, such as tradeoff weighting.

[8] E.g., see von Winterfeldt and Edwards (1986) or Leon (1997). However, there is some evidence that point allocation may result in "peakier" weights (i.e., weights with a greater variance) than rating (Doyle *et al.*, 1997).

[9] Before making the calculations, all weights were normalized by rescaling them so that their sum is 1.0.

[10] In the Seattle case study (Chapter 10), tradeoff weights were also very peaked.

[11] Actually, the full AHP procedure includes steps 1,3,7,8 and 9 of our MCDM process (see Saaty, 1980 for details). AHP represents a philosophy that differs in important ways from traditional decision analysis. AHP's popularity is due mainly to the simplicity of its underlying concepts and its flexibility in problem structuring which allows the user to start from scratch and build up to his or her own decision model in an easy-to-understand fashion. However, the AHP process, especially its scaling procedures, has been criticized for violating a basic principle of rationality: that introduction of an irrelevant alternative should not alter the ranks of the other options (Dyer, 1990). For a recent summary of this and other issues concerning AHP, see Salo and Hamalainen (1997b).

[12] In part because of its popularity, the AHP has often been criticized. The 1 - 9 scale of Table 4.2 has been one target of critics, and several alternative scales have been proposed (e.g., Poyhonen *et al.*, 1997).

[13] *W* is unique up to a scaling constant. Usually, the weights are rescaled so that their sum is 1.

[14] Although using hierarchies reduces the effort somewhat, the assessment task required in AHP can remain burdensome. Proposed ways to alleviate this shortcoming include a user-friendly interface and decision support system (Forman *et al.*, 1996), and the use of incomplete pairwise comparisons (Harker, 1987).

[15] Consider a decision problem in which cost and CO_2 emissions are the only attributes. Let $A_{i\bullet}$ be the worst value of an attribute *i*, and A_i^* be its best, such that $V(A_{i\bullet}) = 0$ and $V(A_i^*) = 1$. If cost and CO_2 are given equal weight, then the following attribute ranges imply that the decision-maker is willing to pay \$5/ton for CO_2 reductions:

$A_{cost\bullet}$ = \$10,000,000/year; A_{cost}^* = \$9,500,000/year;

$A_{CO2\bullet}$ = 500,000 tons/year; A_{CO2}^* = 400,000 tons/year.

That is, equal weights imply that a change of 100,000 tons is worth \$500,000. On the other hand, if the attributes are weighted equally but:

$A_{CO2\bullet}$ = 5000 tons/year; A_{CO2}^* = 4000 tons/year,

then 1000 tons is worth \$500,000, and the implied value is \$500/ton.

[16] Except perhaps as a means of understanding which attributes are most salient in a person's unaided evaluation process. However, we believe that MCDM methods should be used to improve judgments, not imitate them, so weights derived by regression methods should not be used to rank alternatives and make recommendations.

[17] In several experiments, Poyhonen (1998) documents biases that can result from different weighting methods. She concludes that "one way to avoid these problems is to develop more interactive weighting procedures. Different weighting methods should be combined and decision-makers should interactively evaluate the results during the weighting" (*ibid.*, p. 2).

[18] However, this is subject to the caveat made in the introduction: very often people do *not* know what they want, and their views may evolve during a planning evaluation as a result of more information, better understanding, and, most importantly, interaction with others who hold possibly different views. Decision analysis is properly viewed as a procedure for people to develop a coherent set of values rather than as a way to extract pre-existing indifference curves (Payne *et al.*, 1999).

[19] These linear indifference curves assume linear $V_i(A_i)$; note, however, that non-linear $V_i(A_i)$ result in non-linear indifference curves!

[20] In fact, by taking the logarithm of the Power Law, it becomes an additive value function. Thus, if a Power Law accurately represents the preferences of a user, there also exists an additive value function that is an equally valid representation. The $V_i(A_i)$ in that AVF would then be a positive linear transformation of the logarithms of the $V_i(A_i)$ used in the Power Law.

[21] There have been energy planning applications of these methods (e.g., Beccali *et al.*, 1998). For example, ELECTRE III was used to compare eight power sector plans under 15 criteria for the Island of Crete (Georgopoulou *et al.*, 1997). The possibility of using that same method for Hydro Quebec's IRP process was explored by d'Avignon and Saubageau (1996), who considered 18 criteria, including nine social and environmental impacts. Briggs *et al.* (1990) applied PROMETHEE to a nuclear waste management problem, which was also analyzed by Delhaye *et al.*(1991) using ORESTE. Tzeng *et al.* (1992) adopted a combination AHP-PROMETHEE II method to evaluate alternatives for energy-system development in

Taiwan. The criteria were designed to meet technological, social, environmental and economic needs. Fourteen experts from Taiwanese energy agencies were invited to evaluate the alternatives, employing AHP to derive weights. PROMETHEE II was then applied, yielding a multicriteria preference index by which the alternatives were ranked.

[22] An alternative definition of discordance allows X to outrank Y only if $w_i(V_i(A_{iY})-V_i(A_{iX}))$ is no greater than a single threshold q for each and every attribute i. Separate q_i are not defined for the attributes in that case. However, participants in the experiment in Hobbs *et al.* (1992) found the version in the text to be easier to understand.

[23] Technically speaking, the kernel can also include alternatives that outrank each other, creating a "cycle." If no alternatives outside the cycle outrank any of the alternatives in the cycle, then all those options are included in the kernel.

[24] If goal programming is adopted instead, then the MAX would be replaced by the minimization operator, since the overall objective is to minimize distance from the goals. Goal programming under $P = 1$ or infinity can be phrased as linear programs; however, for $P = 2$, a non-linear program results.

[25] See the Seattle case study for an example (Chapter 10), or the review in Hobbs (1986). Salminen *et al.* (1998) give a recent example of the divergence of ELECTRE-type methods and additive value functions in an environmental context.

5 RESOLVING DIFFERENCES (Step 10)

If anyone accuses me of contradicting myself, I reply: Because I have been wrong once, or oftener, I do not aspire to be always wrong
VAUVENARGUES, Reflections and Maxims (1746)

Our agenda is now exhausted. The secretary general is exhausted. All of you are exhausted. I find it comforting that, beginning with our very first day, we find ourselves in such complete unanimity
PAUL HENRI SPAAK, Concluding the First General Assembly of the United Nations (1946)

We know of few, if any, applications of MCDM methods that yield completely consistent results. If more than one method has been used in an application, their results generally diverge. Even if the methods are consistent, it is unlikely that all participants will arrive at the same rankings. Indeed, if they do achieve perfect consistency, the group undoubtedly has been poorly selected. We argued earlier that groups should encompass diverse views, including those of NGOs and likely intervenors in subsequent regulatory procedures. In such groups one should *expect* divergence of results. Finally, a participant applying the same method at the beginning and at the end of the process is very unlikely to arrive at the same result: as noted at the outset, one important purpose of the use of MCDM is to educate, to expose members of a diverse group to each other's views, and to force people to really think about the issues and problems. The more successfully that objective is attained, the more likely it is that a participant will change some of his initial views.

In this chapter, we examine issues involved in resolving disagreements--disagreements among methods that arise when two or more MCDM methods are applied (Section 5.1) and disagreements among individuals (Section 5.2). We believe that differences between methods are to be welcomed as opportunities for gaining insight into one's own priorities and their effects on decisions. We also believe that the purpose of MCDM methods is not to force agreement upon different parties, but to help understand the reasons for disagreement as a starting-point for negotiation. It is sometimes proposed that conflicts in rankings be resolved by methods that mathematically define a single, compromise rank order. Cook *et al.* (1995) summarize several such approaches. However, we believe that the only useful consensus is one that participants assent to after careful thought or negotiation.

5.1 RESOLVING DIFFERENCES AMONG METHODS

Imagine that you have purchased an expensive painting. Before hanging it, you protect it by covering it with a piece of glass. Would it not be disturbing if the glass made more of a difference in what you see than what painting is behind it?

Analogously, you are interested in understanding the preferences of stakeholders concerning energy options. Would it not disturb you if in some comparisons of MCDM methods, *which* method is used to elicit value judgments makes as much or more of a difference in the rankings of options as *who* applies the method? In practice, methods do tend to be consistent for many and sometimes most participants, but for other participants, correlations between results of different techniques are close to zero or even negative. Should it not worry us that choice of method can significantly affect the result?[1]

If different methods yield significantly different results -- obscuring the values lying behind the "glass" of the method -- then we would worry about their relative validity. But if the results converge, then validity differences are less relevant. Considerations such as ease of use can then be used to choose a technique.

5.1.1 Examples of disagreements among methods

For instance, consider some results from our Centerior Energy resource planning study. (Chapters 9 and 10 present additional examples of divergence of method results.) Figure 5.1 shows that the rank order of planning options varied by method. The figure displays the average rank across the 12 participants for each option and for four different methods (lowest rank order = best, highest rank order = worst). It may be seen that some alternatives, such as *d* and *l*, method matters little. Whatever method is used, for example, alternative d received the highest (worst) rank.

Figure 5.1: Average rank by different methods, Centerior case study

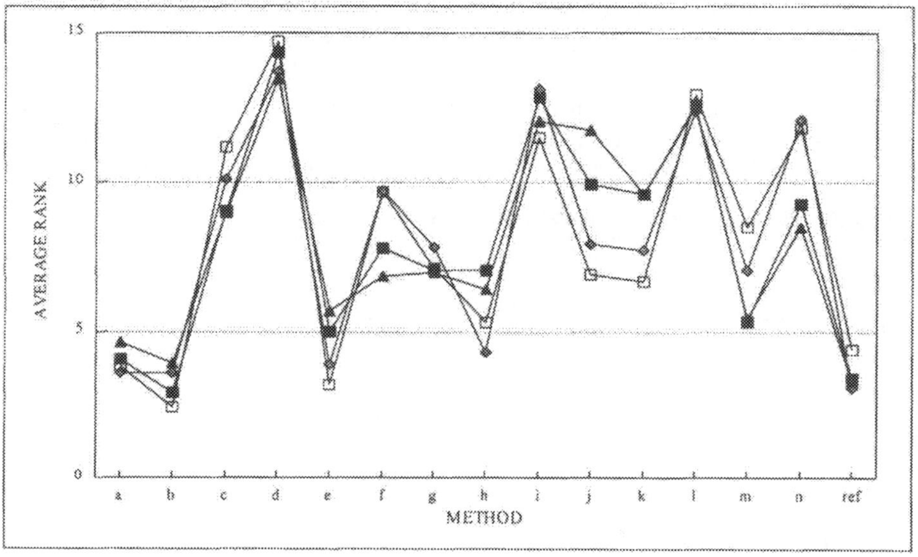

However, for other options -- such as *f, j, k*, and *m* -- the application of different methods results in wide differences in rank. Indeed, the degree of variation among alternatives is not constant. In Figure 5.2 we display these results in a different way, plotting the standard deviation against the average rank. The variation for the best performing plans (*a, ref* and *b*), and for the worst plans (*i, l* and *d*) is much smaller than for those in the middle (only *g*, among the mid-ranked plans, has an unusually low variability).

Figure 5.2: Variability of method differences, Centerior case study

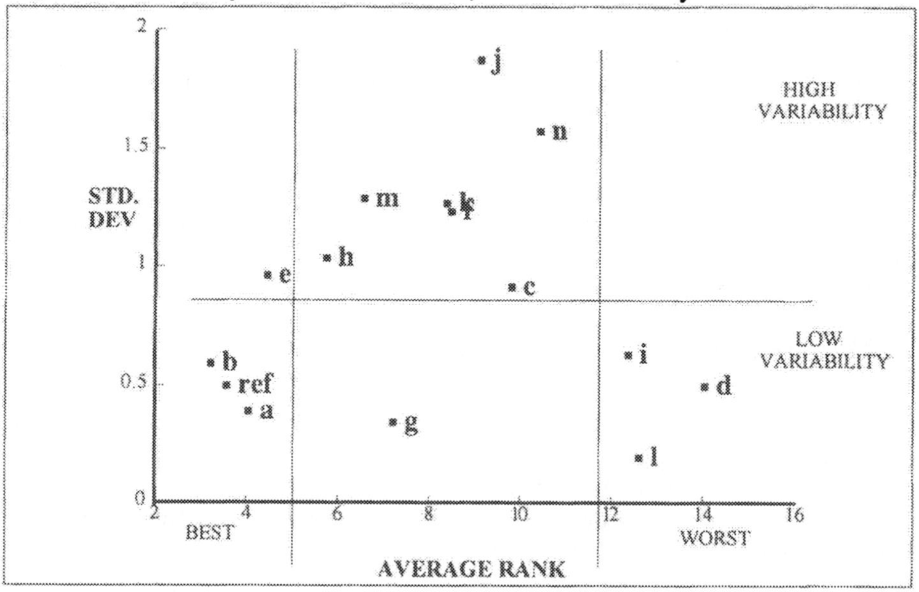

Whenever several methods are used, correlations between rankings of alternatives by the same person can be useful indicators of the extent of intermethod inconsistency. In Table 5.1 we show the between-method correlation coefficients for the Centerior case study.[2] One of the most striking features of these results is the very low correlation between goal programming (GP) methods and additive value functions (AVF), except GP(1,-). For instance, AVF versus GP(5,-) or GP(5,+/-) correlations range from 0.24 to 0.69.

Table 5.1: Between-method correlations, Centerior case study

	holistic	rating H-1 (AVF)	rating H-2 (AVF)	rating NH-1 (AVF)	rating NH-2 (AVF)	swing (AVF)	AHP (AVF)	GP (1,-)	GP (1,+/-)	GP (5,-)	GP (5,+/-)
1 holistic	1	0.58	0.78	0.73	0.58	0.58	0.58	0.76	0.62	0.65	0.4
2 rating H-1		1	*not calculated*		0.82	0.79	0.8	0.73	0.36	0.52	0.26
3 rating H-2			1	0.88	*n.c*	0.83	0.96	0.89	0.64	0.69	0.32
4 rating NH-1				1	*n.c*	0.86	0.91	0.82	0.63	0.62	0.33
5 rating NH-2					1	0.93	0.72	0.63	0.32	0.43	0.28
6 swing						1	0.81	0.71	0.48	0.49	0.28
7 AHP							1	0.78	0.57	0.6	0.38
8 GP(1,-)								1	0.83	0.88	0.59
9 GP(1,+/-)									1	0.8	0.75
10 GP(5,-)										1	0.73
11 GP(5,+/-)											1

Key to rows:

1. Holistic: unaided ranking of methods.
2. Rating H-1: additive value function (AVF) using the first set of hierarchically assessed rating weights chosen at the beginning of the workshop.
3. Rating H-2: AVF using the second set of hierarchical rating weights, chosen at the end of the workshop.
4. Rating NH-1: AVF using the first set of non-hierarchical rating weights.
5. Rating NH-2: AVF using the second set of non-hierarchical rating weights.
6. Swing: AVF with swing weights.
7. AHP: AVF with AHP weights.
8. GP(1,-) : Goal programming with P=1, only bad deviations penalized (Eq. [4.8]).
9. GP(1,+/-): Goal programming with P=1, both good and bad deviations penalized (Eq. [4.7]).
10. GP(5,-): Same as GP(1,-), but with P=5 (which is similar to P=∞, Eq. [4.9]).
11. GP(5,+/-): Same as GP(1,+/-), but with P=5.

At the same time, there are also several methods that give similar results: AHP and the hierarchical rating methods are strongly correlated (0.8 to 0.96): this is a reflection of the fact that both are characterized by relatively "peaky" weights (recall the earlier discussion in Chapter 4).

5.1.2 Resolving differences among methods for specific individuals

The reason why different methods applied by the same person give conflicting conclusions is that people may have general priorities in mind, but have difficulty quantifying those preferences precisely, especially when decisions are unfamiliar and involve important but conflicting values. This vagueness or uncertainty means that the specific value judgments they articulate will be highly "labile" (Fischoff *et al.*, 1980). That is, they depend upon arbitrary aspects of question phrasing and context. People don't know exactly what they want, and as a result, how a method asks them to evaluate options can greatly influence the results. This is, of course, no news to readers of public opinion polls who know that how a question is phrased can swing the results (e.g., "abortion" may be opposed, but not a woman's "right to choose").

Recognizing this reality, we believe that discrepancies between methods are actually a welcome opportunity for reflection and learning (Corner and Buchanan, 1997). Decision analysis should help people gain insight by confronting them with difficult choices and giving them an opportunity to reflect and make their values more coherent. Using more than one method, and then mulling over the differences in results, can help people to think about the tradeoffs in different ways and build confidence in their final judgments (Corner and Buchanan, 1997; Poyhonen, 1998). There is evidence that as a consequence, the results will be more valid representations of people's values (Borcherding *et al.*, 1991; Corner and Buchanan, 1997).

Follow-up interviews are a good way of resolving differences in method results. Sometimes, significant discrepancies prove to be the result of some misunderstanding about how a particular question should be answered. In others, reflection and evolution of thought is the reason for differences, as the *process* brings new information and perspectives to an individual's attention. We conducted follow-up interviews with participants in several case studies, and they proved to be extremely useful.[3] Chapters 9 and 10 summarize some of these interview processes and their results.

It turns out that some individuals find it much easier than others to give quantitative expression to weights. For example, in Figure 5.3 we show the criterion weights expressed by two individuals in the Centerior case study for a variety of different methods. Individual Y expressed weights that are highly consistent, whereas individual X expressed weights that vary significantly across different methods. This does not mean that Y's weights are "better"; it simply means that the choice of method matters more for some individuals than for others.

The outcome of each person's interview should be a set of recommendations that he or she feels comfortable presenting as a starting position for negotiations with others involved in the decision process. The results might include a set of weights and an amalgamation rule that the person feels best represents their preferences. But the evaluations he or she makes of the alternatives do not have to be consistent with those weights or rule. Indeed, if the assessors have thought hard about the choices and alternatives, they are likely to come up with reasons why they do not fully agree with the results of a model. For instance, they might disagree with how the performance of an attribute has been characterized, or they might argue that some important attribute has been omitted.[4] Explicit statements of the recommendations and the reasons for them will be helpful in subsequent discussions with others involved in the decision -- the subject of the next subsection.

Even for a consistent individual (such as Y in Figure 5.3), there are some inconsistencies: for example, the variation in weights for the job loss and emergency power attributes is significantly greater than for the other attributes. This is the sort of point to be clarified in follow-up interviews.

Figure 5.3: Weights by Individuals

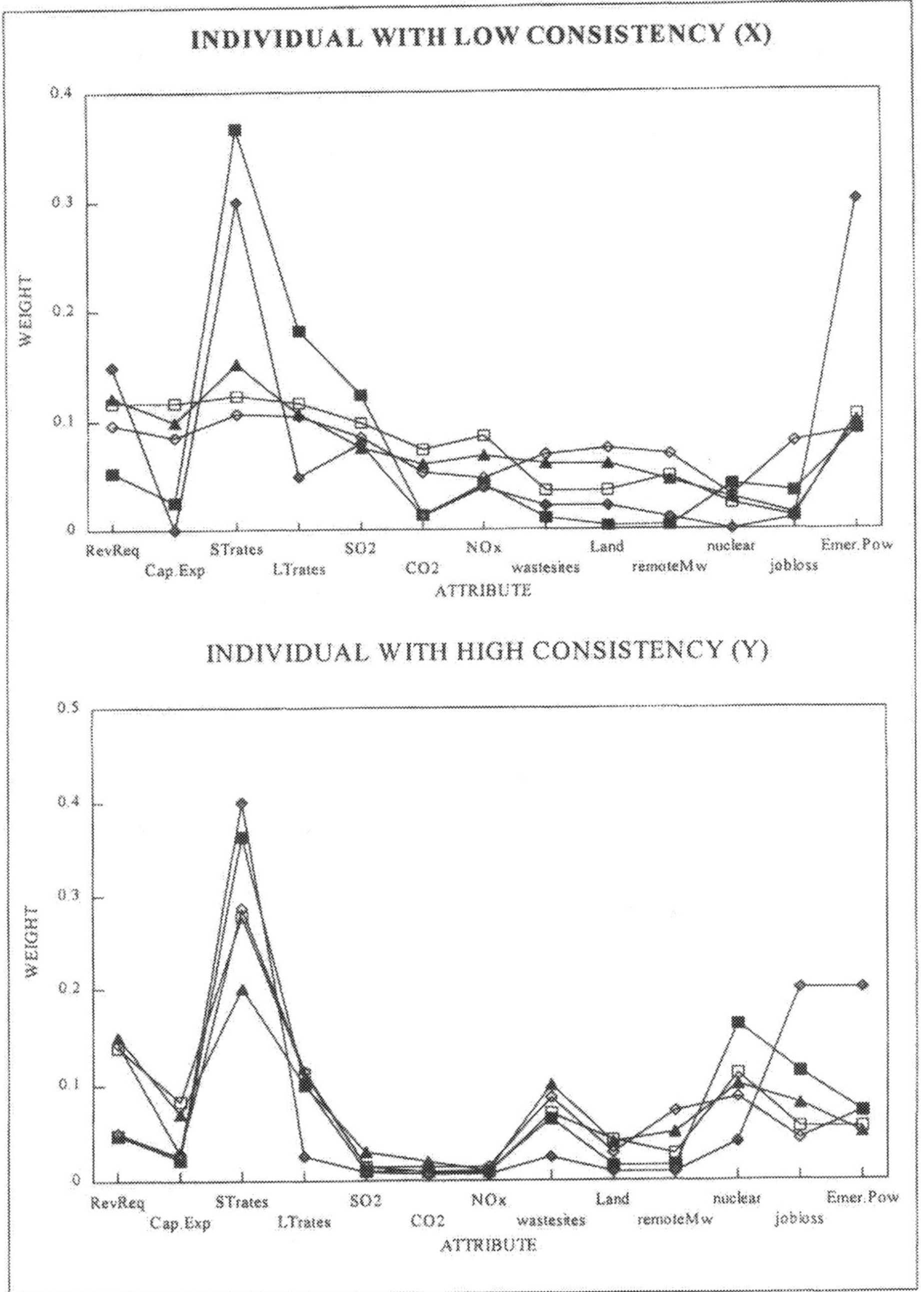

Note: Each set of connected points represents the weights obtained by a particular method (rating, AHP, swing, etc.)

However, the focus of the interview should not be on resolving every discrepancy. Differences in weights and other value judgments matter only if they change the decision.[5] For example, in our BC Gas interviews (Chapter 9), we reviewed with each individual the evaluations of alternatives that resulted from their various weight sets. If no important differences arose in the ultimate recommendations, we did not ask the person to resolve the differences among weights. Only where the recommendations were affected did we explore the reasons for the differences. This was done by examining the quantitative responses people made to each method, and then pointing out how they were inconsistent. For instance, rating weights of 30 and 50 might have been assigned to air quality and cost, respectively, implying a value of \$2000/ton of NO_x. However, when applying the indifference tradeoff method, a value of \$4000/ton of NO_x might have been stated, implying weights of 60 and 50 for the two attributes respectively. The assessors would then be asked to consider those responses; is one more consistent with their preferences than the other? Or does an entirely different response reflect their priorities now?

We found that the process of examining and resolving such conflicts yields additional insight and more confidence in the results. Although applying more than one method and resolving inconsistencies takes more time, we strongly believe that it is worth the additional effort (approximately two hours more per person). Indeed, the participants in the BC Gas study (Chapter 9) reported that the use of more than one method followed by interviews was essential to the study's success.

5.2 RESOLVING DIFFERENCES AMONG INDIVIDUALS

This is a crucial issue because of the importance that is often assigned to achieving consensus among the diverse parties typically involved in energy planning and policy matters. Yet at the same time, one should expect significant differences between individuals in weights, because the weights are expressions of their values: consensus may simply not be achievable. In that case, genuine differences of opinion should not be papered over.

Consider the case of utility resource planning. It is the upper management of the utility or the regulatory commission that decides which tradeoffs are to be made, and what modifications, if any, must be made to the utility's proposal for the future resource mix. If the stakeholder groups that are often involved in resource planning come to a consensus about the resource mix, and if this consensus is consistent with the views of the utility, so much the better. But in many cases, the broad diversity of stakeholder interests reflected in such consultative groups makes consensus very difficult to achieve. Nevertheless, the *process* of attempting to reach a consensus is what is fundamentally important, for understanding *why* consensus is difficult to find provides valuable information to all parties about the values that underlie different viewpoints. What is important is not necessarily that a consensus be reached, but that all viewpoints be aired, and that parties understand each other's points of view. Often, the search for consensus will establish at least some common ground.

Figure 5.4: Average weights by individual

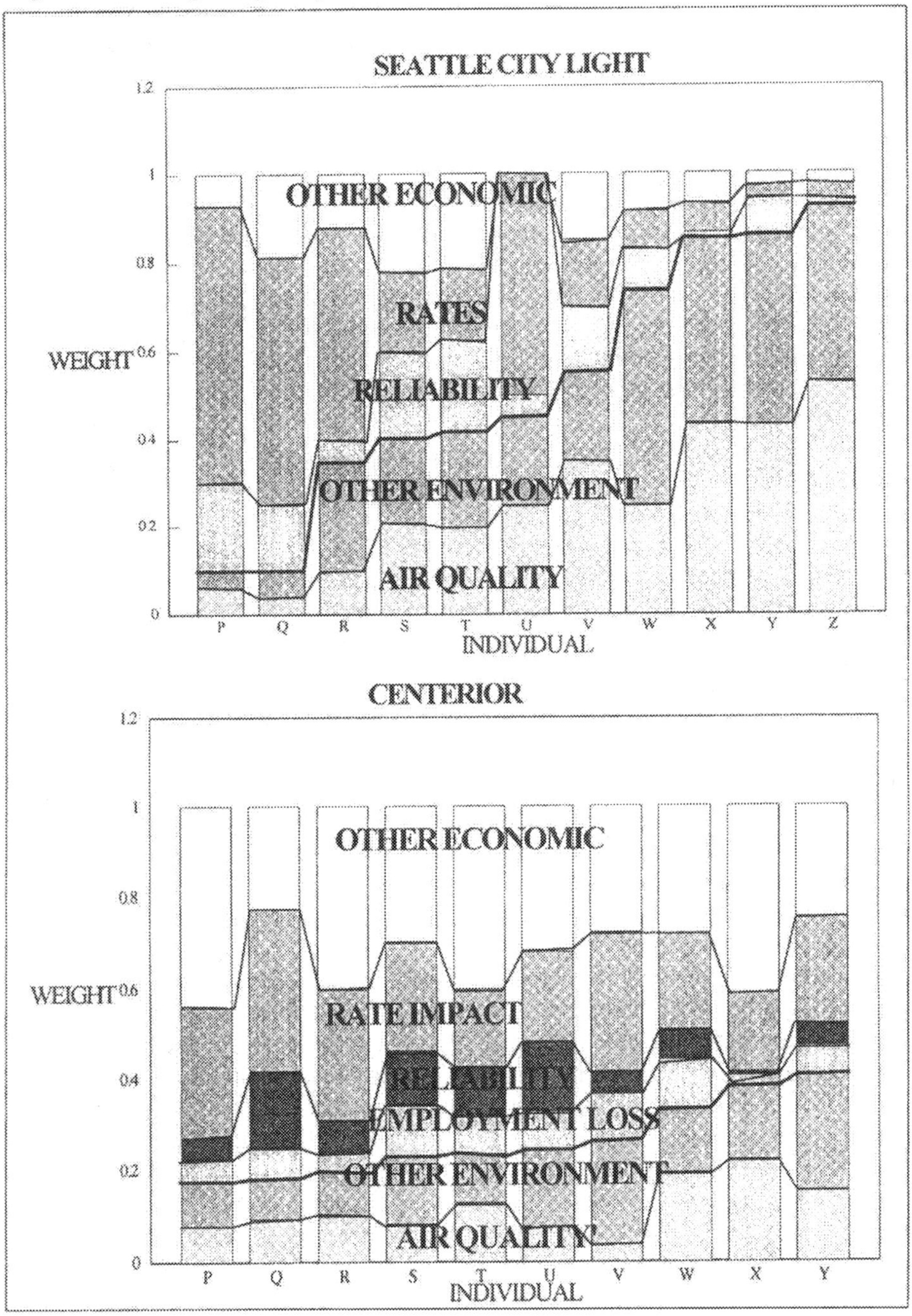

In Figure 5.4 we show the average weights given to the different attributes for selected participants in the Centerior and Seattle case studies. In each case we have sorted the individuals from left to right by the total weight given to environmental factors: in Seattle, this ranges from a minimum of 10% to a maximum of 92%, whereas at Centerior, the range is from a minimum of 18 % to a maximum of 40%. The much greater variation in the Seattle case is simply a reflection of the greater diversity of individuals, which included representatives from a business group and an NGO; meanwhile, no non-utility individuals were present at Centerior. Thus the latter was clearly a more homogenous group in terms of its values. These differences in priorities translated into differences regarding the ranks of alternatives. As shown in the case studies of Chapters 9 and 10, average correlations between alternative rank orders by different individuals were as low as 0.18, depending on the method applied.[6]

However, disagreements in weights do not necessarily imply conflict over which options are best. Brown (1984, 1990), for example, describes two situations where interest groups chose very distinct weights, yet largely agreed on the best option.[7] Whether or not this occurs depends upon the shape of the tradeoff curves (see Section 3.2). If there is a pronounced knee set, and particularly if there are only a few alternatives (or even only one) in that knee set, then such agreement is more likely.

Clearly, where disagreements arise, they are not a matter for resolution by analysts. No amount of massaging of results will change such fundamental differences. Instead, what can be accomplished is two things:

1. *Identify agreement.* Are there plans that everyone (or nearly so) agrees should be in, say, the top five? Are there plans that are clearly undesirable to everyone? Even when weights differ, some alternatives may be high on everyone's lists. Successive elimination (Section 4.3.7) is a mathematical process that can be used to identify what choices are consistent with a range of value judgments supplied by stakeholders (e.g., Hamalainen *et al.*, 1999).

2. *Identify disagreement.* Are there alternatives that some people highly ranked, but others didn't? Because value judgments are explicit in MCDM, it then becomes possible to explore the exact reasons why disagreement occurs. When such disagreements arise because of misunderstanding of alternatives (as occurred in the case study of Chapter 9), or where they can be addressed by redesign of alternatives, further discussion can make agreement more likely. Such discussions can often be made more productive and efficient through structural interaction methods, such as the Nominal Group Technique we apply in Chapter 9. Yet, as that chapter illustrates, some disagreements will usually remain, which can only be resolved by later regulatory or political processes.

Examples of this type of analysis are given in later chapters; Andrews (1992a) and Marttunen and Hamalainen (1995) present others.

Areas of agreement and disagreement can be identified by comparing the rank orders chosen by different people, as in Brown (1984, 1990). Are there any options that are ranked in the top, say, three for each and every person? Are there other alternatives that are not ranked first, second, or third by any person? If there is little consensus, can the positions of the stakeholders be boiled down to two or three different perspectives, as in the BC Hydro case study of Chapter 8?

This type of analysis can be systematized. For instance, the successive elimination method of Section 4.3.7 could be used to delete options that are not found best under any set of weights between those selected by the stakeholders.[8] Hamalainen *et al.* (1999) suggest this approach in an ecological management context. If there is no agreement, cluster analysis or factor analysis could be used to define a few distinct sets of weights or rank orders of options that summarize the positions of the various stakeholders (as in Brown, 1990).

However, we must caution the reader: MCDM methods are not a panacea for dealing with conflict among stakeholders. No one process fits every problem (Yosie and Herbst, 1998). MCDM techniques are just one set of tools that can be useful in some circumstances. First of all, the decision problem must be one for which consensus-based approaches are appropriate (Raab, 1994). There are many situations in which this is not the case: for instance, if the managers have sufficient information and public acceptance is not critical to effective implementation.[9]

Second, if a consensus-based approach is adopted, competent facilitators must be engaged who possess excellent communication and negotiation skills. Their overriding objectives are to help participants to view the process as fair and one in which everyone has something to gain (Susskind and Cruikshank, 1987). Two tasks that are critical to achieving these objectives are for participants to clearly understand both their goals and those of the other parties, and for participants' goals to help guide the creation of new alternatives that can better achieve those goals. MCDM methods can assist in both of these tasks (Gregory and Keeney, 1994; Keeney, 1992; Maguire and Boiney, 1994; Keeney and Raiffa, 1991). However, for MCDM to be useful, the group must have sufficient time to provide thoughtful responses and to digest the results. Further, the facilitator must understand the methods and their limitations so that the approach can be adapted, iterated, or even thrown out as the process evolves and the participants' needs change. We hope that the explanations and applications of MCDM methods provided in this book will help practitioners gain that understanding.

NOTES

1 Some of this literature is summarized in Chapters 9 and 10. Those chapters present two experiments often showing large disagreements among methods. Some of the differences between methods represent predictable biases. For instance, non-hierarchical rating tends to yield "flatter" or more uniform weights than hierarchical weighting. As another example, indifference tradeoff weighting often leads to a higher weight for the attribute than the assessor is asked to vary (e.g., if one is asked to state an amount of money that one is willing to sacrifice for a fixed improvement in some other attribute).

2 These correlations are calculated as follows. For each participant, the Spearman correlation is calculated between the two sets of alternative ranks (16 options each), resulting from two different methods (such as additive value functions with rating weights versus goal programming, P=1). The table's entries are the averages of the correlations across the participants.

3 The questionnaires used in these follow-up studies are contained as an Annex in Hobbs and Meier (1994b).

4 Chapter 9 gives examples of such rationales.

5 Successive elimination (Section 4.3.7) provides a formal approach for determining what decisions are consistent with all sets of weights or other value judgments within a given range.

6 In Chapter 10 we note that similar disagreements appear in numerous other MCDM studies.

7 In these cases, the US Bureau of Reclamation successfully used rating and weighting methods to solicit public input on water project design. Although interest groups chose greatly different weights in both studies, the top-ranked alternatives were the same for all groups.

8 This can be done as follows. Let w_{is} be the value of weight i chosen by stakeholder s. Let λ_s be a variable to be chosen by the optimization model that represents how much emphasis is to be placed on stakeholders's values. (This λ_s is not to be confused with the eigenvalue λ used by the Analytic Hierarchy Process.) The following constraints would then be added to the math programs of Section 4.3.7:

$$w_i - \Sigma_s \lambda_s w_{is} = 0 \text{ for all attributes } i$$

$$\Sigma_s \lambda_s = 1$$

$$\lambda_s \geq 0 \text{ for all stakeholders } s$$

9 The Vroom-Yetton (1973) model is a popular approach for determining what type of public involvement approach, if any, is appropriate. The recommended approach is determined by the answer to a series of dichotomous questions (e.g., "Does the manager have sufficient information" and "Is conflict likely within the relevant public?"). Daniels *et al.* (1996) discuss how the Vroom-Yetton model can be used in environmental management.

6 AN ILLUSTRATIVE NUMERICAL EXAMPLE

JACKIE (very slowly): "take tube A and apply to bracket D"
VICTORIA: "reading it slower does not make it easier to do"
VICTORIA WOOD, Mens Sana in Thingummy Doodah (1990)

"Why," said the Dodo, "the best way to explain it is to do it."
LEWIS CARROLL, Alice's Adventures in Wonderland (1865)

The case studies in the remainder of this book are detailed applications of the methods introduced in Chapters 2 through 5. However, those introductions skim over many of the computational procedures. Therefore, in this section we present detailed illustrations of several of the procedures through the vehicle of a simple problem.

Consider the following hypothetical energy supply problem for a rural area of a developing country. There are four alternative plans: A, B, C, D. Alternative A corresponds to no action being taken. There are three attributes: *cost* (as present worth, in $1,000,000); *wetlands lost* to construction (in acres); and *population served*. The performance of each alternative on each attribute is given in Table 6.1.

Table 6.1: Performance A_{ij} of four energy alternatives on three attributes

	attribute *i*		
alternative *j*	*cost*	*wetlands lost*	*population served*
A	0	0	0
B	20	100	15,000
C	30	400	25,000
D	25	120	15,000

6.1 EXCLUSIONARY SCREENING (STEP 4)

As an example of this method, which was presented in Section 3.1, assume that exclusionary cutoffs are set at a maximum of 300 acres for *wetlands lost* and a minimum of 10,000 *population served.* As Figure 6.1 shows, alternative A would then be unacceptable (because it would supply no one), as would plan C (as it would cause 400 acres of *wetlands* to be lost). Plans B and D survive the screen, even though D is dominated. As an example of an implied tradeoff, the cutoff of 10,000 people implies that serving 15,000 people (as the acceptable plan B does) is worth at least $20,000,000 and 100 acres of *wetlands* (B's incremental loss compared to A).

Figure 6.1: Exclusionary screening

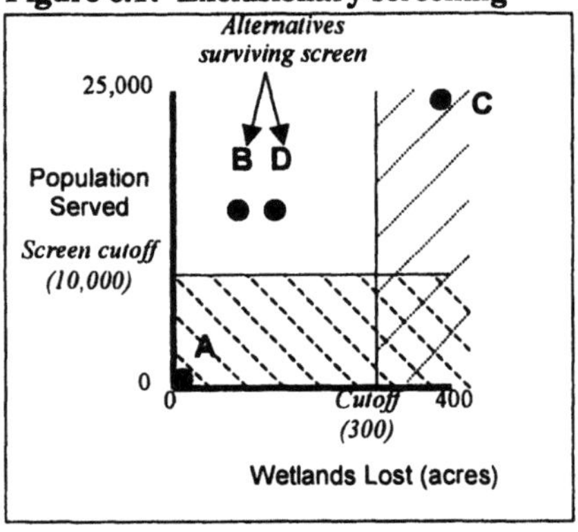

6.2 DOMINANCE ANALYSIS (STEP 6)

This method was described in Section 3.3. As an example of its application, refer to Table 6.1: it may be seen that alternative B dominates D, because B is at least as good in all attributes and is strictly better in *cost* and *wetlands*. Thus, D can be dropped from further consideration if the dominance principle is applied.

6.3 SCALING, WEIGHTING, AND AMALGAMATION (STEPS 7 - 9)

These processes are illustrated here for the case of additive value functions. The first task is to scale the three attributes -- that is, to define single attribute value functions $V_i(A_j)$ (as in Figure 4.1). For instance, one could judgmentally assign values between zero and one for each level of the attribute (as suggested in Section 4.1.1). Zero is assigned to the worst level, and one to the best level. Other levels get values in between. For example, in this case one might specify:

cost:	$V_{cost}(0) = 1$	$V_{cost}(20) = 0.2$	$V_{cost}(30) = 0$
wetlands:	$V_{wet}(0) = 1$	$V_{wet}(100) = 0.3$	$V_{wet}(400) = 0$
population served:	$V_{pop}(0) = 0$	$V_{pop}(15000) = 0.5$	$V_{pop}(25000) = 1$

To determine values of the value functions for other levels of the attributes, linear interpolation could be used, as in Figure 6.2. Alternatively, one could solve for the values of the parameters *a*, *b*, and *c* in the exponential function

$$V_i(A_i) = a + be^{cA_i}$$

that result in the function passing through each of the three points.[1] In the case of V_{pop}, this procedure would yield:

$$V_{pop}(pop) = -0.78394 + 0.78394e^{0.3289pop}$$

Figure 6.2: Piecewise linear single attribute value functions

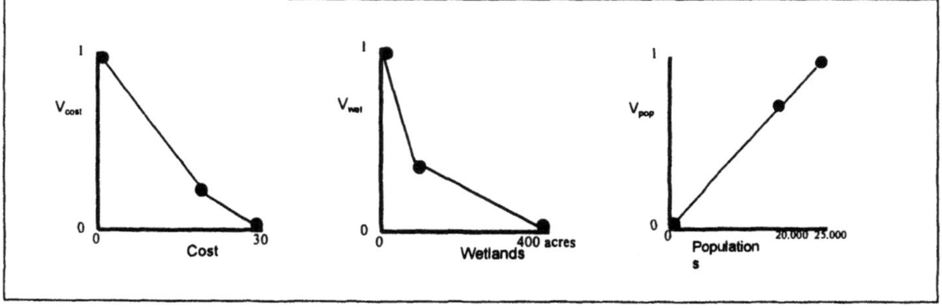

The next step is to assign weights. We consider here the case of rating (Section 4.2.3); other methods are applied below. When rating, one might directly assign the following weights to the three attributes in the problem here: $w_{cost} = 40$, $w_{wet} = 20$; $w_{pop} = 100$. Often it is convenient to divide each weight by the sum of the weights to achieve "normalized" weights: $w_{cost} = 0.25$; $w_{wet} = 0.125$; and $w_{pop} = 0.625$. This operation does not affect the ranks of plans.

Since an additive value function is being used (Section 4.3.1), the third step is to weight and sum the three value functions. This yields the following worths of the first three alternatives: $TV_A = 0.375$; $TV_B = 0.4$; and $TV_C = 0.625$. For instance, TV_B was calculated as follows:

$$TV_B = 0.25*0.2 + 0.125*0.3 + 0.625*0.5 = 0.4$$

C is "best", because it has the highest value. If option D were also examined, its value would be $TV_D = 0.3725$ if exponential single attribute value functions are used. Note that this is less than B's value; this is necessarily so, since B dominates D.

6.4 INDIFFERENCE TRADEOFF WEIGHTING (STEP 8)

Now consider how we can determine weights by instead asking assessors for tradeoffs they are willing to make (Section 4.2.6). The attributes are traded off two at a time. First we consider *cost* and *wetlands*. A user can trade these off by finding two alternatives that are equally attractive but differ in those two attributes. Say that it is decided that the following two hypothetical alternatives are equivalent:

Option F: *cost* = 30, *wetlands* = 0, *population* = 25,000
Option G: *cost* = 20, *wetlands* = 400, *population* = 25,000.

Thus, the assessor is willing to let 400 acres of *wetlands* be destroyed if that would save 10 units of *cost*. Figure 6.3 portrays this graphically. Then *wetlands* and *population* are traded off. For instance, let it be decided that the following two hypothetical alternatives are equally desirable:

Option G: *cost* = 20, *wetlands* = 400, *population* = 25,000
Option H: *cost* = 20, *wetlands* = 0, *population* = 15,000.

Figure 6.3: Indifference tradeoff between alternatives F and G

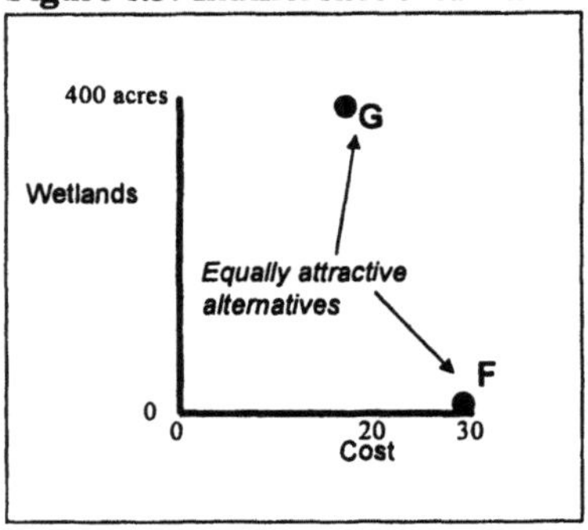

Therefore, a loss of 400 acres of *wetlands* is found to be as undesirable as a decrease of 10,000 *population served*. Based on the value functions chosen in Section 6.3, these tradeoff judgments imply that the weights are: w_{cost} = 0.625; w_{wet} = 0.125; w_{pop} = 0.25. That is, those weights will make the values of F and G equal and the values of G and H equal. For instance, the ratio of w_{cost} to w_{wet} was obtained by setting the overall value of F equal to G in an additive value function:

$$TV_F = w_{cost}V_{cost}(30) + w_{wet}V_{wet}(0) \quad + w_{pop}V_{pop}(25{,}000)$$
$$= TV_G = w_{cost}V_{cost}(20) + w_{wet}V_{wet}(400) + w_{pop}V_{pop}(25{,}000)$$

Some simplification and rearrangement yields:

$$w_{cost}/w_{wet} = [V_{wet}(0) - V_{wet}(400)]/[V_{cost}(20) - V_{cost}(30)] = [1-0]/[0.2-0] = 5$$

which is a restatement of Eq. [4.5]. Thus, *cost*'s weight should be five times that of *wetlands*, because the assessor is only willing to give up a little bit of *cost* in order to improve *wetlands* from its worst value to its best value. A similar analysis of the alternatives G and H yields a ratio of 0.5 for w_{wet}/w_{pop}.

Given these two ratios, and the stipulation that the weights add to 1, the above weights (0.625, 0.125, 0.25) result. No other weights that sum to 1 will equate the values of these plans.

These weights are then used to evaluate the alternatives using the additive function. This gives $TV_A = 0.75$; $TV_B = 0.2875$; and $TV_C = 0.25$. A is now "best," whereas C was preferred under rating weights of Section 6.2. Chapters 9 and 10 show that in practice, as well as in this simple example, rating and indifference tradeoff assessments can yield different weights and, ultimately, different decisions.

6.5 AHP WEIGHTING (STEP 8)

In the Analytic Hierarchy Process (AHP), the user chooses weights by comparing attributes or groups of attributes two at a time, assessing the ratios of their "importance" (Section 4.2.4). These ratios are then used to compute weights of individual attributes and a measure of consistency of the user's assessments. In the simple example, it is necessary to determine the weights of the three attributes: *cost* (w_{cost}), *wetlands lost* (w_{wet}), and *population supplied* (w_{pop}) by specifying the "relative importance" of one pair of attributes at a time. For example, after considering the ranges of feasible values of all attributes, suppose the assessor views *cost* as being 2 times as important as *wetlands lost* and 1/3 times as important as *population supplied*, while stating that *wetlands lost* is about 1/6 times as important as *population supplied*. This implies that $w_{cost}/w_{wet} = 2$, $w_{cost}/w_{pop} = 1/3$, and $w_{wet}/w_{pop} = 1/6$. The results of the assessment can be summarized in the form of the matrix R shown in Table 6.2.

Table 6.2: Summary of pairwise judgments: the consistent case

	cost	*wetlands*	*population*
cost	1	2	1/3
wetlands	1/2	1	1/6
population	3	6	1

Obviously, the diagonal part of the matrix assumes that an attribute is relatively as important as itself, and the lower triangular part of the matrix (below the diagonal) assumes that the relative importance reciprocates whenever the order of attributes is reversed. These two assumptions ensure some degree of consistency in judgments. A set of weights that is consistent with the above judgments is $w_{wet} = 0.111$, $w_{cost} = 0.222$, and $w_{pop} = 0.667$ (or any set of numbers satisfying the proportions w_{wet}: w_{cost}: w_{pop} = 1:2:6).

In practice, however, such tasks are made difficult because pairwise judgments are rarely completely consistent. Inconsistencies make it impossible to define a set of weights that will satisfy all judgments simultaneously. For example, in comparing *wetlands* and *population*, the user may assess the former to be 1/4 times (instead of 1/6) as important as the latter. The resulting matrix of assessment is shown in Table 6.3.

Table 6.3: Summary of pairwise judgments: the inconsistent case

	cost	*wetlands*	*population*
cost	1	2	1/3
wetlands	1/2	1	1/4
population	3	4	1

Solving Eq. [4.5] yields a set of weights that is, in one sense, most consistent with these judgments. The resulting normalized weights are 0.2385, 0.1365, and 0.635.

6.6 GOAL PROGRAMMING AMALGAMATION (STEP 9)

Goal programming was summarized in Section 4.3.2. For scaling purposes, goal programming usually assumes linear value functions. Thus (Figure 6.4):

cost:	$V_{cost}(0) = 1$	$V_{cost}(20) = 0.33$	$V_{cost}(30) = 0$
wetlands:	$V_{wet}(0) = 1$	$V_{wet}(100) = 0.75$	$V_{wet}(400) = 0$
population:	$V_{pop}(0) = 0$	$V_{pop}(15000) = 0.6$	$V_{pop}(25000) = 1$

Suppose the user sets goals of 5 for *cost*, 50 for *wetlands lost*, and 20,000 for *population* to be supplied. The linear value function assumption then results in $G_{cost} = 0.833$, $G_{wet} = 0.875$, and $G_{pop} = 0.8$ (Figure 6.4). Suppose also that the planner views the *cost* and *wetlands lost* attributes to be equally important and each is viewed as being a little more than one half as important as the *population* attribute. Accordingly, 0.26, 0.26, and 0.48 might be used as a set of normalized weights for the three attributes, respectively.

Figure 6.4: Single attribute value functions and goals used in goal programming

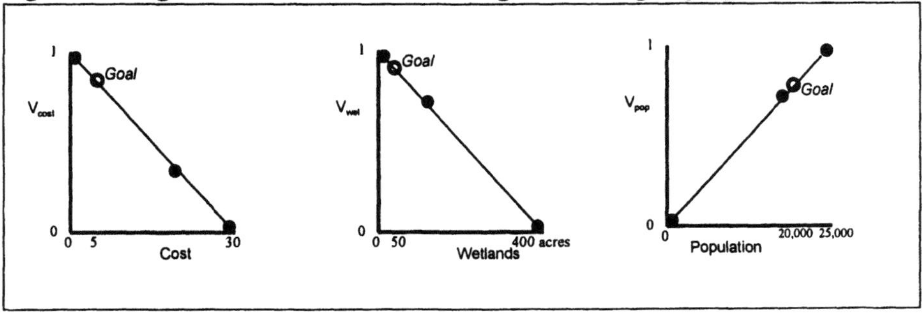

Therefore, the distance from goals for plan A is (Eq.[4.7]):

$$= [(0.26*|0.833-1|)^P + (0.26*|0.875-1|)^P + (0.48*|0.8-0|)^P]^{1/P}$$
$$= 0.460 \quad \text{when } P = 1$$
$$= 0.387 \quad \text{when } P = 2$$
$$= 0.384 \quad \text{when } P = \infty$$

Computing the distance from goals for all plans and each value of *P* yields the results shown in Table 6.4. The table shows that plan B is ranked best and plan D second best for all values of *P*; i.e., B has the lowest distance from the goals, while D is second closest. Plan A is ranked better than plan C for the "city block" and "Euclidean" distances, but worse for the "minimax" distance.

Table 6.4: Results of Goal Programming Calculations under $P = 1,2, \infty$

	distance metric		
	$P = 1$ "city block"	$P = 2$ "Euclidean"	$P = \infty$ "minimax"
A	0.460	0.387	max{0.043,0.032,0.384}=0.384
B	0.260	0.165	max{0.130,0.032,0.096}=0.130
C	0.540	0.329	max{0.217,0.228,0.096}=0.228
D	0.315	0.203	max{0.173,0.045,0.096}=0.173

6.7 SINGLE AND MULTI-ATTRIBUTE UTILITY FUNCTIONS (STEPS 7-9)

We now illustrate the application of scaling, weighting, and amalgamation methods that capture decision maker attitudes towards risk (Sections 4.1.2, 4.2.7, and 4.3.4). First, one determines the single attribute utility functions using gamble questions. For *cost*, we start by defining the utility of the extreme values as $U_{cost}(30) = 0$ (worst) and $U_{cost}(0) = 1$ (best). Then assume that it is decided that the following two hypothetical alternatives are equally desirable:

"Sure Thing": *cost* = 20
"Gamble": 0.5 chance of *cost* = 0; 0.5 chance of *cost* = 30

Because the expected utility of the gamble is 0.5 (equal to $0.5U_{cost}(0) + 0.5U_{cost}(30) = 0.5*1 + 0.5*0$), $U_{cost}(20)$ must also equal 0.5. We could then fit an exponential utility function to these three points, as we did with the deterministic scaling functions. Alternatively, we could obtain another point on the utility function by finding a value of *cost* which would be just as valuable as a 50:50 gamble between 20 and 30, whose expected utility would be 0.25. Additional points can be estimated in a similar manner.

Continuing to the other attributes, assume that for *wetlands* one is indifferent between:

"Sure Thing": *wetlands* = 100
"Gamble": 0.5 chance of *wetlands* = 0; 0.5 chance of *wetlands* = 400

implying that $U_{wet}(wetlands{=}100) = 0.5$. Similarly, for *population*, one might be indifferent between:

"Sure Thing": *population* = 15,000
"Gamble": 0.5 chance of *population* = 0; 0.5 chance of *population* = 25,000

implying that $U_{pop}(population{=}15{,}000) = 0.5$.

As a result of the above calculations, we have the following single attribute utilities for the options A, B, and C, based on the attribute values of Table 5.2.

Option A: $U_{cost}=1.0$, $U_{wet}=1.0$, $U_{pop}=0.0$
Option B: $U_{cost}=0.5$, $U_{wet}=0.5$, $U_{pop}=0.5$
Option C: $U_{cost}=0.0$, $U_{wet}=0.0$, $U_{pop}=1.0$

We can ignore D, since it is dominated by B.

Once the single attribute utility functions are chosen, then the weights must be determined. The first step is to rank the attributes in importance. This is done by determining which attribute is most important to move from its worst to best possible level. Say that the assessor determines that it is most important to improve *population* from 0 to 25,000. The next most important attribute to improve from worst to best is (from 30 to 0). The least important attribute is thus *wetlands*.

The particular values of the weights are determined by finding a probability w_i in a "Gamble" among the best and worst possible alternatives (having an overall utility of 1 and 0, respectively) that makes one indifferent between the gamble and a "Sure Thing" in which only one of the attributes i is at its best level. Say that for *population*, one decides that the following options are indifferent:

"Sure Thing": *cost* =30, *wetlands* = 400; *population* = 25,000
(only *population* is best, others are at their worst levels)

"Gamble": 0.6 chance of *cost*=0, *wetlands*=0, *population*=25,000 (all best);
0.4 chance of *cost*=30, *wetlands* = 400, *population* = 0 (all worst)

This implies that the weight for *population* is 0.6, which can be shown as follows. Assume that the multiplicative utility function Eq. [4.11] applies.[2] Then the utility of the sure thing is:

$$\{[1 + Kw_{cost}U_{cost}(30)][1 + Kw_{wet}U_{wet}(400)][1 + Kw_{pop}U_{pop}(25{,}000)]-1\}/K$$

which after substitution simplifies to just w_{pop}. Meanwhile, the expected utility of the gamble is:

$$0.6\{[1 + Kw_{cost}U_{cost}(0)][1 + Kw_{wet}U_{wet}(0)][1 + Kw_{pop}U_{pop}(25{,}000)]-1\}/K$$
$$+ 0.4\{[1 + Kw_{cost}U_{cost}(30)][1 + Kw_{wet}U_{wet}(400)][1 + Kw_{pop}U_{pop}(0)]-1\}/K$$

By definition, the utilities of the best and worst outcomes in the gamble are 1 and 0, respectively, so its expected utility is 0.6. Equating the expected utility of the sure thing and gamble results in $W_{pop} = 0.6$, as stated.

To obtain the other two weights, we could use the ratios resulting from the indifference tradeoff questions (Section 6.5), which would imply that $W_{cost} = 1.5$ and $W_{wet} = 0.3$. There must be an inconsistency here, however. This is because those results imply an infeasible probability of 1.5 (> 1) for the gamble that would be used to set *cost*'s weight. Such inconsistent results are common in utility assessments because of the difficulty people have handling probabilities. Here, we instead use the gamble method to directly assess the other weights. For *cost*, assume indifference between:

"Sure Thing": *cost*=0, *wetlands* = 400; *population* = 0
(only *cost* is best, others are at their worst levels)

"Gamble": 0.25 chance of *cost*=0, *wetlands*=0, *population*=25,000 (all best);
0.75 chance of *cost*=30, *wetlands* = 400, *population* = 0 (all worst)

implying that the weight for *cost* is 0.25. Finally, for *wetlands*, one is indifferent between

"Sure Thing": *cost*=30, *wetlands* = 0; *population* = 25,000
(only *wetlands* is best, others are at their worst levels)

"Gamble": 0.15 chance of *cost*=0, *wetlands*=0, *population*=25,000 (all best);
0.85 chance of *cost*=30, *wetlands* = 400, *population* = 0 (all worst)

implying that the weight for *wetlands* is 0.15. Using w_{pop}= 0.6, then the three weights sum to 1.0, implying that the additive utility function is appropriate. If instead the weights summed to a number other than 1.0, then the multiplicative form should instead be used.[3]

With these weights and single attribute utility functions, one finds that the resulting utility of each alternative is: *U(A)* = 0.4; *U(B)* = 0.5; *U(C)* = 0.6. Thus, C is viewed as best.

To illustrate the computation of expected utility when attribute levels are random, assume for the moment that plan B has a 50% chance of a *cost* of 20 and an identical chance of a *cost* of 25. Assume further that if 20 is realized, then B's overall utility is 0.5, as just mentioned, but if instead 25 occurs, its utility is instead 0.45. Then the expected utility of B would be 0.5*(0.5) + 0.5*(0.45), or 0.475, and this value would then be compared with similarly calculated expected utilities for the other options.

6.8 AMALGAMATION BY ELECTRE (STEP 9)

We apply ELECTRE (Section 4.3.6) to the three alternatives A, B, and C. Assume that a value of 0.501 is given for the "voting" ("concordance") parameter p. This is equivalent to requiring that an alternative X win a "majority vote" of the attributes in order to outrank another option Y. Assume that the following rating weights have been assigned to each attribute: w_{cost} = 50, w_{wet} = 25, and w_{pop} = 90, which when normalized yield w_{cost} = 0.303, w_{wet} = 0.152, and w_{pop} = 0.545. Further, say that the maximum difference ("discordance") that will be tolerated for *cost* is q_{cost} = 25; for *population*, no more than a difference of q_{pop} = 8000 will be permitted. However, *wetlands* are relatively unimportant, and no difference in level of *wetlands* would result in the veto of an alternative -- so the discordance threshold q_{wet} for that attribute is set equal to the range of the entire attribute (400 acres).

Executing ELECTRE, it is found that the concordance indices *C(k,j)* between each possible pair of alternatives *k, j* are:

		Alternative *j*		
		A	B	C
	A	----	0.455	0.455
Alternative *k*	B	0.545	----	0.455
	C	0.545	0.545	----

Recall that *C(k,j)* is the "vote" on the proposition that *k* outranks *j*; it represents the sum of the normalized weights for the attributes in which *k* is better than *j*. For instance, A is superior to B in *cost* and *wetlands*, whose weights sum to 0.455.

We now turn to the discordance indices. For each comparison of plans, the number of attributes *D(k,j)* that violate the maximum tolerable amount that alternative *j* can be better than *k* are:

		Alternative *j*		
		A	B	C
	A		1	1
Alternative *k*	B	0		1
	C	1	0	

For example, the lower left entry of 1 shows that A is better than C by more than the tolerable amount in one attribute, *cost*; that is enough to veto the proposition that C outranks A. The following outranking relationships result from the above analysis:

		Alternative *j*		
		A	B	C
	A		0	0
Alternative *k*	B	1		0
	C	0	1	

where a "1" in the *k*-th row and *j*-th column indicates that *k* outranks *j*. These relationships can be shown graphically (Figure 6.5).

Figure 6.5: ELECTRE outranking relationships for a simple example

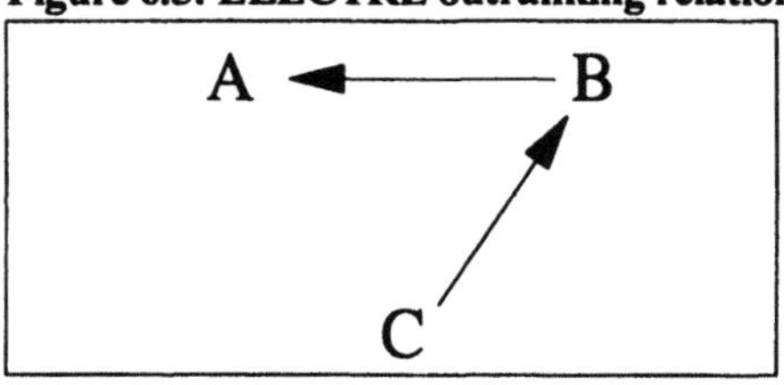

An option *k* outranks *j* if $C(k,j) > p = 0.501$ (*k* wins the "vote") and $D(k,j)=0$ (there is no strong "dissent" to the proposition that *k* outranks *j*). This analysis shows that C is the only option that is not outranked by any other option, and that it outranks B (as the above two tables show that $C(\text{C,B}) = 0.545 > 0.501$, and $D(\text{C,B})=0$). Alternative B in turn outranks A. Hence, C appears to be best, but A has advantages that make it difficult to compare to C.

6.9 SUCCESSIVE ELIMINATION (STEP 9)

As an illustration of this method, the second version presented in Section 4.3.7 is applied to just alternatives A, B, and C. The values of the attribute value functions are assumed to be known with perfect precision, and are equal to the values assumed in Section 6.3. Say that the constraints on the imprecisely known weights are:

$$2w_{wet} \leq w_{cost} \leq 3w_{wet}$$

$$0.5w_{pop} \leq w_{wet} \leq w_{pop}$$

$$w_{cost} + w_{wet} + w_{pop} = 1$$

Then for each pair of alternatives, two linear programs are set up, as explained in Section 4.3.7. For instance, in comparing A vs. B, the following linear program attempts to determine if A could be better than B for some set of weights, under the assumption that the value functions of Figure 6.2 are accurate:

$$
\begin{aligned}
MAX\ d_{AB} &= TV_A^* - TV_{B^*} \\
&= [w_{cost}V_{cost}(0) + w_{wet}V_{wet}(0) + w_{pop}V_{pop}(0)] \\
&\quad - [w_{cost}V_{cost}(20) + w_{wet}V_{wet}(100) + w_{pop}V_{pop}(25{,}000)] \\
&= [w_{cost}(1\text{-}0.2) + w_{wet}(1\text{-}0.3) + w_{pop}(0\text{-}0.5)]
\end{aligned}
$$

subject to the constraints:

$$2\,w_{wet} \leq w_{cost} \leq 3w_{wet}$$

$$0.5w_{pop} \leq w_{wet} \leq w_{pop}$$

$$w_{cost} + w_{wet} + w_{pop} = 1$$

$$w_{cost}, w_{wet}, w_{pop} \geq 0$$

The weights resulting from solving this optimization problem are w_{cost} = 0.6, w_{wet} = 0.2, and w_{pop} = 0.2. The most possible weight has been put upon *cost* and *population*, in which A excels. The objective d_{AB} is 0.52. Thus, there exist feasible weights that cause A to score better than B; B therefore cannot eliminate A. On the other hand, if instead d_{BA} was being calculated, the objective function of the linear program would become:

$$
\begin{aligned}
MAX\ \ d_{BA} &= TV_B^* - TV_{A^*} \\
&= [w_{cost}V_{cost}(20) + w_{wet}V_{wet}(100) + w_{pop}V_{pop}(15{,}000)] \\
&\quad - [w_{cost}V_{cost}(0) + w_{wet}V_{wet}(0) + w_{pop}V_{pop}(0)] \\
&= [w_{cost}(0.2\text{-}1) + w_{wet}(0.3\text{-}1) + w_{pop}(0.5\text{-}0)]
\end{aligned}
$$

The weights maximizing this expression under the above constraints are w_{cost} = 0.4, w_{wet} = 0.2, and w_{pop} = 0.4. These weights place the highest possible emphasis on *population*, which is B's strong suit. The resulting d_{BA} is -0.26; so even under the most favorable weights, B cannot outscore A, and can be eliminated from further consideration.

Table 6.5: Solutions to example application of successive elimination

			optimal values of weights			
plan j	plan k	d_{jk}	w_{cost}	w_{wet}	w_{pop}	conclusion
A	B	0.52	0.6	0.2	0.2	
B	A	-0.26	0.4	0.2	0.4	drop B
A	C	0.6	0.6	0.2	0.2	
C	A	-0.2	0.4	0.2	0.4	drop C
B	C	0.08	0.6	0.2	0.2	
C	B	0.06	0.4	0.2	0.4	

Table 6.5 shows the results of the computations for all pairs of alternatives. It reveals that A eliminates both B and C under those weight constraints; i.e., there is no feasible set of weights that can make A inferior to either B or C. However, neither B nor C eliminate each other. Therefore, if we wanted to know which alternative is second best, it would be necessary to further tighten the constraints upon the weights.

NOTES

[1] If the three points lie on a straight line, then the function fitted would instead be $V_i(A_i) = a + bA_i$.

[2] The same result would occur with the additive function.

[3] For instance, if $w_{cost} = 0.4$, $w_{wet} = 0.2$, $w_{pop} = 0.6$, then the multiplicative form is appropriate, and K would equal -0.4796. The utilities of the alternatives, as calculated by the multiplicative function, are $TV_A = 0.562$, $TV_B = 0.549$, and $TV_C = 0.6$. C is still best, but now A is second best.

7 A REVIEW OF MCDM APPLICATIONS IN ENERGY PLANNING AND POLICY

Few things are harder to put up with than the annoyance of a good example
MARK TWAIN, Puddn'head Wilson (1894)

Judge not, that ye be not judged
NEW TESTAMENT, Matthew 7:1

There are literally thousands of MCDM applications in a very wide range of fields.[1] In energy planning and policy alone, applications number in the hundreds. A complete review of this literature would be both tedious and necessarily superficial. Rather, we review a representative sample of MCDM applications to a range of energy operations, planning and policy problems. In each problem area we review one or two applications in some detail, with briefer references to other examples; the primary emphasis in this review is to examine how these applications have adhered to the principles set forth in Chapters 2 to 5.

We invite the reader to look over each of these case studies even if the particular problem addressed is not of immediate interest. Each study illustrates different points about the theory and practice of MCDM, elaborating upon the necessarily sketchy method overviews of previous chapters. The applications include environmental impact assessment, resource bidding, integrated resource planning, transmission system design, generation expansion, reporting of corporate environmental performance, dispatch, national energy policy planning, and energy planning in developing countries.

In Chapters 8-10, we present three in-depth case studies of the use of MCDM in energy supply and demand-side planning. These applications present details of theory and application not possible in the short summaries of this Chapter. Issues we discuss include stakeholder involvement and alternative development (Chapter 8), use of multiple value elicitation methods to measure reliability and user confidence (Chapter 9), and comparisons of the results and validity of alternative weighting and amalgamation methods (Chapter 10).

7.1 MCDM METHODS IN ENVIRONMENTAL IMPACT ASSESSMENT

An environmental impact statement for an energy facility may describe several dozen impacts for each of several siting or design options. To summarize impacts and facilitate comparisons of the options, it is often helpful to boil down the impacts to one or a handful of overall indices (McDaniels, 1995).

MCDM methods have been used to derive such indicators of environmental quality or impact since the early 1970s (Rau and Wooten, 1980; Hobbs, 1985). In one of the earliest applications of this type, Dee *et al.* (1973) derived an additive value function-based index of the environmental impact of water resources projects for use by the US Bureau of Reclamation. Montana Power Co. (1991) used a similar index to evaluate new generation resources. Such indices have also been created to summarize corporate and even national environmental performance (Miakisz, 1995; Hope *et al.*, 1991).

Qualitative impact comparison methods have long been a part of environmental impact statements (EIS), and range from checklists to more complex impact matrices. In fact, such methods have all the essential ingredients of an MCDM method: attributes, usually represented by the columns of the matrix; a set of discrete alternatives, usually represented by the rows of the matrix; and a value function $V_i(A_i)$ that translates the attributes into one scale of impact. In such impact matrices, this value function is often defined on a simple discrete scale and displayed in shades of grey. Two examples of this application are presented below -- one qualitative, the other based on additive value functions.

7.1.1 Puget Sound Area Reliability Plan EIS: A greyscale impact matrix

The strategy impact comparison matrix developed for the EIS for the Puget Sound Area Reliability Plan (BPA, 1992) is a good example of the qualitative approach, and is illustrated in Figure 7.1. Four levels of impact are distinguished: high, moderate, low, and minimal. These impact designations were developed by group consensus in a series of discussions by the team responsible for the EIS. The major problem with such an approach is perhaps not in the matrix itself -- for it may indeed be the case that a typical group of experienced environmental analysts can relatively easily reach a consensus about the environmental desirability of alternatives -- but in how the environmental rankings are integrated into the overall decision. In this particular study, seven evaluation factors were used: the six shown in Table 7.1, and an environmental attribute. In this table, the strategy that ranks best in each attribute (i.e., each row) is shown in bold, and strategy 2 was selected as the preferred alternative overall.

The environmental rankings are not included in this table. Yet from the environmental analysis, strategy 2 was only second best, and the preferred *environmental* alternative was strategy 3. There is no discussion at all as to whether the additional environmental impacts between strategies 2 and 3 were considered inconsequential.

Figure 7.1: The greyscale impact comparison matrix

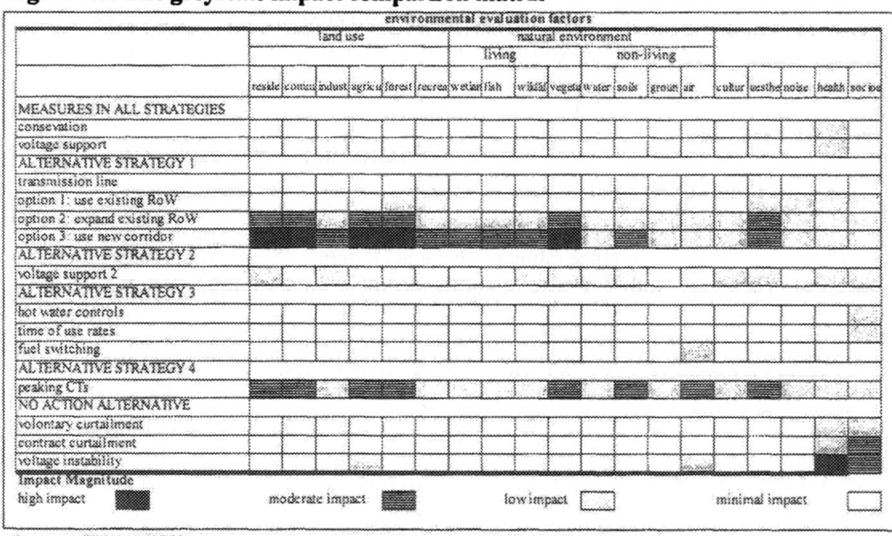

Source: BPA (1992)

Table 7.1: The economic and technical evaluation summary

evaluation factors	strategy 1	strategy 2*	strategy 3	strategy 4
	alternative strategies			
net present value, $million	67	**105**	128	39
sensitivity to load growth, $million	196	126	**84**	102
near-term revenue requirements, $million	50	25	50	**20**
long term revenue requirements, $million	75	**40**	110	105
deliverability (1=high, 4=low)	1.6	**1.5**	2.0	1.7
reliability (1=high, 4=low)	**2.0**	**2.0**	**2.0**	**2.0**

Source: BPA (1992). *=preferred alternative. **Bold** indicates best entry in row.

Thus, tradeoffs are entirely implicit, and nowhere in the EIS -- which runs to several hundred pages -- does the term tradeoff even appear. Indeed, the "preferred" alternative was selected by a five-utility Plan Steering

Committee, albeit with advice from a citizen review panel. Although the EIS emphasized steps 1-3 of our general MCDM process (alternative and impact characterization), steps 4-10 (tradeoff description, screening, and value judgments) were described in a most cursory manner. In sum, it is extremely difficult to ascertain whether and how the environmental ranking influenced the actual decision.

7.1.2 A US Federal Energy Regulatory Commission approach: Additive value functions

The US Federal Energy Regulatory Commission (FERC) has developed an environmental impact index to be applied to natural gas pipeline proposals (Stewart and Horowitz, 1991). The intention of the system is to fulfill the 1970 National Environmental Policy Act's mandate to "*utilize a systematic, interdisciplinary approach which will ensure the integrated use of the natural and social sciences and the environmental design arts in planning and decision making.*"[2] The system includes 64 environmental attributes (Table 7.2). These attributes are combined into an environmental index using additive value functions.

A unique aspect of this system is how the weights for the attributes were derived. The weight w_i for attribute i is calculated as follows (Stewart and Horowitz, 1991):

$$w_i = \sum_{k=1,..9} wd_k R_{ik}$$

where R_{ik} = the importance rating of attribute i in terms of importance dimension k. The nine importance dimensions are:

- regional human environment (short, intermediate, and long-term)
- local human environment (short, medium, and long-term)
- natural environment (short, medium and long-term)

wd_k = weight for importance dimension k, independent of the attribute i.

Thus, the overall importance of an attribute is the weighted sum of nine dimensions of importance. The weights wd_k were initially assessed by the Analytic Hierarchy Process (AHP, Section 4.2.4). They were then adjusted so that the resulting w_i for a subset of 12 attributes were highly correlated with a set of attribute ranks directly chosen by two experts. Most of the resulting weight (62.5%) is assigned to natural environment, with the remainder split between regional and local human environment. In each of those categories, long-term impact receives 57% of the weight, intermediate impact gets 29%, and short-term effect is assigned the rest.

The R_{ik} are assessed for each of the 64 criteria using the following verbal scale: severe (all impacts likely to be considered unacceptable by FERC); considerable (some impacts unacceptable); moderate (impacts acceptable after mitigation and natural recovery); slight (impacts eliminated by mitigation and natural recovery); and none (no change in environmental quality). The verbal assessments for each criterion are then converted into an ordinal scale and multiplied by the wd_k's to obtain the criterion's overall weight w_i. As an example, attribute $i=1$ (potable water) is judged to have "considerable" impact upon the short-term local human environment and a "moderate" impact upon the short-term natural environment. Its effect on the other seven dimensions is either "slight" or "none."

Table 7.2 : Environmental criteria in the FERC system

Water use and quality	22. areas of high cultural resource sensitivity	42. locally important, passive recreation areas crossed by new pipeline
1. potable water supplies	23. areas of medium cultural sensitivity	43. national parks crossed by pipelin
2. municipal watersheds crossed by pipeline construction	**Geology**	44. state parks crossed by pipeline loop
3. toxic sediments at river or stream crossings	24. active faults crossed by the pipeline	45. state parks crossed by new pipeline
4. major water crossings	25. areas with the potential for seismic soil liquefaction	46. disturbance to native American religious sites
5. minor stream crossings	26. crossing of karst terrain	48. temporary RoW in orchards and vineyards
6. crossings of lakes or reservoirs	27. areas of potential landslide	49. cranberry bogs crossed during pipeline construction
Vegetation and wildlife	**Soils**	50. construction of above- ground facilities in flood plains
7. privately owned upland forest temporarily converted to open land for new or loop pipeline	28 installing new pipeline in farmland	51. inactive surface mines crossed by pipeline
8. privately owned upland forest permanently converted to open land	29. installing pipeline loop in farmland	52. new pipeline RoW near essential public facilities
9. forested or scrub-shrub wetlands converted to open wetland	30. soil with high erosion hazard crossed by pipeline	53. construction across hazardous waste sites
10. herbaceous wetlands cleared during construction	31. soil with moderate erosion hazard	54. crossing of national and state trail
11. significant fisheries	32. prime farmland converted to non-farm use	55. compressor stations etc. within 1 mile of scenic area
12. high-quality & cold-water fisheries	33. pipeline across annually cultivated fields with stony or clayey subsoil.	56. new pipeline R-o-W within 1 mil of scenic areas
13. significant habitat crossings	**Land use, recreation & aesthetics**	57. pipeline construction across federal or state wild and scenic rivers
14. national or state wildlife refuges crossed	34. non-parallel RoW	58. sand dunes disrupted by construction
15. national or state forest crossed	35. new pipeline RoW within 50 ft of existing or planned residences	59. commercial trawling or shellfish areas crossed
16. federally listed or proposed threatened or endangered species	36. compressor stations, etc. within 500 ft of residential areas	60. commercial anchorages and navigation channels crossed
17. critical habitat of federally listed or proposed threatened/endangered species	37. construction RoW within 50 ft of homes	61. military reservations crossed by pipeline
18. state-listed or proposed threatened, endangered or rare species	38. natural or wild area crossed by pipeline	**Air and noise quality**
cultural resources	39. natural or wild areas crossed by pipeline loop	62. NOx emissions from compressor stations
19. listed or eligible resources on the National Register of Historic Places (NRHP) in Right of Way (RoW)	40. active recreation areas crossed by pipeline	63. noise levels at noise sensitive are
20. listed National Historical Landmark (NHL) in RoW	41. locally important, passive recreation area crossed by pipeline loop	64. abandonment and/or removal of pipeline facilities potentially contaminated with PCBs.
21. cultural resources recommended as eligible for the NHL or NRHP.		

Source: Stewart and Horowitz (1991), Appendix, pp. 130-132.

The main advantage of FERC's approach to weighting is that it provides a systematic way of assessing the importance of a very large number of criteria. Further, the authors made an effort to validate their system by comparing the weights with a ranking of attributes that experts chose directly. The major disadvantage of this system is that it appears that the ranges of the attributes were not explicitly considered when assessing weights. For instance, whether potable water impacts are important or not should depends on the amount of water involved. As another example, "long-term", "intermediate term," and "short-term" are vaguely defined, making the task of assessing the wd_k and R_{ik} ambiguous and difficult to replicate. Importance judgments made in the absence of information on the ranges of attribute values are vacuous, and fundamentally arbitrary.

7.2 MCDM METHODS IN RESOURCE BIDDING SYSTEMS

Over the past few years, what in effect are MCDM systems using additive value functions (AVF) have come into fairly wide use for capacity bidding systems -- one example of which, albeit imperfect, has already been presented in Annex 2.1. Often such systems are referred to as "rating" or "scaled scoring" systems. Mintzer *et al. (1990)* provide a typical rationale:

> *"...scaled scoring systems are easy to use, facilitate comparisons of environmental impacts measured in different units, and are more comprehensive than economic analysis ... using weighted scaled scores, technologies (or a combination of technologies contained in an energy futures scenario) can be ranked from the least to the highest cost on the basis of all economic, environmental and social costs that can be quantified."*

These are impressive claims, but, unfortunately, many of the applications have paid little attention to methodological issues, with the consequence that the validity of the results is uncertain. The procedure used by the State of New York is a typical case, and illustrates some of the issues.

A first general problem is consistency in weighting of attributes (Step 8). In some bidding programs approved by the New York Public Utility Commission (NYPUC), the bid prices are adjusted to reflect externalities -- for example SO_2 emissions have been valued at \$832/ton, and CO_2 at \$1.1/ton: valuations that are based on control or mitigation costs. In others, the bidding programs require the use of so-called "environmental scoring forms." Table 7.3 presents such an example as used by the Niagara Mohawk utility in New York: there the "weight" assigned to SO_2 is 7, and to carbon dioxide 3.[3] The claim is made that the scoring forms are supposed to reflect "the environmental perceptions in each utility service area" (Putta, 1990); yet it seems fairly clear that "weights" can be manipulated in such a way as to produce almost any desired result.

Table 7.3: The environmental scoring form used by Niagara Mohawk

	Weight w	Points P 0	1	2	3	4	5	score $w \times P$
air emissions								
SO_x (lbs/MWh)	7	>6	4.0-6.0	2.5-3.9	1.5-2.4	0.5-1.4	<0.5	___
NO_x (lbs/MWh)	16	>6					<0.1	___
CO_2 (lbs/MWh)	3	>1500	1050-1500	650-1049	250-649	100-249	<100	___
TSP (lbs/MWh)	1	>0.3	0.2-0.3	0.1-0.19	0.05-0.099	0.01-0.049	<0.01	___
water effects								
cooling water flow[a]	1	80-100	60-79	40-59	20-39	5-19	<5	___
fish protection	1	none		operational restrictions		provided	no public water used	___
water quality [b]	1	A or better	B	C+	C	D	no water use	___
Land effects								
average required (acres/MW)	1	0.3-0.5	0.2-0.29	0.1-0.19	0.05-0.09	0.01-0.05	<0.01	___
terrestrial	1	unique[d]		rural or low density suburban		industrial area	no land used	___
visual aesthetics	1	highly visible		within existing developed area		not visible from public roads		___
transmission	2	new OH >5 miles	new OH 1-5 miles	new UG >5 miles	new UG 1-5 miles	use existing facilities	energy conserva-tion	___
noise(L_{eq}-back ground L_{90})	2	5-10			0-4.9		<0	___
solid waste disposal (lbs/MWh)	2	>300	200-300	100-199	50-99	10-49	<10	___
solid waste as fuel (% of BTUs)	1	0	1-30	31-50	51-80	81-90	91-100	___
fuel delivery	1	new RR spur	truck & existing RR	new pipeline	barge	use existing pipeline	no fuel use	___
km from sensitive receptor area[c]	1	<10	10-39	40-69	70-100	>100	energy conserva-tion	___
							total score (210 max) =	___

Notes: a. annual intake as % of lake volume; b. NYS water quality classification of receiving water; c. per 6 NYCRR 225.1.1.(b)(4); d. unique ecological or historical value

Source: Reprinted from *The Electricity Journal*, Vol. 6(3), S. Putta, "Weighing Externalities in New York State," pp. 42-47, Copyright 1990, with permission from Elsevier Science.

Another problem, that of validity in scaling (Step 7), is illustrated in Figure 7.2: the scaling function is arbitrary, and is unlikely to have interval properties. It is in fact quite common to encounter systems in which the physical attribute is converted into a step function corresponding to some number of points on a scale -- these are sometimes called "binned ranges."[4] However, in the absence of some specific threshold effects that would justify some deviation from the presumption of linearity, much better would be to maintain the original physical attribute values, and simply normalize to a 0-1 scale.[5] The important point here is that transformations ought not to be arbitrary.

Figure 7.2: The attribute scaling function for CO_2 emissions in the environmental scoring system for Niagara Mohawk (New York)

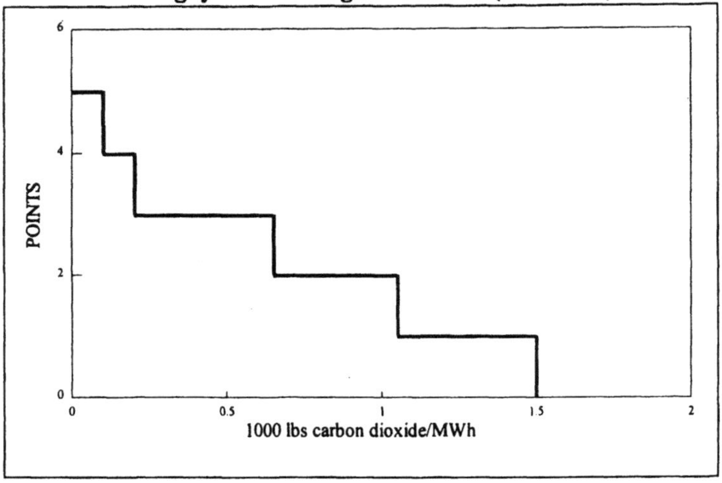

Problems of attribute selection (Step 1) abound. For example, exactly what is the environmental consequence expressed by "land use" in acres (or acres/MW)? As we noted in Section 2.1.1, differences in market value of the land should be reflected in the cost attribute, and the opportunity costs of lost production (say in the case of agricultural land) are readily monetized. But if the attribute reflects just non-market values, these are already captured in the attribute "terrestrial", in which sites of unique historical or ecological value are penalized. Finally, the "transmission" criterion may also double count land area consumed. In short, it is unclear why acreage has any relevance.

There are similar problems with the attribute "fuel delivery." The scale used, ranging from "new railroad spur" to "no fuel use," corresponds closely to the costs of the fuel supply connection, which should be incorporated in the system cost attribute. If there are particular environmental issues associated with the construction of such new transportation links, then those should be included in the corresponding environmental attribute. In any event, the requirement for preference independence is unlikely to be present; one might be more motivated to put a higher weight on the "fuel delivery" criterion if barge transport took place in relatively pristine rivers than if the water is instead very polluted.

7.3 MCDM METHODS IN TRANSMISSION AND DISTRIBUTION SYSTEM PLANNING

T&D planning involves tradeoffs among system cost, reliability, and environmental considerations (e.g., Atanackovic *et al.*, 1998). Although MCDM methods have been applied less frequently in this area than to other aspects of utility planning, the following two examples illustrate how such methods can help the T&D planner. One focuses on tradeoffs between cost and reliability in system design, and the other looks at cost-environment tradeoffs in transmission line routing.

7.3.1 Analyzing cost-reliability tradeoffs

Tradeoffs between cost and reliability are important in all aspects of power system design and operations. They are particularly critical in T&D design, yet are rarely examined explicitly. Reasons for this include engineers' lack of training in probability and reliability, and the absence of accepted methodologies for estimating the reliability of transmission and distribution networks. Generally, standards based upon accepted engineering practice are used to determine when, where, and what type of circuits need to be added. AC load flow models are often used to simulate system performance under a set of system contingencies to see if standards are violated. For example, would voltages remain within acceptable limits if a given line were to fail under the proposed system design?

However, we expect greater acceptance in the future of models for exploring cost-reliability tradeoffs. Ramirez-Rosado and Adams (1991) show one way of doing this for a distribution design problem. Below, we fit their method into the 10-step procedure of Figure 2.1.

Step 1: attribute selection

Two criteria are considered:

1. The sum over time and busses of weighted deviations of bus voltages from a target voltage (20 kV in this case), where the weights can depend on bus location and time of the deviation. This is similar in spirit to the goal programming amalgamation method of Section 4.3, and
2. The cost of line construction.

Step 2: definition of options

The options considered are possible combinations of lines connecting seven load points with two substations, and the resulting power flows and voltages. The feasible configurations are defined by a set of constraints in a non-linear mathematical program. This approach has the advantage that it permits the planner to implicitly consider a very large number of alternatives, lowering the likelihood that very good solutions will be overlooked.

Step 3: quantifying impacts

Impacts of each configuration upon the two criteria are automatically calculated by the mathematical program, which expresses the two criteria as mathematical functions of the decision variables (line locations and bus voltages).

Step 4: screening

Alternatives are not explicitly screened. Instead, a few configurations are directly chosen by two amalgamation methods in Step 9.

Step 5: tradeoff curves

Tradeoff curves are not explicitly developed, but could easily be obtaining by weighting and summing the two criteria, minimizing that sum, plotting the resulting solution, and then modifying the weights and repeating the procedure. However, in Figure 7.3, we display some of the non-dominated solutions that were generated using two different amalgamation methods, described below. A more careful analysis would have generated more points to give the planner a fuller picture of the options that are available.

Figure 7.3: Voltage penalty vs. line cost

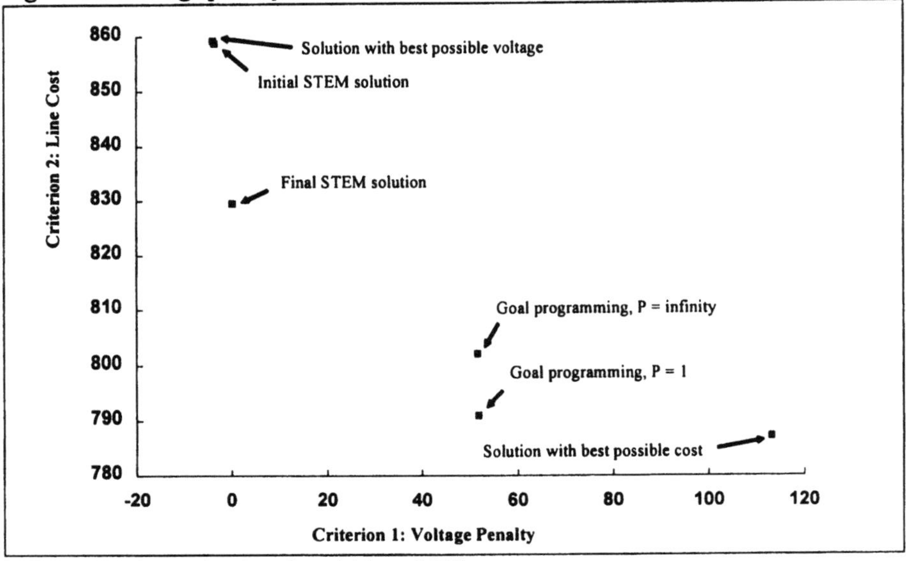

Based on data in Ramirez-Rosado and Adams (1991)

Examination of the figure reveals that only two of the six plans are truly competitive: those labeled "Final STEM Solution," and "Goal Programming, P=1." (The "STEM" method is described in Step 9, below). These two solutions would make up the knee set. Although the other solutions are undominated, they would certainly not be in the knee set because their advantages are very weak. For instance, the "Goal Programming, P = infinity" point is much worse in cost than the "P=1" point but has only a negligibly better voltage penalty.

Step 6: dominance analysis

No dominance analysis is performed. If the mathematical program is correctly formulated, only non-dominated designs will be generated (see Annex 3.1).

Step 7: scaling

The only scaling performed is as follows:

- Voltages above 19 kV receive a small bonus (in the form of a negative penalty) in the first criterion.
- Voltages below 19 kV but above 18.6 kV are penalized more heavily per unit of deviation.
- Constraints in the mathematical program prohibit any bus voltage from falling below 18.6 kV.

Step 8: weighting

Weights are used by the two amalgamation methods applied in Step 9. They appeared to be arbitrary, as no justification is given for them. In an actual application, planners would want to give careful consideration to the monetary value of additional reliability when choosing weights. There is, of course, a large and growing literature on the economic value of power system reliability upon which such consideration could be based (Tollefson *et al.*, 1994).

Step 9: amalgamation

Two amalgamation methods are applied:

- STEM, an iterative mathematical programming method that generates a solution; asks if the planner likes it; and then, if the planner is dissatisfied, generates another solution based on the criterion that the planner wishes to improve.
- Goal programming (Section 4.3.2), where the goals are set equal to the best-possible values of the two criteria (-4 for voltage deviations, 787 for cost). Two values of the exponent P were considered: 1 and ∞. As we point out in Section 4.3.2, if $P=1$ and if goals are set equal to the best-possible values, then this method yields the same result as minimizing the weighted sum of the two criteria (i.e., the additive value function, Section 4.3.1). On the other hand, if $P=\infty$, then results are very sensitive to the goals. As the goals were arbitrarily set to the best-possible values of each attribute, the resulting solution is also arbitrary.

In both cases, non-linear programming must be used to find the feasible configuration that has the best value. The values of the two criteria for the solutions from these methods are shown in Figure 7.3. The two goal programming solutions are very similar, but STEM yields a much more expensive system with a better voltage profile.

Step 10: resolving differences

Not addressed in the published study.

Conclusion

The main advantage of the above MCDM approach is that its mathematical programming formulation allows a huge number of alternatives to be implicitly screened and evaluated. The disadvantage of the implementation is that Steps 8 and 9 seem arbitrary, with no justification given for the weights or amalgamation methods chosen. Without such a justification, it is difficult to be confident that the supposedly "best" solution yielded by a method really is most preferred by the planner or stakeholder. However, since there are just two objectives here, it would be a simple matter to generate a wider range of options to be shown in the tradeoff plot, and presented to the responsible decision-makers who could then directly decide which points are most acceptable. This is the approach we recommend for problems with just two or three criteria.

7.3.2 Transmission Line Routing

Weedy (1989) reviews various methods for delineating corridors for new transmission facilities. MCDM methods have found use since the 1970s (e.g., Economides and Sharifi, 1978; Guidinger, 1987), in which a route is chosen to minimize a weighted sum of environmental, cost, and other impacts. Below we summarize how this method is typically applied in terms of the 10 steps of the general MCDM procedure.

Step 1: attribute selection

The criteria may number in the dozens, and can be grouped as follows:

- damage to natural ecosystems
- conflict with existing and proposed land uses
- potential for sharing of right of ways

- impact on culturally significant features
- visual impact
- potential exposure to electromagnetic fields (EMF), perhaps measured with a proxy such as population density within a certain distance, or population inversely weighted by distance
- cost
- transmission losses

The most salient environmental issue today is EMF (Gulliver and Vito, 1993; DeCicco *et al.*, 1993), although aesthetics are also important.

Step 2: definition of options

Options may be a few discrete routes predefined by planners. But in network routing methods, they are defined as possible combinations of segments; for instance, Figure 7.4 shows a network with dozens of possible routes between nodes 2 and 39.

Figure 7.4: Network structure showing transmission links between nodes 2 and 39

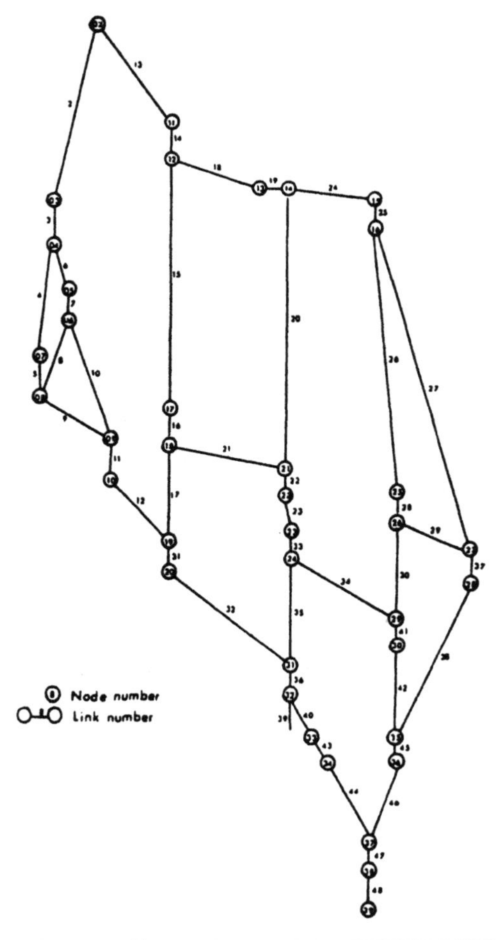

Reprinted from *Electric Power Systems Research*, Vol. 16(3), B.M. Weedy, "Environmental Aspects of Route Selection for Overhead Lines in the USA," p. 222, Copyright 1989, with permission from Elsevier Science.

Step 3: quantifying impacts
Impacts are first quantified by tallying the environmental, financial, and other effects of a line of the prescribed voltage for each segment. Then the total impact of the entire route is defined as the sum of the impacts of the constituent segments.

Step 4: screening
Screening occurs in the selection of the segments and connections to be considered; for instance, no segments might be allowed through areas with dense populations.

Step 5: tradeoff curves
Tradeoff curves are rarely generated by network methods, mainly because there are so many criteria. However, it might be possible to define a single environmental index that could be plotted against cost; the challenge would be to define an index that could be interpreted meaningfully (as in the example of Section 7.1).

Step 6: dominance analysis
Dominance analysis is not explicitly performed. Instead, an amalgamation method is used to generate one or more routes (sequence of segments); these routes will be nondominated if additive value functions are used.

Step 7: scaling
Scaling is usually linear in network models. For instance, if segment A has twice as many meters of wetland to be traversed as segment B, then segment A would be penalized twice as much. For reasons discussed later in this section, such scales need to be ratio scaled, unlike the mere interval scales appropriate to most MCDM applications.

Step 8: weighting
The various criteria for each segment are assigned weights. All the considerations we outlined in Section 4.2 concerning appropriate choices of weights apply here: i.e., weights should be ratio scaled and represent the willingness of decision-makers to trade off one criterion for another.

Step 9: amalgamation
A sequence of segments minimizing the weighted sum of criteria scores is chosen, usually by a linear programming or network solution method. This is the additive value function approach, but with a difference: because the number of segments is not the same for all routes, the individual criteria scales must be *ratio scaled*, not merely interval scaled. The reason for this can be illustrated by an example. Say there are just two routes between nodes 2 and 39. The first route traverses segments A and B, and the second route consists only of one segment, C. The criteria must have a non-arbitrary zero point (be ratio scaled), or else the ranking of the two routes could be changed by altering the arbitrary zero point. Consider, for example, the following scores for environmental impact for the three segments:

$$I_A = 2;\ I_B = 4;\ I_C = 7$$

The first route would then have the lowest total impact score (2+4=6, versus 7 for the second route). But if these impacts were only interval scaled, we could add an arbitrary constant to each, say 2. Then the new impact scores would be:

$$I_A = 4;\ I_B = 6;\ I_C = 9$$

and now the second route would have the lower score (9, versus 4+6=10 for the first route). This demonstrates that the criteria scales cannot have an arbitrary zero point when routes can consist of different numbers of segments.[6]

Step 10: resolving differences
Not addressed in the published study.

Conclusion

The advantage of the network approach to routing is that it can consider many possible routes. Its disadvantages are:

- with many criteria, it is difficult to plot and visualize tradeoffs
- additive value functions are the most appropriate amalgamation method, but criteria must be measured on a ratio scale.

7.4 MCDM METHODS IN SUPPLY CAPACITY PLANNING

We describe two basic approaches to multicriteria analysis of generation plans. One approach contrasts a few discrete plans and uses amalgamation methods to rank them. The other incorporates an MCDM method within a linear or dynamic programming model for capacity expansion. MCDM methods can also be used for detailed design of individual power or cogeneration plants (Balestieri and Correia, 1997), and for evaluating emissions compliance plans for generating systems (Srinivasan and Tettamanzi, 1997).

7.4.1. Ranking Discrete Supply Plans

Traditional supply-side planning consists of defining alternative supply plans that meet future demands at a given reliability, and then choosing the plan that minimizes the present worth of cost for a given set of cost assumptions. Today, supply planning is more sophisticated in at least three ways. Firstly, planners consider several criteria other than cost. Secondly, they explicitly recognize uncertainty in evaluating plans. And thirdly, they recognize market opportunities for buying and selling power. To handle these complications, multicriteria methods have been used to evaluate alternative supply plans.

Lootsma *et al.* (1990) use such an approach to evaluate electricity supply plans for the Netherlands. Rankings were made by each of 18 experts from universities, government ministries, and energy companies. This application is summarized below:

Step 1: attribute selection

The nine criteria considered are: average cost (cents/kWh); total cost; total SO_2, NO_x, and CO_2 emissions; domestic natural gas reserves; profits of the domestic natural gas industry; percentage of total energy supply by gas and oil; and installed nuclear capacity.

Step 2: definition of options

Three general strategies are compared: nuclear, coal, and gas.

Step 3: quantifying impacts

The performance of each strategy on each criterion is estimated for the years 2000 and 2010. Further, three scenarios are considered in each year:[7]

- low oil prices, GNP, and electricity demand
- medium, oil prices, GNP, and electricity demand
- high oil prices, GNP, and electricity demand.

Step 4: screening
No screening procedures are described in the case study.

Step 5: tradeoff curves
The authors explicitly reject using tradeoff curves or other devices to display the tradeoffs, asserting that decision-makers would be confused by so much information and would not be able to make a reasoned choice. Instead, the authors emphasize Steps 7, 8, and 9 of the MCDM process. We believe, however, that tradeoff analysis is a useful complement to weighting and amalgamation, because it can help make decision-makers more confident in the values they choose for weights, and in the resulting rankings.

Step 6: dominance analysis
No dominance analysis is performed.

Step 7: scaling
An unusual scaling method is used in which decision-makers are asked to perform pairwise comparisons of the strategies for each criterion.[8] For example, the total SO_2 emissions of the nuclear strategy in the year 2000 under the low scenario would be 180,000 tons, while the coal strategy would yield 190,000 tons. The decision-maker is asked which of the two values he prefers, and whether the preference was absent (score of 0), weak (2), strong (4), or very strong (6).

A scale or value function $V_i(A_{ij})$ for the attribute is then derived from those judgments as follows. First, the authors assume that perception is logarithmic; that is, statements of relative value of two alternatives are proportional to the logarithm of the ratio of their values. For instance, if a person states that A is twice as preferred as B, but four times as preferred as C, then the actual value ratios are e^2 for A:B and e^4 for A:C (where e is 2.7128 . . .). The rationale for this is the finding by psychologists that this is how people tend to compare physical magnitudes, such as brightness of light or loudness of sound (Stevens, 1966). This means that $V_i(A_{ij})$ is of the form:

$$V_i(A_{ij}) = exp\{c\ A_{ij}\}$$

where $exp\{x\}$ is e raised to the power x, and c is an arbitrary constant. Unfortunately, the choice of this arbitrary constant actually changes the ranks of alternatives, a result that should understandably disturb the decision-makers.

The A_{ij}'s in this study were not the original criteria scores (e.g., tons of SO_2), but rather form a scale that is statistically estimated from the pairwise preference judgments. The A_{ij} are the values that result in the best (least squares) fit to the responses to all the pairwise comparisons.

Step 8: weighting
Weights are chosen by a procedure similar to Step 7, where each participant is asked to compare two criteria at a time, stating which is more preferred (more important) and to what extent.

Step 9: amalgamation
The resulting scales and ranks are combined by the Power Law:

$$TV_j = \prod V_i(A_{ij})^{w_i}$$

where A_{ij} is the score on criterion i for strategy j, and w_i is that criterion's weight. As explained in Section 4.3.3, the Power Law method assumes logarithmic perceptions of relative importance of criteria. The score for each strategy is then averaged over the three scenarios by a complicated procedure that allows the participants to give the probability of each scenario as a range, rather than a single point. For instance, someone could say that the probability of the low scenario was between 0.2 and 0.4. This is a useful feature of this method, as decision-makers are often more comfortable stating ranges than precise values.

Rankings are then produced for each of the 18 experts. For a value of 0.5 for the arbitrary constant *c*, ten experts rank the nuclear strategy as best. Gas is ranked first by six experts, while coal is ranked best only by one of the experts. The experts examined the scores, and their sensitivity to the choice of *c*. Several of the experts expressed their dislike of the dependence of the results on *c*.

For comparison purposes, additive value functions (AVFs) are also applied with the same scales and weights. Table 7.4 summarizes a set of strategy scores for each scenario and for two different assumed values of *c* in the exponential equation. For example, the nuclear strategy receives an AVF score of 34 under the medium scenario if $c = 0.5$. The sensitivity of the results to the arbitrary constant *c* is obvious: nuclear and coal are tied under the Power Law in the medium scenario if $c = 0.5$, but nuclear is far inferior if $c = 1$. The AVF ranks depend even more strongly upon *c*. The table also shows that the Power Law yields very different rank orders from AVFs. Although the top option is usually the same, the relative ranking of coal and nuclear is often switched.[9]

Table 7.4: Netherlands supply planning: Power Law vs. additive value function scores for each strategy under different scenarios and different scaling constants *c*

Oil Price/ GNP/ Demand Scenario	Amalgamation Method	Constant $c = 0.5$			Constant $c = 1$		
		Nuclear	Coal	Gas	Nuclear	Coal	Gas
Low	Power Law	29	34	37	7	43	50
	AVF	33	32	34	21	38	41
Medium	Power Law	31	31	39	9	34	57
	AVF	34	31	36	21	36	43
High	Power Law	30	33	37	7	39	54
	AVF	35	31	35	21	37	42

Source: Lootsma *et al.* (1990)

Step 10: resolving differences

Not addressed in the published study.

Conclusion

A strength of this MCDM application was a careful laying out of the consequences of each strategy under several scenarios at different times in the future. Another strength was the attempt to measure the preferences of a diverse group of people. The major disadvantages of the methodology used in this case were the complexity and opaqueness of the scaling, weighting, and amalgamation procedures, and the dependence of the results on an arbitrary constant. A complex procedure such as the Power Law can be justified only if there is evidence that it is more valid than the simpler AVF method. However, there was apparently no attempt to confirm whether the participants' preferences conformed to the logarithmic model, or if the calculated weights were consistent with tradeoffs that people were willing to make. We believe that use of a model with unclear assumptions and complex computational procedures is less likely to be accepted by decision-makers and the public.

7.4.2. Multiple criteria analysis of supply plans using linear and dynamic programming

As an alternative to considering a handful of predefined plans, many utilities use linear and dynamic programming models to sort through a large number of possible sequences of supply additions (Turvey and Anderson, 1977). Most applications of such models have emphasized the minimization of the present worth of cost. However, environmental and other non-cost objectives can be added to such models to allow the planner to generate alternatives representing a range of tradeoffs. For instance, Kavrakoglu and Kiziltan (1983) and Quaddus and Goh (1985) modified linear programming capacity expansion models in this manner (as we do in our Seattle case study, Chapter 10). The World Bank has also applied MCDM to electric sector planning in Costa Rica and Hungary (Crousillat and Merrill, 1992). Below we summarize the dynamic programming-based method of Yang and Chen (1989) that they applied to the Taiwan Power Company system.

Step 1: attribute selection

The selected attributes are thirteen in number, and include:

- present worth of generation cost
- environmental impact (SO_2, NO_x, and particulate emissions, in parts per million in the stack gas multiplied by the number of generating units times the number of years each unit is in service)
- risk of plant disaster (measured as the cumulative capacity, in MW-years, of each fuel type: nuclear, hydro, coal, oil, and gas)
- fuel import vulnerability (measured as the MWh production by nuclear, coal, gas, and oil)

As we pointed out earlier in Chapter 2, total emissions are, at best, only a rough indication of environmental impact, as their effect depends on plant location. For instance, a plant located so that prevailing winds blow emissions out to sea might have minimal impact, even if emissions are large. The definitions and rationales of the last two groups of attributes is unclear. For instance, given a fixed nuclear and hydroelectric power capacity, why would additional oil capacity be considered to "increase the risk of plant disaster"?

Step 2: definition of options

The alternatives considered are different capacity expansion paths that could meet assumed demand growth through the year 2005. The alternative paths are implicitly defined by the dynamic programming network, which describes which additions can be made in which years.

Step 3: quantifying impacts

Impacts are automatically quantified by the dynamic program. Production costing runs for each year determine the amount of fuel cost, emissions, and "fuel import vulnerability." The capacity additions along the chosen path determine the capacity costs and "risk of plant disaster."

Step 4: screening

The only screening that is performed occurs when the dynamic programming network is set up. In defining the network, some combinations of generating plants will be excluded in order to limit the size of the problem; the excluded combinations will generally be those that the planner believes are unlikely to be economic (e.g., all baseload plants) or will fail to meet reliability criteria.

Step 5: tradeoff curves

Tradeoff curves could be generated by solving the dynamic program under several different weight sets. This is what Yang and Chen (1989) do; Figure 7.5 shows an example in which solutions from two different weight sets are compared. Present worth of cost, ppm of SO_2, and total MWh generated from oil are shown. Perhaps because there are 13 criteria, the authors do not present explicit tradeoff plots.[10]

Step 6: dominance analysis

No explicit dominance analysis is performed, but the dynamic program guarantees that a non-dominated solution will result from minimizing a weighted sum of the criteria.

Step 7: scaling

Linear scales are used for all attributes.

Step 8: weighting

The Analytic Hierarchy Process (see Section 4.2.4) is applied by three different groups: nuclear engineers, power engineers, and energy economists. Each group has six people. First, the participants compare the four general categories of attributes against each other using AHP's 1/9 to 9 scale. Then the attributes within the risk category are compared with each other. The four import attributes are also compared. Finally, the authors distribute each group's environmental weight among the three pollutants in proportion to estimates of the external costs they impose on the public. In addition, a poll was taken of members of the public to determine a fourth set of weights. The disadvantages of the criteria as defined in Step 1 become apparent when the AHP questions are phrased. For example, how can someone respond to the following question meaningfully?

> *Is gas 1/9, 1/7, 1/5, 1/3, 1, 3, 5, 7, or 9 times as important as coal in terms of the "plant disastrous risk" objective?*

This question is too vague to be significant. Yet somehow people responded; for instance, the nuclear engineers, on average, found that gas is 80% as important as coal in terms of risk. This may simply mean that people will supply numbers when asked, even if they signify nothing.

Step 9: amalgamation

Amalgamation is done four times by additive value functions, once for each group's weights. The amalgamation is performed by defining the weighted sum of the criteria as a single-objective function to be minimized by the dynamic program.

Step 10: resolving differences

As shown in Figure 7.5, the study addresses directly the fact that groups holding different values will arrive at rather different results.

Conclusion

The advantage of this approach to generation planning is that it allows some tradeoffs among various criteria to be considered. For instance, a comparison of Figures 7.5a and 7.5b show that the latter solution's higher weight on "plant disastrous risk" would result in much less reliance upon nuclear power. Under Yang and Chen's (1989) assumptions, this decrease in risk would cost over five billion dollars.

However, the use of AHP's ambiguous questions in this context means that we suspect that the weights do not truly represent the values of the groups. If a recommendation for a plan is to be based on this type of methodology, each person should apply more than one weighting method as a check to ensure that the weights are valid, as is done in Chapters 8 through 10 below. Preferably, a method that forces users to explicitly consider the units of the various criteria should be applied.

Figure 7.5: Comparison of plans

(a) Weights: Cost=0.12; Environment=0.16; Risk=0.21; Import=0.51
(nuclear engineers group)

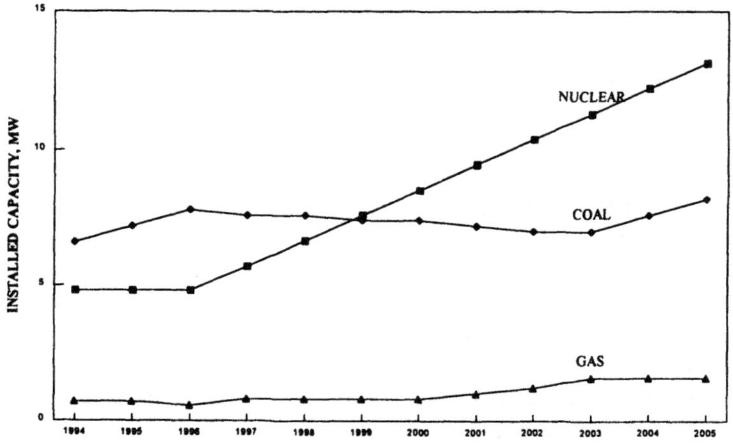

Cost: $40x10^9$; SO_2: 9400 ppm-generating unit-year; Nuclear Risk: 103,000 MW-yr
Oil (import vulnerability): 18,500 GWh

(b) Weights: Cost=0.16; Environment=0.13; Risk=0.53; Import=0.18
(energy economists group)

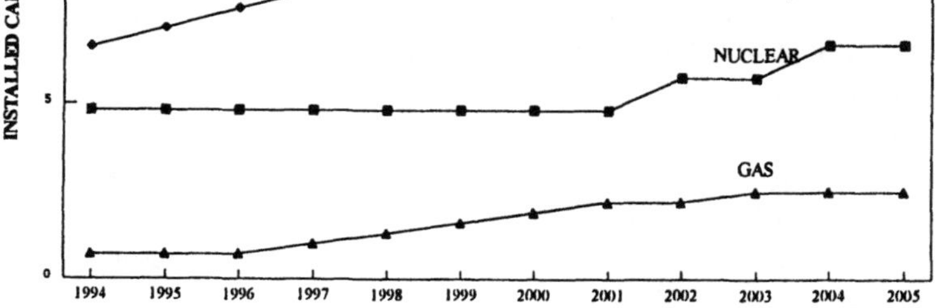

Cost: $45.2x10^9$; SO_2: 10,900 ppm-generating unit-year; Nuclear Risk: 59,000 MW-yr
Oil (import vulnerability): 20,500 GWh

Source: Yang and Chen (1989)

7.5 MCDM METHODS IN RESOURCE PLANNING

Integrated resource planning considers not only the supply options included in the studies of the last section, but also demand-side management (DSM). As in supply studies, MCDM methods can be used in two general ways:

1. to rank a set of predefined plans, and
2. to create objective functions for linear programming or dynamic programming models. The models can then be used to generate tradeoff curves and to select a plan that represents a acceptable compromise among the objectives.

As an example of the mathematical programming approach, the dynamic programming package PROVIEW/PROSCREEN, which has been adopted by over 60 utilities, has the capability of generating tradeoff curves and optimizing a mix of objectives (Energy Management Associates, 1993). Cincinnati Gas & Electric used that model to compare several different strategies for complying with the SO_2 control program of the 1990 US Clean Air Act (Bloemer and Knue, 1993). In that study, a strategy is defined as a commitment of resources (DSM, supply expansion, fuel choice, and emissions control) for the first few years of the planning period, followed by later resource commitments. Those commitments can depend on what scenario of loads and prices for fuel, SO_2 allowance prices, and control equipment is realized. The strategies are ranked in terms of the sum of their expected costs and emissions, which were equally weighted. The expectation (probability weighted average) is taken over the possible scenarios, each assumed to have the same probability.

A typical application of MCDM in IRP is the one used by a Wisconsin utility to rank four alternative resource plans (Hanson *et al.*, 1991).[11] The 10 steps of MCDM analysis are as follows:

Step 1: attribute selection

Nine attributes are selected (Table 7.5). "AFUDC" indicates interest costs accumulated during plant construction; its ratio to net earnings is considered to be an indicator of "earnings quality." One quarrel we have with the list of attributes is that conservation and coal are given as *criteria*. They are more properly thought of as *alternatives*; it is more appropriate to define attributes that describe what it is about those options that planners like or dislike (e.g., emissions, public relations, etc.). There is a danger of double counting, in that conservation might be given a high weight because one wants to avoid the pollution that results from fossil fuel use, but at the same time high weights are also directly given to SO_2 and solid waste.[12]

Step 2: definition of options

Four alternatives are considered:

1. Retire four existing 130 MW coal units and do not implement DSM; as a result, a new coal unit will be required relatively soon.
2. Retrofit those four units with atmospheric fluidized bed combustion (AFBC) and implement 150 MW of DSM.
3. Same as 2, except implement 445 MW of DSM.
4. Retrofit two units with AFBC and implement 445 MW of DSM

Table 7.5: Summary of attribute values

	alternative:	1	2	3	4
1	present worth (customer cost), 1986 $million	16.6	16.49	16.92	16.86
2	levelized rates, 1986 cents/kwh	6.16	6.28	6.54	6.52
3	AFUDC/net income (average %)	15.7	10.5	6.3	9.2
4	SO_2 emissions (average tons/year)	76.87	74.43	70.42	71.88
5	solid waste (1000 tons/year)	1,924	1,288	1,168	908
6	new employment (jobs created)	178	206	395	375
7	conservation (MW)	0	150	445	445
8	baseload plant (in-service date)	1995	2006	2010	2008
9	coal consumed (million tons)	174	162	154	151

Source: Hanson *et al.* (1991), p.41

Step 3: quantifying impacts

Production costing and financial models are used to quantify the attributes for each alternative (Table 7.5).

Step 4: screening

No screening procedures are discussed by the authors, although presumably some screening approach is used, for example to determine the amounts and types of DSM in each plan.

Step 5: tradeoff curves

With so few alternatives and so many criteria, it is probably easiest to present tradeoffs in the form of a table, rather than graphically. Table 7.5 shows several interesting tradeoffs. Plan 1 has the cheapest rates, but Plan 2 has the lowest present worth of costs (due to its inclusion of relatively cheap DSM). The higher cost DSM plans (3 and 4) yield lower emissions and coal use, more employment, and less AFUDC. Clearly, the relative importance of the attributes will make a difference in the ranking of the plans.

Step 6: dominance analysis

Each of the four plans is non-dominated.

Step 7: scaling

Scaling is accomplished using the risky utility function approach of Keeney and Raiffa (1976). Seven utility managers and outside experts were interviewed to obtain single attribute utility functions, along with the weights in Step 8. The authors report that there are considerable differences in the shape of the utility functions for the attributes chosen by different people -- consistent with what we find in the Seattle case study (Chapter 10).[13]

Step 8: weighting

The authors say they used the methods of Keeney and Raiffa (1976) to weight the attributes, implying that the indifference tradeoff method in combination with the lottery method was probably used. The weights differ greatly among the participants. For example, one person put most of his or her weight on levelized rates, while non-economic criteria (numbers 4-9) were given most of the weight by a majority of the other participants.

Step 9: amalgamation

Either an additive or multiplicative utility function is used to rank the alternatives. As a result, four of the seven experts rank plan 2 as best. Plan 4 is chosen by two experts, while the expert who placed most of the weight on rates picks plan 1. The heavy weighting of non-economic criteria was the reason for those results; if instead zero weight is assigned to those attributes, then plan 1 would be found best by six of the seven participants.

Step 10: resolving differences

Not addressed in the published study.

Conclusion

This study had the advantage of using a relatively transparent amalgamation method. Different value judgments, such as the expressed willingness to trade off attributes, can be easily translated into changes in scales, weights, and plan ranks. Such sensitivity analyses are much more difficult with, for example, the complex MCDM method used by Lootsma *et al.* (1990).

We have three concerns about the Hansen *et al.* (1991) study. First, as we mentioned, there is obvious double counting in the attributes, resulting in a bias towards conservation. Second, a greater range of alternatives should have been explored, perhaps with the aid of a linear or dynamic programming capacity expansion model. For example, an amount of DSM between the extremes of 120 MW and 445 MW might be preferred by most of the experts.

Our final concern is that the ranks should have been presented to the experts to see if they agreed with them. Such feedback is useful for two reasons:

1. If the decision-makers agree with the rankings, this provides evidence that the method is valid and trustworthy.
2. If the decision-makers initially disagree with the ranks, then the reasons for the disagreement can be explored. There are two possible outcomes of this exploration. One is that the users adjust their scales and weights to better reflect their values. As we noted in Chapter 4, scales and weights are often inconsistent and unreliable because the decision-makers' values themselves are not fully coherent (Fischoff *et al.*, 1980). The advantage of an MCDM procedure with these types of checks is that it helps decision-makers to think hard and make their values more consistent.[14] The other possible outcome is, that upon considering the reasons for the MCDM method's ranking, the decision maker gains some insight into the tradeoffs and decides that the ranking is reasonable after all.

The advantages of feedback are explored more fully in Chapters 5, 9, and 10.

7.6 MCDM APPLICATIONS IN GENERATION SYSTEM OPERATIONS

There are several objectives that a power system operator might consider, including fuel cost, system security, minimization of use of load controls, and emissions. For instance, Maekawa *et al.* (1992) analyze the problem of trading off fuel cost, transmission line overloads, and automatic frequency control (AFC) generator margins (the difference between generator output and its upper and lower limits) in real-time dispatch. They use an amalgamation method based on fuzzy set theory to choose an optimal dispatch schedule. The scales in this method are termed "fuzzy set membership functions", which are then combined using an amalgamation method that the authors do not describe. We question this application for the same reason that we questioned the distribution planning method of Ramirez-Rosado and Adams

(1991): with only three objectives, we believe that decision-makers would appreciate being presented with a range of possible dispatch schedules in the form of tradeoff curves. A single solution resulting from a complex MCDM method would (deservedly) be distrusted.

Figures 7.6 and 7.7 present examples of such tradeoff curves. In Figure 7.6 non-dominated strategies for real-time operation of a power distribution system have been generated for two objectives (Muslu and Anderson, 1989):

1. minimization of fuel cost (in $1000/day) (attribute X in Figure 7.6).
2. minimization of curtailment of air conditioning loads that are participating in load control program (in MWh/day) (attribute Y in Figure 7.6).

Curtailing load lowers costs, but also results in customer complaints. Presented with the information in this curve, an operator can decide what level of curtailment represents a reasonable compromise between the objectives.

Figure 7.6: Reliability-fuel cost tradeoff curve

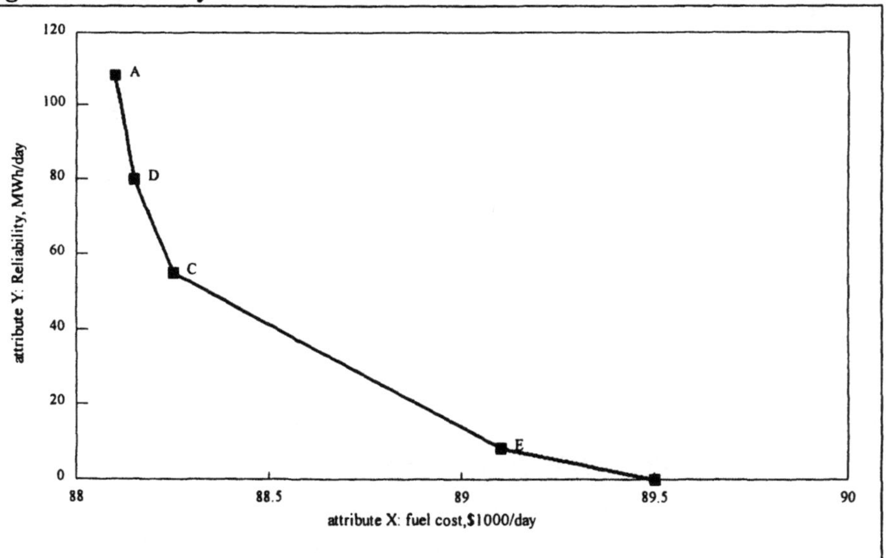

Source: Reprinted from *Electric Power Systems Research*, Vol. 16(1), M. Muslu and M.D. Anderson, "A Multiobjective Optimization Method for Electrical Power Distribution Systems with Load Control Alternatives," pp. 1-9, Copyright 1989, with permission from Elsevier Science

Although in the past emissions have not been considered in operating power systems, emissions dispatch is increasingly viewed as a cost-effective strategy for decreasing the environmental impacts of power systems. "Emissions dispatch" is defined as the operation of a power system in such a way that emissions are lowered at some increase in cost. Several models have been developed for exploring tradeoffs between emissions and costs in dispatch (Bernow *et al., 1991*; Talaq *et al.*, 1994; Chen and Chen, 1997).

As an example of emissions dispatch, Figure 7.7 shows tradeoffs between SO_2 emissions and cost for the state of Ohio, averaged over the year

(Heslin and Hobbs, 1989). On the lower right is the least-cost dispatch point, where fuel costs are minimized without regard to emissions. By shifting the plant merit order and fuel sources, emissions can be decreased, but at a cost.[15]

Figure 7.7: Ohio emissions-cost tradeoff curves, with and without fuel switching

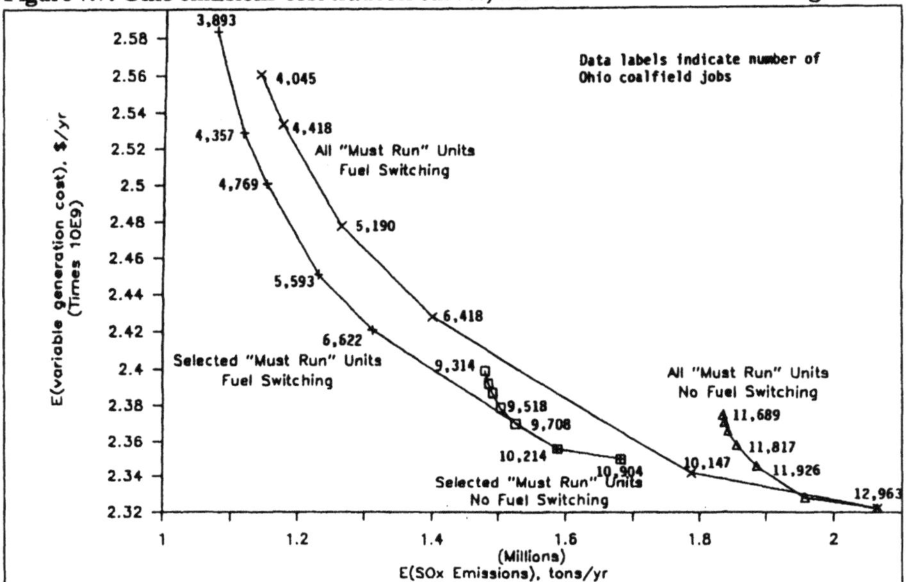

Source: "A Multiobjective Production Costing System for Emissions Dispatch and Fuel Switching," by J.S. Heslin and B.F. Hobbs which appeared in the *IEEE Transactions on Power Systems*, Vol. 4(3), 536-542, Copyright 1989, IEEE

7.7 USING MCDM TO MEASURE CORPORATE ENVIRONMENTAL PERFORMANCE

Corporate accounting systems have two audiences: internal management and external stakeholders. A central focus of *management* accounting is upon *projected* costs and benefits of decision alternatives. These costs and benefits may have multiple dimensions, including environmental effects. MCDM methods can help managers organize this information and use it to make decisions. Indeed, most of the applications of MCDM described in this chapter are of this sort: to provide information on tradeoffs to decision-makers.[16]

On the other hand, *financial* accounting reports are directed at stakeholders external to the firm. They can include the firm's owners and lenders, financial markets, consumers, regulators, labor unions, and environmental and social activist groups. The focus of financial information is to document *past* performance of the entire firm or its subdivisions using standard methods a standard format. The intent is to help stakeholders make meaningful comparisons of different firms, or to track performance of a single firm over time.

Within the last decade, stakeholders have begun to demand that information beyond the standard financial statements (income statement, balance sheet, and cash flow) be provided. A suite of "key performance indicators" (KPIs) covering a wide range of concerns is desired. For instance, environmentally conscious consumers have pressured corporations to publish information on issues such as air toxics emissions, rain forest exploitation, and recycling efforts. In an effort to begin standardization of these environmental reports, accounting groups and NGOs have begun to publish reports on "best practices" (e.g., Epstein, 1996; Ditz and Ranganathan, 1997). As we mentioned in Chapter 1, this is a particularly important issue in restructured electricity markets in which some sellers position themselves (honestly or not) as "green" producers of power.

But because "environment" -- like "social impacts" and other broad categories of concerns -- is an umbrella for a wide range of diverse impacts, there is a need for KPIs that combine these impacts into one or a few summary indicators. Like the environmental impact indicators of Section 7.1, value judgments are inescapable in constructing such aggregate KPIs. For instance, Northern Telecom's Environmental Performance Index (EPI) assigns 25% of its weight to compliance indices (e.g., fines), 12.5% to resource consumption (water, energy), 12.5% to site remediation, and 50% to fifteen categories of air, water, and land releases (NORTEL, 1996). Thus, these KPIs are similar to the environmental impact indices of Section 7.1, in that they use MCDM methods to aggregate a wide range of indices. Furthermore, the methods of Chapters 2 through 5 can be used to develop them.

Below, we summarize how Niagara Mohawk Power Corporation applied the ten steps of MCDM analysis in the construction of their EPI (Miakisz, 1995).

Step 1: attribute selection

Three categories of attributes are defined: emissions/wastes (SO_2, NO_x , CO_2, paper waste recycled, re-use of combustion wastes, non-PCB hazardous wastes disposed, herbicide impacts, and heavy metal discharges); compliance (notices of violation, fines, water discharge permit exceedences; excess opacity, and spills of oil, PCBs, and chemicals); and environmental enhancements (expenditures on outdoor recreation, fish & wildlife habitat, environmental restoration, and land protection). The air attributes in the first category can be criticized because they are measured on a weight per MWh produced basis, rather than tonnage; our complaint is that health and ecological impacts are related to total emissions, not rates per MWh. Another criticism can be directed at the third category of attributes, as they represent actions rather than outcomes. Ideally, indicators of resource use or quality (e.g., hectares of high-quality habitat, visitor-days) ought to be used instead.

On the other hand, this list fortunately avoids the obvious double counting that plagues a separate environmental benchmarking system that Niagara Mohawk and several other US utilities are developing (Miakisz, 1995). The latter system includes as attributes both emissions and a set of alternatives designed to lower emissions (DSM, heat rate efficiency, and alternate fuel vehicles). Counting both the means (DSM, heat rate, alternate fuels) and the ends (emissions reductions) will overweight emissions relative to other environmental effects (such as chemical spills and herbicides) that are unaffected by those means. Careful accounting of emissions effects alone could avoid this double counting by, for instance, including the net region-wide emissions reductions that result from using alternate fuel vehicles rather than gasoline vehicles (as BC Gas does in the case study of Chapter 10).

Step 2: definition of options

Because the purpose of an EPI is to summarize annual corporate performance for external stakeholders, no alternatives are defined. However, the intent is that utility management will consider effects upon the EPI when making decisions with environmental implications.

Step 3: quantifying impacts

Unlike most of the applications of MCDM in this chapter, impacts are not quantified by models or expert judgment. This is because the other applications focus on *future* effects of hypothetical alternatives. Instead, the EPI peers backward in time, using accounting systems to track *past* emissions, spills, compliance, and expenditures.

Steps 4-6: screening, tradeoff curves, and dominance analysis

Because the purpose is to document past corporate performance, rather than advantages of alternative courses of action, none of these steps are executed.

Step 7: scaling

Emission/waste attributes are scaled linearly by noting the percentage increase or decrease in each, relative to a pre-1992 benchmark. Desirable changes are given positive values (e.g., increases in re-use of combustion wastes, decreases in NO_x), while undesired changes are defined as being negative. In contrast, a -2, -1, 0, +1, and +2 scale is used for the compliance attributes. Zero represents no change in performance relative to the benchmark. The disadvantage of this coarse scale is that it can hide significant changes from year to year. For the final set of attributes, each \$200,000 of expenditure on environmental enhancement is assigned one point.

Step 8: weighting

Miakisz (1995) does not describe which method was used to assign the specific weights for the attributes. However, we can summarize some of the results. For instance, NO_x is given the most weight among the emission/waste attributes, twice that of SO_2, and five times that of CO_2. Because of the existence of a tradable emissions allowance system for SO_2 in the US, it could be argued that SO_2 should receive zero weight, because national emissions are capped. Therefore, if Niagara Mohawk reduces SO_2 emissions, it will save money, but if the emissions allowances are used elsewhere, national emissions are unchanged (Hobbs, 1994). Meanwhile, among the compliance attributes, notices of violations and fines each received 25% of the compliance weight. The other attributes in that category received 10% of the weight.

The relative weights assigned the three general categories are apparently based upon "the relative importance of (the) category, magnitude of investment required to obtain pollutant reductions, and so forth" (Miakisz, 1995). However, if the overall index is supposed to measure how the environmental impacts of the corporation change over time, magnitude of investment should be irrelevant to the EPI; only *results* should matter.

Step 9: amalgamation

An additive value function is used, undoubtedly (and justifiably) for simplicity. The results show a steady improvement from 1992 to 1994, with most of the improvement due to emissions changes.

Step 10: resolving differences

One would expect that there would be differences of opinion within corporate management and between different groups of external stakeholders concerning the relative importance of the various attributes. However, how they were resolved, or even whether they were addressed explicitly, is not discussed in the documentation.

Conclusion

The use of KPIs such as the above environmental performance indicator provides important information that is lacking in traditional financial statements. But if the resulting KPIs are to be meaningful rather than misleading, care should be taken to ensure that attribute sets represent all important concerns without double-counting or other distortions, and that the weighting systems are transparent and justified. Since weights represent value judgments about which managers and stakeholders will naturally disagree, KPIs should be documented so that users can alter the weights to see if conclusions about the firm's performance are robust. This can be facilitated by making the information and calculation procedures available on web sites, as has Northern Telecom (NORTEL, 1996). It is also desirable that different utilities use the same attribute definitions to facilitate comparison among companies; this is the goal of the multi-utility environmental benchmarking system now under development by a number of US utilities.

7.8 MCDM METHODS IN NATIONAL ENERGY POLICY

Energy policy making is inherently multi-objective in nature, as Jones' (1989) review of 12 policy reports shows. MCDM methods can help governments evaluate energy sector plans and policies. For instance, an expert committee appointed by the Government of Finland evaluated a range of policies using a formal multicriteria procedure (Hamalainen and Karjalainen, 1992). Criteria considered included a range of health and environmental risks, in addition to economic and political factors. The Analytic Hierarchy Process was used to select weights. Earlier, the Parliament of Finland applied a similar procedure (Hamalainen, 1988). Keeney *et al.* (1987) present another application of MCDM methods to national energy policy.

As another example, Jones *et al.* (1990) asked 25 individuals in the United Kingdom to choose weights for 15 different policy criteria. Examples of the criteria include supply security, cost, contribution to the greenhouse effect, diversity, acid rain, and capital requirements. Weights are chosen by the rating method on a 1-100 scale. These weights are then used to evaluate five distinct energy strategies through the year 2010:

1. free market in energy
2. nuclear power emphasized
3. a mix of coal and nuclear, plus some DSM
4. an emphasis on DSM, with some additional fossil fuel supplies
5. primarily DSM and renewable energy.

The policy application by Karni *et al.* (1992) to Israeli electricity pricing is particularly interesting because they use more than one MCDM method. We summarize the steps they went through below.

Step 1: attribute selection
Fifteen criteria are considered, grouped into six categories: reliability, cost, pricing efficiency, inflationary effects, social equity, and external goals (public acceptance, effects on national budget, and infrastructure).

Step 2: definition of options
Five pricing policies are compared, each with a different theme:

1. average cost-based pricing with no reference to marginal cost
2. price based on short run marginal cost
3. price based on long run marginal cost
4. conservationist pricing, where prices are set higher than cost for some customers
5. current policy, where prices are subsidized for some customers.

Step 3: quantifying impacts
The value of each of the 15 attributes for each policy is determined by panels of experts in the relevant fields. These are all determined on a subjective 1-5 scale, where 5 represents the best possible outcome and 1 stands for the worst.

Steps 4,5, and 6:
Not performed.

Step 7: scaling
The subjective scales created in Step 3 are directly used in the amalgamation.

Step 8: weighting
Nineteen policy makers, representatives of industry, and other stakeholders each choose two sets of weights for the six overall categories of criteria. One set results from direct allocation of 100 points in proportion to each category's importance. The other set is chosen by the Analytic Hierarchy Process (AHP). The two weighting methods yield very similar category weights; the major difference is that many participants directly assigned zero weight to the last three categories of criteria. However, the eigenvector calculation procedure used by AHP always gave those categories non-zero (yet relatively small) weights.

Each person's 15 individual attribute weights are then calculated by allocating his or her category weight among the attributes in the category, using ratios chosen by panels of experts in relevant policy areas. One possible rationale for such a procedure is that policy-makers should set priorities for general goals, while experts are better able to prioritize within categories. Another rationale is that policy makers have neither the time nor patience to weight a large number of criteria.

Step 9: amalgamation
Additive value functions (AVF) are used to rank the five policies. In addition, each stakeholder directly and holistically ranks the five policies in terms of overall attractiveness. Some consistent patterns of results emerge from this analysis. Sixteen of the nineteen experts rank the current policy last by both AVF and holistic methods. This is true even though the current policy is best in four of the 15 attributes. Furthermore, 17 of the experts rank short-term marginal cost pricing as best by AVF.

However, there are also inconsistencies. Only six of the experts prefer short-term marginal cost pricing when asked directly. More would rather choose the long-term approach. The correlations between the holistic and AVF rankings are actually negative for two of the stakeholders, and are 0.5 or lower for seven of the stakeholders. Unfortunately, the authors apparently did not confront the stakeholders with the inconsistencies between the two ranking approaches. However, the authors do suggest the following possible causes for the disagreements between the AVF and holistic ranks:

- The stakeholders disagreed with experts' assessments of the performance of each policy on the criteria (Step 3)
- Stakeholders use a definition for "importance" when choosing weights that disagrees with their willingness to trade off the attributes when directly ranking options[17]
- When directly ranking options, the stakeholders consider latent attributes, such as implementability, that were excluded from the list (Step 1).

Step 10: resolving differences

Not addressed in the published study. However, with the rankings differing between both methods and who applied them, this step should have been undertaken.

Conclusions

This case study has an important advantage over others we have reviewed in this chapter; namely the use of more than one method. Such comparisons promote learning and confidence in the results. For instance, because the point allocation and AHP methods give similar results, the authors can have more confidence in the validity of the weights.[18] However, the differences between the direct holistic rankings of alternatives and the AVF rankings indicated that there might be problems with the MCDM assessment, and that there is a need for further discussion. Unfortunately, as in the other cases in this chapter, it seems that the stakeholders were not asked to resolve the disagreements.

7.9 POWER SECTOR APPLICATIONS OF MCDM METHODS IN DEVELOPING COUNTRIES

In this concluding section of our review we examine the application of MCDM in developing countries. With electricity growth rates very much higher than in the developed countries -- typically 7-10% per year -- and with an increasing awareness of environmental issues,[19] the inclusion of environmental externalities into power sector planning in these countries has received growing attention. In particular, institutions such as the World Bank have been grappling with the issue of how environmental factors should be incorporated at the planning stage, because the shortcomings of waiting until project implementation before environmental issues are addressed in a project EIS are all too apparent (Meier and Munasinghe, 1994; World Bank, 1998a).[20] Moreover, in Asia in particular there is a growing awareness of the greenhouse gas emission problem, given that India and China are expected to add over 50,000 MW of coal fired capacity just over the next decade (World Bank, 1996). We review two representative areas of application of MCDM methods: power plant siting, and the definition of appropriate environmental attributes.[21]

7.9.1 Power plant siting

With US consulting firms active in the fast-growing countries of Asia, it comes as no surprise that the techniques used in the US for power plant siting in the 1960s and 1970s are now being applied to studies in these new markets. For example, in Table 7.6 we show the AVF impact score used in a siting study for a gas-fired thermal plant on the Island of Kalimantan, Indonesia. The ratings for each criterion (here the *rows* of the matrix) were on a discrete step-function scale (similar to that used for ranking generation bids in New York, described in Section 7.2), with the ratings -3, -2, -1, 0,+1, +2, and +3.

Table 7.6: The rating matrix for the Tanjung Batu gas-fired thermal plant, Indonesia

		Gunung Lok		Tanjung Batu		Jongkung		Loa Duri	
	weight	impact rating	impact score	impact rating	impact score	impact rating	impact score	impact rating	impact score
I. Air quality	10								
Surrounding the project site	4	0	0	0	0	0	0	0	0
Industrial areas (power consumer)	6	3	18	3	18	3	18	3	18
II. Water Quality	10								
Mahakam River	6	-1	-6	-2	-12	-2	-12	-3	-18
Surface and groundwater	4	0	0	0	0	0	0	0	0
III. Water Quality	10								
Mahakam River	6	-1	-6	-1	-6	-1	-6	-1	-6
Surface and groundwater	4	0	0	0	0	0	0	0	0
IV. Biological resources	18								
endangered species	6	0	0	0	0	0	0	0	0
wild animals	6	0	0	0	0	0	0	0	0
economic biota	6	-1	-6	-1	-6	-1	-6	-1	-6
V. Land use	24								
Productive cultivated land	6	-1	-6	-2	-12	-3	-18	-3	-18
protection forest	6	-1	-6	0	0	0	0	0	0
human settlements	6	0	0	0	0	0	0	0	0
facilities and infrastructure	6	0	0	0	0	0	0	0	0
VI. Socieconomic	18								
number of population	6	0	0	-1	-6	-1	-6	-2	-12
job opportunities	6	2	12	2	12	2	12	2	12
local economic activities	6	6	1	6	1	2	12	2	12
VII. Social Cultures	10								
social institutions	6	0	0	0	0	-1	-6	-2	12
historical heritage	4	0	0	0	0	0	0	0	0
Total	100		6		-6		-24		-54

Inspection of this matrix does not inspire much confidence in the methodology, much less in its results. For nine of the 18 criteria, all sites scored zero. For another four criteria, the scores are also identical for all four sites. There is no explanation in the text why there are two identical "water quality" criteria (scored differently): one may speculate that one of them in fact refers to water "quantity." It is very hard to avoid the conclusion that the entire exercise was little more than an afterthought, with its numerical precision quite specious.[22]

Many of the concerns about validity of past MCDM applications noted in Chapters 2 through 4 are also illustrated by the siting studies undertaken in Sri Lanka in the mid 1980s in an attempt to find a suitable location for a first coal-fired power plant (with an ultimate capacity of 900 MW). A preliminary screen identified Trincomalee Bay on the east coast as the best location, because of the superb natural deepwater harbor that would permit coal imports at relatively low cost.[23] Within this general area, seven sites were identified as possible candidates, and AVF used to rank them: the results are shown in Table 7.7. The most favorable site was assigned rank number 7, the least favorable rank number 1 (Black and Veatch, 1988a). The ranks were determined by "*. . . specialists knowledgeable about each of these siting factors*", and consensus weights were estimated by "*several siting specialists*". The final scores were calculated as follows: "*...The rank numbers were multiplied by the factor weights for each factor and the products summed to produce a weighted rank total for each site.*" Unhappily this procedure violates most of the principles necessary for method validity,[24] particularly the need to avoid algebraic manipulation of ordinal numbers (ranks).

Table 7.7: Site rankings in the Trincomalee study

site	coal unloading	transmission	site	generation facilities	supporting infrastructure	land use environment	overall ranking
"consensus" weights:	25%	10%	15%	20%	10%	20%	
site 1	2	1	2	1	4	7	2.7
site 2a-1	7	3	3	5	6	4.5	5
site 2a-2	6	3	4	7	6	6	5.6
site 2a-3	5	3	6	6	6	4.5	5.2
site 3a-1	4	5.5	5	3	2.5	2	3.8
site 3a-2	3	5.5	7	4	2.5	2	3.6
site 4	1	7	1	2	1	2	2

Source: Black & Veatch International (1988a).

This application illustrates another problem in the application of such methods. Five of the six attributes used here are in fact based on costs: indeed, we are told "*. . . capital costs and expected costs of operation were generally used as a basis for ranking the sites.*" If in fact the purpose of the exercise is to examine the tradeoff between "land use and environmental factors on the one hand," and factors related to system costs on the other, then everything related to cost should be subsumed into a single cost attribute.

The Trincomalee EIS proved highly controversial, and in the end the necessary permit from the Coast Conservation Department was denied on grounds of unacceptable impacts of thermal discharges. In a subsequent more broader siting study, AVF was again used to rank sites (Black and Veatch, 1988b). In Table 7.8 we show, as an example, the "air quality scores." On the 30-point scale used for air quality, 30 is the best, 0 is the worst. The basis for the scoring is described only in very general terms as "number of sensitive receptors and the exposure risk of significant receptors". Even as *relative* indications, these scores are very ambiguous. Finally, it is unclear whether the scores are comparable across the various generation types. One can only speculate, therefore, why a coal plant at Sapugaskanda would have an air quality score of 11, while a diesel plant is scored at 13.

The most serious problem with such valuations is not that expert judgment may necessarily be incorrect, but that the relationship between the score and possible mitigation actions is not demonstrable. In other words, there is no way of subsequently performing a sensitivity analysis that might explore the use of alternative fuels, alternative pollution control strategies, or of alternative technologies. Indeed, as we shall note below, actual calculations of the impacts associated with emissions at these sites proved to show differences an order of magnitude greater than differences among the relative scores shown in this table.

Table 7.8: Air quality scores at potential Sri Lankan power plant sites

site ID	site	combined cycle	oil	coal	diesel
NC-1	Mundal East	21	20	20	20
NC-3	Bolawata	21	20	20	20
NC-5	Ekala	21	20	20	20
C-1	Mahawetta				7
C-3	Kelanitissa	6			5
C-6	Peliyagoda	11	6	6	11
C-7	Sugathasua	6			5
C-8	Yakbedda	11			11
C-10	Sapugaskanda	13	11	11	13
C-11	Siyambalape	13	11	11	13
C-12	Biyagama	13	11	11	13
G-3	Bope				13
G-6	Armitage Hill				13
G-7	Haliwate				13
G-8	Wakawella	13	9	10	13
G-9	Bataduwa	13	9		13
M-1	Kudawella East	30	28	29	27
M-16	Weligama	30	28	29	27

Source: Black & Veatch International (1988b)

7.9.2 Incorporating environmental concerns in power sector planning

With the World Bank under heavy criticism for its involvement in power projects that have major environmental impacts, and with project-level environmental assessments under attack for being *ex post facto* justifications for decisions already made,[25] the Environment Department of the Bank has been examining the use of MCDM methods to better incorporate environmental concerns at the planning stage. Many of the major environmental impacts of power projects in tropical countries are particularly difficult to monetize, perhaps most notably so in the case of the loss of bio-diversity associated with inundation of forests at major hydro projects. The case study of Sri Lanka (Meier and Munasinghe, 1994) is the first application of MCDM methods to power sector environmental work at the World Bank.

The details of the methodology and the results of this study are described elsewhere.[26] Rather what is important and relevant to MCDM practice is the way in which the attributes were defined, and the importance placed upon location (as already described in Annex 2.2 in the case of biodiversity).

Some care was taken to be sure that attributes reflected *actual impacts*, rather than measures of regulatory convenience. Moreover, attributes were sparingly selected, on grounds that their proliferation would needlessly clog the analysis, and cloud the presentation of essential conclusions to decision-makers. Each attribute was associated with an important concern of national importance; in addition to the six environmental attributes shown in Table 7.9, the analysis also included present value of system cost, levelized tariff, and reliability (expressed as unserved energy).

Table 7.9: Summary of environmental attributes

attribute	units	impact
emissions of CO_2	[1000 tons]	global warming
population exposure to air pollutants	[person-ug/m^3/yr]	human health impacts
biodiversity index	[index]	diminution of biodiversity; impact on habitat of endemic species (see Annex 2.2).
increase in surface temperature> 1°C	[ha]	ecosystem impacts from thermal plumes
employment		discounted incremental employment
emissions of acid rain precursors (SO2 and NO_x)	[1000t/y]	potential for acid rain damages

Source: Meier and Munasinghe (1994)

Rather than using SO_2, NO_x and/or particulate emissions each as separate attributes, the study derived a health impact index that reflected the population exposure to these pollutants. This was defined as

$$X = \sum C_j P_j$$

where X = cumulative population exposure to the incremental ambient concentration attributable to the power plant

P_j = population in the j-th grid square

C_j = incremental average annual concentration in the j-th square, attributable to th emissions from the source in question. These were calculated using a a simple Gaussian plume model for a set of wind direction and wind speeds

The health index had the useful property that its magnitude was location specific, and resulted in several important policy conclusions. For example, it was shown that it really did not matter very much whether base-load coal plants sited in remote locations and with relatively tall stacks were equipped with FGD systems; far more important to reduction of population impacts was NO_x and SO_2 control at urban diesel plants with low stacks that burn relatively high sulfur residual fuel oil.

A similar conclusion was reached by a more recent study of the Indian State of Bihar (SCADA, 1998). Population exposure to emissions from urban self-generation were found to be 50-100 times greater than to emissions at large coal-fired stations (even though the latter were not in compliance with emission standards for particulates and NO_x).

Another interesting finding concerned the difference between calculated values of health impacts and the air quality scoring system used in earlier siting studies. The earlier air quality scores, on a scale of 1 to 30, for a coal plant located on the South Coast, and a diesel plant in Colombo, were 29 and 11, respectively, a ratio of about 2.7:1. In contrast, actual calculation of impacts showed that the two differ by a factor of at least 11:1.

NOTES

[1] See, e.g., the surveys by Chankong and Haimes (1983); Cohon (1978); Goicoechea *et al.* (1982); Hobbs (1979); Loucks and Somlyody (1988); and Zeleny (1982). For a review of energy sector applications, see Huang *et al.* (1995).

[2] The same Act spurred utilities applying for nuclear plant operating licenses from the US Nuclear Regulatory Commission to use multicriteria approaches to compare alternative plant sites (Rowe *et al.*, 1979)

[3] At first glance the relative weights (SO_2 to CO_2) of 7 to 3 appear inconsistent with the NY-PUC valuations. However, when one takes into consideration the scales used, the ratios are comparable. This can be shown as follows. In the scoring form of Table 7.3, a reduction of 4 lb SO_2 (say from 5 lb/MWh, which lies in the middle of the range that scores 1 point, to 1 lb/MWh, for the mid-value of the range that scores 4 points) is worth 7 x 3 = 21 points (equal to the weight for SO_2 [7] times the range of the scale [3 points] corresponding to 7 lb). Thus, one pound of SO_2 is worth 21/4 =5.25 points. Meanwhile a similar calculation reveals that decreasing CO_2 by 1100 lb (from 1,275 lb/MWh, the mid-value of the range that scores 1 point, to 175 lb/MWh, in the middle of the range scoring 4 points) is worth 3 x 3 = 9 points (CO_2's weight of 3 multiplied by the five points corresponding to 1,100 lbs). Therefore 5.25 points on the SO_2 scale will be equivalent to 1100 x 5.25/9 = 672. Thus 1lb of SO_2 is worth as much as 672 lb of CO_2. This is broadly consistent with the NYPSC evaluation of $1.1/ton of CO_2 and $832/ton of SO_2 (a ratio of 1:756)

[4] Goldman *et al.* (1992) provide a detailed evaluation of binned v. linear scales (also using the Niagara Mohawk system as an example), and conclude that this may introduce serious distortions when used in bidding systems.

[5] If uncertainty in bidder performance is explicitly quantified as probability distributions (which has apparently never been done), then the scaling might be non-linear due to risk attitudes (see Section 4.1 on utility functions).

[6] The same requirement for ratio scales also exists for any problem in which individual projects are rated, but the number of such projects to be selected is not known ahead of time. This is, for example, a problem in energy R&D planning, where the number of research proposals to be funded can vary (e.g., GRI, 1994).

[7] These scenarios are logically similar to the three scenarios considered in the Centerior case study (low stress, reference, and high stress), as described in Section 2.2. The actual rule used by Lootsma *et al.* (1990) is somewhat more complex, but is equivalent to the one we present.

[8] The Analytic Hierarchy Process involves a similar procedure. Its weighting approach is summarized in Section 4.2.4.

[9] However, Lootsma *et al.* (1990) state that they have found elsewhere that which amalgamation method is applied seems to make little difference when there are five or more alternatives.

[10] PROVIEW/PROSCREEN, another dynamic programming-based expansion package, allows the user to define several different resource portfolios for the last year of the time horizon, and then uses tables to display the tradeoffs among their costs, rates, emissions, and other criteria (Energy Management Associates, 1993).

[11] See also Chapters 8-10 for more detailed case studies of BC Hydro, BC Gas, and Seattle City Light, and Hobbs and Meier (1994a) for a case study of Centerior Energy.

[12] This problem also arose in our BC Gas and Seattle studies, Chapters 9 and 10.

[13] In the Seattle case we found that some people think that more conservation is undesirable (has lower utility) because it may be less reliable, while others believe that more conservation has higher utility because of the cost savings and experience that results. (Note that this indicates there is double counting between the conservation and cost attributes!)

[14] Of course, there is the danger that a user will arbitrarily adjust the parameters until the results agree with his or her prejudices about which option is best.

[15] However, under the US Clean Air Act, SO_2 emissions are subject to an allowance system which effectively converts those emissions into just another cost. All a system operator needs to do is multiply each generator's emissions by the appropriate allowance price, add that cost to the generator's fuel expense, and then use traditional least-cost dispatching algorithms. Treating SO_2 as a separate attribute would constitute double-counting (see Section 2.1).

[16] Managerial accountants are increasingly turning to MCDM methods to evaluate investments, particularly when stakeholder input is desired and environmental consequences are involved (Earl and Clift, 1998).

[17] In Section 4.2, we argue that this is likely for weighting methods that ask directly for weights because they leave the definition of attribute "importance" ambiguous.

[18] Yet both methods are direct weighting methods: therefore their results may have converged because in each case the user was using the same -- though *incorrect* -- definition of attribute importance. A better check on the validity of the weights would have been a set of tradeoff judgments, as in the indifference tradeoff method (see Section 4.2 and Chapter 9).

[19] As evidenced, for example, by the controversy over the World Bank's involvement in the Narmada hydro projects in India, which would displace 170,000 people in order to provide irrigation to 2 million hectares.

[20] Largely at the insistence of the World Bank and other international financial institutions, developing countries have had to demonstrate that facilities proposed for financing are in the "least cost" plan. In practice this has meant the application of sophisticated capacity expansion optimization models, such as the WASP model (Jenkins and Joy, 1974). The objective function in such models is the minimization of the present value of system costs for a given demand forecast for some given level of system reliability.

Perhaps the most important objection to the use of models such as WASP is the difficulty they have dealing with the increasing uncertainties currently faced by the sector, ranging from fluctuations in the prices of imported fuels to the impacts of construction cost overruns and droughts. The so-called "least cost" plan may in fact be optimal only for a relatively narrow band of external circumstances assumed, and may deviate from optimality for most other combinations of external factors (Meier and Munasinghe, 2000).

Given these kinds of difficulties, in the early 1990s the World Bank began to explore the use of MCDM in electricity supply planning, notably the studies by Crousillat (1989) and Crousillat and Merrill (1992) for Hungary and Costa Rica, and Meier and Munasinghe (1994) for Sri Lanka. These involved tradeoff analysis to identify decision sets or "short lists" (using dominance analysis techniques described in Chapter 2), and subsequent risk analysis to measure robustness, quantify exposure, and develop hedges against uncertainties.

[21] Integrated resource planning is not a term that is in general use in developing countries, although the need to go beyond the traditional emphasis on supply side expansion to include efficiency improvement and demand-side management is receiving increasing attention (see e.g., Nadel *et al.* (1991) or Rao (1998) for DSM studies of India).

[22] With large areas of the forest of Kalimantan having been destroyed in the great fires of 1996, 1997 and 1998, one might be forgiven in believing that even had the worst site been chosen by this exercise, it would not have mattered very much. Indeed, might say the cynics, elevating to great importance an environmental impact assessment exercise for a power plant, whilst ignoring the real villains of Kalimantan's environmental destruction (the corrupt logging companies owned by cronies of the former President), lacks all common sense.

[23] Sri Lanka has no fossil resources of its own, and with few remaining hydro sites, must look to thermal baseload generation by the end of the 1990s to meet strongly growing electricity demand (expected to be around 7% per year).

[24] Even more unhappily, the procedure was justified as follows: "...*Weighted ranking systems are commonly used in engineering circles in the United States to evaluate factors other than costs: these ranking methods are particularly applicable to siting studies, since at least one of the major siting factors cannot be evaluated with costs. We have prepared many siting studies involving different geographical states in the United States which involved use of this method.*"

[25] The EIS prepared for the proposed coal plant at Trincomalee in Sri Lanka provoked almost as much controversy over the *process* as it did about actual impacts, with comments such as "*. . . it is therefore evident that the entire feasibility study process, terminating with the EIA, has followed a policy known as 'decide-announce-defend' . . . the EIS is an ex-post facto document seeking to environmentally justify a project which was formulated without any environmental considerations whatsoever . . . this EIA is a post planning exercise, whereas it ought to have been an ongoing exercise from the inception of the project*" (Meier and Munasinghe, 1994).

[26] The study made extensive use of the dominance and tradeoff curve analysis concepts discussed in Chapter 3.

8 MCDM AT BC HYDRO: THE 1995 PLAN

Insofar as it represents a genuine reconciliation of differences, a consensus is a fine thing; insofar as it represents a concealment of differences, it is a miscarriage of democratic procedure
J. WILLIAM FULBRIGHT, Speech, US Senate (October 12, 1965)

The truth is never simple, and rarely pure
OSCAR WILDE, The Importance of Being Earnest (1895)

In this chapter, we describe the use of MCDM at BC Hydro to assist the preparation of the 1995 Integrated Electricity Plan (BC Hydro, 1995a).[1] A number of the highlights of this study have already been presented in previous chapters to illustrate specific methodological points. Here the emphasis is on *process* rather than algebra: how and why things were done is the focus of this Chapter, not the details of the outcome (which are well described in the Plan itself).[2] Particular emphasis is placed on the processes involved in Step 1 (attribute definition).

8.1 STRUCTURING THE PROCESS

The MCDM process conducted by BC Hydro was undertaken at the insistence of the British Columbia Utilities Commission (BCUC), which in a November 1994 decision (BCUC, 1994) ordered:[3]

The Commission directs BC Hydro to institute with the new IRP consultative committee a multi-attribute tradeoff analysis (MATA) for the purposes of portfolio development and selection.[4]

The Commission noted further:

..the Commission wishes to indicate its agreement with the position of the Energy Coalition that there is no contradiction between a stakeholder committee making recommendations on a complete range of IRP issues, and the utility then taking these recommendations into account in the final preparation of its IRP and offering an explanation in those instances where the stakeholder recommendations are not followed. The value of the collaborative is not that "the people" decide, but that the utility is made accountable for its decisions . . .The Consultative Committee should try for consensus. Where consensus cannot be realized, members may write dissenting opinions.

The Commission then went on to set a June 30 deadline for the submission of the 1995 resource plan, and directed that the Consultative Committee be in place by the end of 1994. The Committee held its first meeting on January 26, 1995.

As we noted in Chapter 2, deciding *who* will participate in an MCDM process is the starting point. Who is included, and who is excluded, may well determine what criteria are chosen for the process; which options should be examined; and what will be the probable outcome. In this case, the membership was largely determined by the BCUC.[5] The most controversial aspect of the Committee's membership was the participation of the Independent Power Producers Association (which was expressly ordered by the BCUC): some of the customer group and environmental stakeholders felt this was inappropriate in light of its interest in the outcome.[6]

Table 8.1: The Consultative Committee

Interest Represented	Participating organizations
Residential customers	Public Interest Advocacy Centre Kootenay-Okanagan Electricity Consumers Association
Industrial customers	Council of Forest Industries Mining Association of British Columbia
Regional representatives	Prince George Region Development Corporation Kootenay Regional Advisory Group Greater Vancouver Regional District
Environmental interests	BC Energy Coalition Slocan Valley Watershed Alliance
Independent power producers	Independent Power Association of British Columbia
Natural resources dependent sector	United Fishermen and Allied Workers' Union
BC Hydro	BC Hydro, Power Supply Division BC Hydro, Customer Services Division

An important step was the appointment of an independent moderator, who chaired the many meetings of the Committee, and successfully facilitated the process which began in an atmosphere of distrust between some Committee members and BC Hydro,[7] but which ended, if not with a consensus on the details of the plan, at least with the recognition that the process had provided an opportunity for stakeholder interests to be heard and considered.[8]

Neither BC Hydro nor any of the members of the Committee had any experience in MCDM, and a request for proposals for consultants to assist the Committee was one of its first decisions. However, by the time the consultants were engaged in late March 1995, the Committee already had four meetings, and had almost concluded its discussion of planning objectives. Moreover, at this time there was still an expectation that the deliberations of the Committee would conclude in late May, and that an Electricity Plan could be completed to meet the BCUC deadline of June 30.[9]

It was quickly apparent that a three-month period of time was insufficient to complete the essential deliberations of the Consultative Committee, much less to conduct a multi-attribute tradeoff analysis (MATA) that met appropriate standards of rigor. Even with the extension to September 30, 1995 that was granted by BCUC in early June, the time available was extremely short, particularly in view of the fact that this was BC Hydro's first experience in preparing an Electricity Plan in consultation with a stakeholder group. Most utilities in North America prepare resource plans on a two-year cycle, and set aside at least a nine-month period for analysis and Consultative Committee deliberations. Moreover, in an optimal schedule, most of the preparatory work in setting up the databases and models for MATA should be done *before* the first meeting of the Consultative Committee: in the BC Hydro case, this all had to be done in the months of April through June 1995, during the meeting period itself.[10] Table 8.2 shows meeting schedule for the MATA.

Table 8.2: The meeting schedule

	Consultative Committee	BC Hydro Managers Group
March 23/24	Introduction to MATA Links between objectives and attributes	
April 6/7	Trial runs of the model Resource screening and weighting exercise	
May 4/5	Scenario development Tradeoff analysis for DSM	Introduction to MATA
May 25		Preliminary screening
June 8/9	First portfolio building	
June 20 - June 27	Individual meetings with members of the Consultative Committee to review weights	
June 27/28	Small group meetings: three groups construct the second round portfolios X, Y and Z	First portfolio building
July 20/21	Further refinement of portfolios	
July 24		Revised portfolios Discussion of major issues
August 16/17	Final portfolios constructed BCH, F1, F2F	
August 22		Risk assessment Conclusions

We could of course have chosen a very simplistic approach to MATA that might have satisfied the letter of the wording of the BCUC directive about methodology, and perhaps met the original deadline. But such an approach would have been in keeping neither with the *spirit* of the directive, nor with BC Hydro's desire to prepare an electricity plan that provided meaningful input from the Consultative Committee, and that met the standards of technical excellence expected of one of North America's leading utilities.

We considered it important that the same process of MATA, and of portfolio construction, that was followed by the Consultative Committee was also conducted in-house by BC Hydro. If BC Hydro's senior managers were to understand how and why the Consultative Committee came to its conclusions, it was important that they fully understood the MATA process itself. To this end a group of senior BC Hydro managers was convened, which from March to August conducted a parallel exercise of portfolio building and tradeoff analysis. Although the timing of this group's work was some weeks behind that of the Consultative Committee in its early stages, information was exchanged between the two sets of deliberations as appropriate.

It should not be surprising that the Consultative Committee reflected a much more diverse set of interests, as reflected in their weights, than the BC Hydro managers. On the Consultative Committee the extremes were reflected by individual C, who gave 85% of the total weight to cost factors, and by individual Y, who gave no weight at all to costs (Figure 8.1, top). Nevertheless, even in the BC Hydro managers' case (Figure 8.1, bottom), there were several members who gave a preponderance of weight to environmental factors (although none gave zero weight to cost).

8.2 OBJECTIVES AND ATTRIBUTES (STEP 1)

By the time the consultant was engaged to advise on the MATA process, BC Hydro and the Consultative Committee had already spent considerable time discussing IRP objectives, and by the end of March 1995 had come to closure on a list of 38 objectives, or 66 when subheadings to the main objectives are included. A partial list of these objectives, which shortly thereafter were finalized by both the Consultative Committee and BC Hydro's own Steering Committee, is shown in the first column of Table 8.3.

The relevance of such a broad and extensive list of objectives to the actual planning process was unclear. Moreover, in drawing up this list, it was evident that little attention was given to precisely how the objectives would be used in the actual decision-process. The problems include:

- Much of the list reads like a checklist for project-level environmental impact assessment (EIA), rather than a list of objectives capable of translation into attributes for use in MATA. Indeed, there is little distinction between issues appropriate for project level EIA, such as "Minimize aesthetic impacts of facilities," and issues that *are* appropriate for planning, such as "Minimize impacts on global climate from greenhouse gas emissions."

- In any event, not all 66 objectives and sub-objectives can possibly have equal weight. The list gives no suggestion of any prioritization.
- For objectives to be considered in a meaningful way in MATA, they have to be capable of quantitative expression. One has to be able to measure to what extent that objective is being met. For example, the objective "Minimize personal disruption" may be germane to the management of a construction project. But how does one measure performance on this objective in the analysis that is conducted for a 20-year plan?

Figure 8.1: Comparison of weights: BC Hydro managers and Consultative Committee

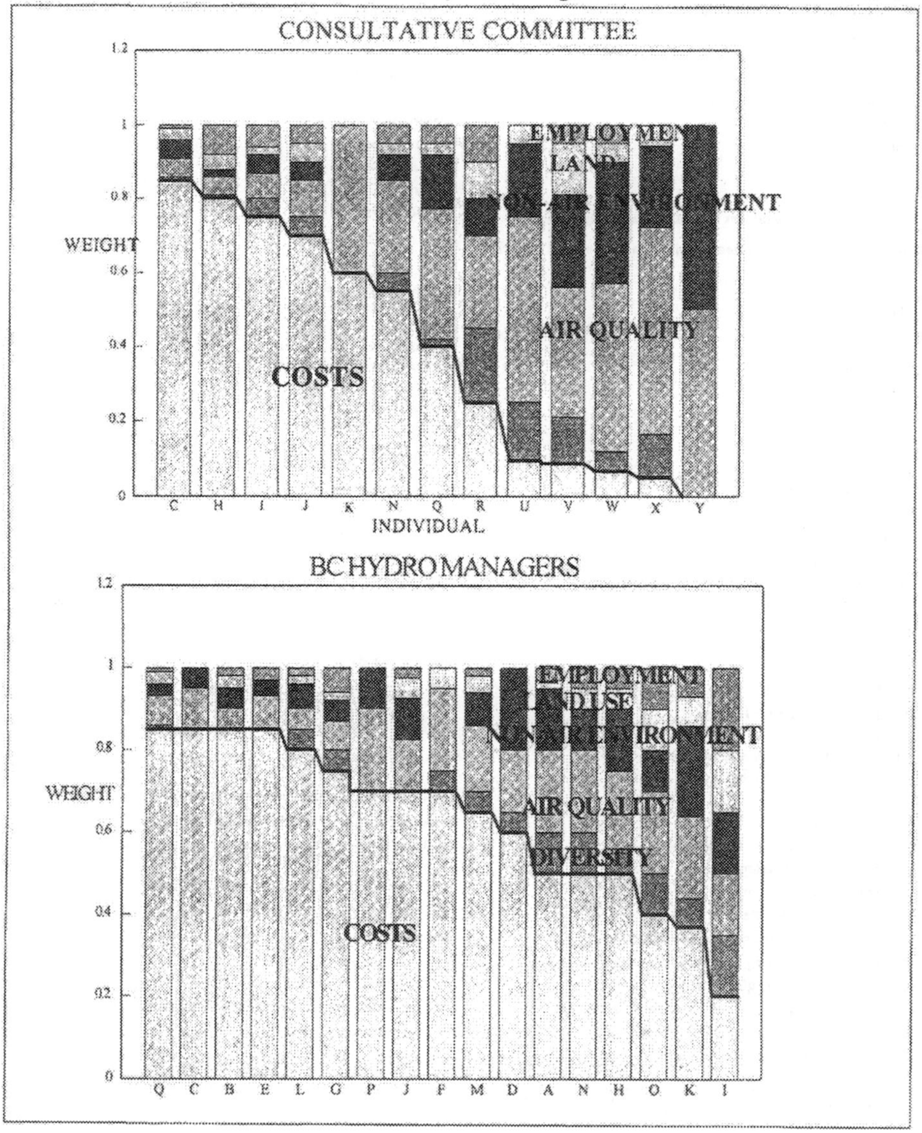

Table 8.3: Objectives for the plan [partial list]

IRP objective	MATA treatment
Minimize costs of electrical services to consumers	
minimize financial costs of meeting electric service needs	attribute [rate impact]
minimize adverse impacts of required capital expenditures by consumers	attribute [system cost, includes DSM customer costs]
avoid rate shock	attribute [rate impact]
minimize costs to other government agencies	attribute [system cost]
Provide high quality customer service	
meet electric service needs	
secure energy supply	portfolio issue [must meet reliability constraints]
secure capacity supply	portfolio issue [must meet reliability constraints]
provide reliable service	
minimize service outages	portfolio issue [must meet reliability constraints]
minimize service restoration time	portfolio issue [must meet reliability constraints]
provide high power quality	
minimize voltage changes	portfolio issue [must meet reliability constraints]
minimize frequency changes	portfolio issue [must meet reliability constraints]
Provide flexible service to adapt to changes:	
changing demands	scenario [load forecasts]
changing technologies	resource option
changing regulations	scenario
changing values	scenario
Provide choice to consumers regarding electricity services	attribute [diversity]
Minimize adverse and promote positive environmental impacts	
on land resources, including:	
flora	attribute [species persistence]
fauna	attribute [species persistence]
soils	EIA
ecosystems	attribute [terrestrial ecosystems]
biodiversity	attribute [species persistence]
on water resources, including:	
flora	attribute [species persistence]
fauna	attribute [species persistence]
ecosystems	attribute [species persistence]
biodiversity	attribute [species persistence]
on air resources: visibility	attribute [particulates]
on local climate	EIA
on global climate from GHG emissions	attribute [GHG emissions]
on wilderness preservation	attribute [terrestrial ecosystems]
on natural aesthetics	EIA
Minimize health and safety impacts	
early deaths	attribute [pwae]
sicknesses	attribute [pwae]
injuries	EIA
Optimize socioeconomic impacts	
minimize displacement of residents	EIA
minimize aesthetic impacts of facilities	EIA
minimize personal disruption	EIA
compact communities	resource option [community energy planning]
complete communities	resource option [community energy planning]
support regional economic development	attribute [employment]
foster sustainable development	
foster community viability	
Ensure fair and equitable treatment	
among customers bearing costs of new programs	attribute [diversity]
among current and future generations regarding resource availability	sensitivity analysis [discount rate]
among nations regarding global energy consumption	attribute [GHG emissions]
among firms and individuals providing electrical services	corporate, plus separate discussion of consistent treatment of IPPs
regarding evaluation of electrical service options	attribute [diversity]
of aboriginal peoples	corporate, EIA, attribute [terrestrial ecosystems], in part by recognizing aboriginal land claims

Abbreviations: pwae = population-weighted air emissions
EIA = issue for project level environmental impact assessment

Indeed, the list as presented is an almost textbook example of the consequences of the desire to be all inclusive, and of the reluctance of those participating in the definition of objectives to eliminate issues irrelevant to planning decisions that some may nevertheless consider important.[11]

In an ideal situation, each objective ought to have a corresponding quantifiable attribute that is used in the MATA process. Nevertheless, if one wishes to construct a process in which the links between objectives and the final result is to be transparent, one has to work with objectives that can be quantified at a level consistent with the data likely to be available at the planning stage. If indeed all the attributes are quantified and decision relevant, structuring them hierarchically offers a way to retain focus (see Section 4.2.3).

But even if all 38 objectives could be quantified, it is unclear that one would want to do so. Making tradeoffs among too many attributes bogs down the analysis, tends to obscure the really important issues, and tends to produce "flat" rankings that differentiate little between options. As noted in an earlier chapter, few individuals have the ability to deal rationally with more than seven or eight objectives simultaneously. Making tradeoffs between four or five clearly articulated and well-defined objectives is difficult enough.

These problems were taken up at the first two meetings with the MCDM consultant. It was first agreed that it was indeed unnecessary for each of the listed objectives to have a separate attribute. However, a showing did need to be made that the overall process was sensitive to each of the listed objectives, and that in one way or another these objectives needed to be incorporated into BC Hydro's decision-making. Those objectives that would have no effect on decisions could be dropped:

1. Some of the "objectives" are more appropriately treated as resource options for portfolio construction. For example, the best way to ensure that adequate attention is given to the objective "promote implementation of appropriate new and existing technology" is to consider portfolios that include significant levels of alternative technologies.
2. Several of the "objectives" could be combined. For example, objectives "keep industry and business electricity costs competitive with other jurisdictions", "avoid rate shock," and "minimize financial costs of meeting electrical service needs," are all captured by an attribute for rates.
3. Some of the "objectives" were best treated in sensitivity analysis. For example, objective "provide flexible service to adapt to changes: changing demands" is best captured by the MATA by examining the robustness of portfolios to alternative load forecasts.
4. Some "objectives" have more relevance as general corporate goals. For example, objective "ensure fair and equitable treatment among firms and individuals providing electrical services" is a matter for general attention of BC Hydro's management to ensure a fair and transparent procurement policy.

The result of this process of mapping the individual objectives into the different components of the MATA process is shown in the second column of Table 8.3. Each objective maps into one of the following categories:

- reflected in an attribute
- treated as a sensitivity analysis issue
- treated as a portfolio building/ resource option issue
- flagged as needing appropriate attention at the project-level stage (EIA)
- treated by sensitivity analysis
- flagged as a general corporate objective.

From this process then emerged the list of attributes summarized in Table 8.4.[12]

Table 8.4: MATA attributes

attribute	units of measurement	primary concern
system cost	present value of system costs	long-term economic efficiency
rates	percentage change in rates over reference case	impact on customers
diversity	index	flexibility, risk management
upstream air emissions	million tons of GHG emissions	impacts of gas exploration, development and extraction
particulate emissions (see Box 2.2 for further description)	tonnes	rural air quality
greenhouse gas emissions	tonnes	climate change
population-weighted air emissions	person-tonnes	health risk
species persistence	# of affected red-listed species	biodiversity
terrestrial ecosystems	hectares lost	biodiversity
land use	hectares	community impact
employment	person-years of employment	employment generation

8.3 DEFINING THE ALTERNATIVES (STEP 2)

BCUC was quite specific about the characteristics of the MATA process to be used: "...the MATA should be conducted along the lines of that conducted by West Kootenay Power Company[13] as part of the development of its IRP" (BCUC, 1994). Yet almost at the very first meeting of the Committee it was clear that most members would not be satisfied with such a process (see Box 8.1), because most of its members wanted to examine resource portfolios, not just rank individual resources.

Box 8.1: "Rating-weighting" as used by West Kootenay Power

The 1995 West Kootenay Power IRP (1995) used an approach to MATA known as "rating-weighting," a variant of the additive value function (Section 4.3.1). The process began by characterizing the attributes of the individual resources available to the utility. The resources were then rated on a scale of 1 to 10 (or 1 to 100) by members of the West Kootenay Consultative Committee. The ratings for some resources followed directly from the measurement scale used (such as GHG emissions per kWh, or cost per kWh). For others, subjective scales were used (such as "high," "medium," and "low," then translated to a 1 to 10 scale)

Weights were then elicited from Consultative Committee members, and applied to the resource ratings. Notional "representative weight sets" were also used: one set of weights assigned 100% to the financial attributes, and none to others; another set put equal weight on all attributes, and so on.

Each resource *j* is then ranked according to the weighted score S_j :

$$S_j = \Sigma_i w_i R_{ij}$$

where w_i is the weight for the *i*-th attribute, and R_{ij} is the rating score for resource *j* on attribute *i*. The rating scores are were normalized such that the best-rated option was given a value 1, the worst-rated a value of 0, so that the above formula results in the best resource having the highest score.

The figure below outlines the approach: this may be contrasted with the portfolio analysis approach of Figure 8.2 (p. 169, below).

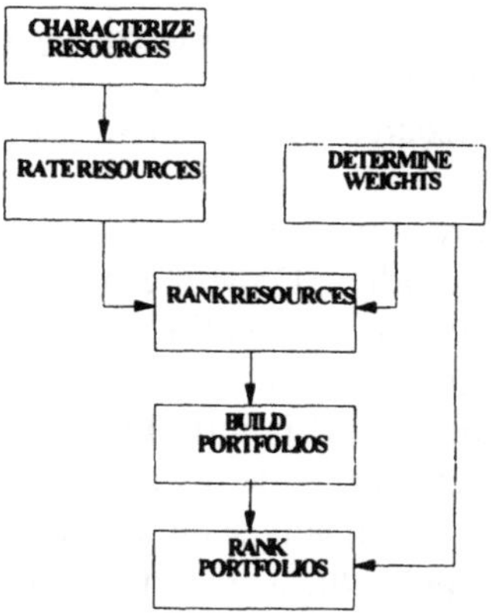

Although widely used, and despite the obvious merit of simplicity, there are many reasons why such simple rating-weighting schemes are unsatisfactory:

1. <u>How one rates a resource should be a function of timing</u>. For example, how one rates the desirability of woodwaste plants depends upon *when* they might be brought into service: if they cannot be brought into service within 2 to 3 years, then 1) they may no longer be available, since some other use has been made of the woodwaste resource and/or 2) a woodwaste plant may be competitive with gas plants in some years, but not in others (as a result of anticipated changes in future gas prices).

2. Use of a single set of "average" weights, even if a sensitivity analysis is conducted on those weights, may be inappropriate. Indeed, several members of BC Hydro's Consultative Committee took strong objection to the use of a weight set based on averages of individual weights of the members of the Committee. It was held that such an average simply reflected the particular composition of the Committee, and did not necessarily reflect the range of weights of the public at large.
3. Use of a rating-weighting scheme tends to lead to a mechanistic "black box" process, rather than one in which the focus of Consultative Committee deliberations is on an understanding of the tradeoffs to be faced, and on consensus building based on structured group discussions.
4. The performance of a resource depends upon the system context. For example, the unit cost and environmental performance of a thermal resource is a function of how it is dispatched. Yet how a unit is dispatched depends upon what other units are available in the system at a given point in time. In other words it may be hard to rate the performance of an individual resource without the context in which that resource operates. Moreover, it is not so much the cost of an individual unit that matters, as the cost of the portfolio as a whole. For example (as we shall see below), the value of DSM is much higher to a high-cost portfolio with significant levels of relatively expensive alternative technology, than it is to a portfolio based largely on low cost gas resources.
5. Some important attributes defy characterization on an individual resource basis. For example, one cannot reasonably estimate the rate impact of an individual resource without running a financial model for a complete portfolio.

The difficulty of rating resources in isolation of a portfolio is exemplified by DSM. In BC Hydro's earlier studies, DSM measures were assigned zero impact for the environmental attributes.[14] Yet the environmental impact of DSM depends upon which supply-side resources are displaced; indeed, most often, the impacts are positive. When DSM is introduced into a portfolio, the most likely result is that fewer new supply-side resources are required, leading to a reduction in attribute values for impacts such as land use; and, if fewer thermal resources need to be dispatched, a reduction of air emissions. This is simply not reflected in stand-alone resource characterizations, and distorts the results of rating-weighting schemes of individual resources.[15]

However, the most important methodological objection to simple rating-weighting schemes is that the link between the resulting ranked list of resources and the actual portfolios is vague. Suppose that resources "A", "B", "C" and "D" are the top ranked resources, in order ("A" the best, "B" the second best, etc.). It does not follow that the order in which the utility should bring them into its planning portfolio is the same -- in part because some resources may only be available at specific points in time, in part because "A", "B" and "C" may be completely different in nature. Resource "A" might be a capacity only, supply-side additions such as Revelstoke 5 and 6 hydro -- while "B" may be some set of demand side management programs. The result is that the rationale for constructing a portfolio at the end of the process becomes very difficult to demonstrate -- yet it is precisely this accountability of showing the link between decisions and impacts, that BC Hydro was directed to demonstrate in its planning.[16]

Figure 8.2: The portfolio analysis approach to MATA

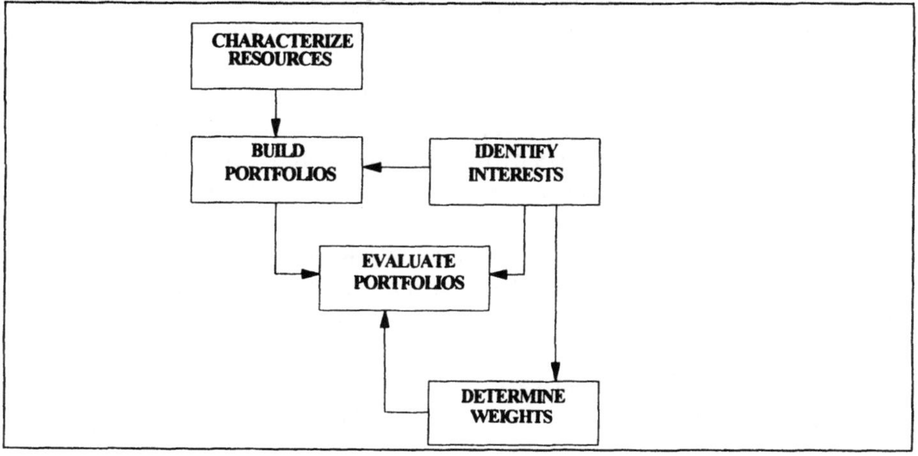

The alternative of Portfolio Analysis (Figure 8.2) is unquestionably more complex and time consuming, and one which requires a more sophisticated -- yet necessary -- modeling effort. The main features of this approach are:

1. Resources are evaluated in the context of an overall portfolio, which means that *timing* of a resource is one of its intrinsic characteristics. Even screening of resources is done in the context of a portfolio; to this end the process begins with a "reference portfolio," typically the previous year's planning portfolio. It it always *portfolios* that are assessed and compared, not individual resources.
2. It is the attribute values of the portfolio, not those of individual resources, that matter. For example, it is the total GHG emissions of the portfolio, over some planning horizon, that is of primary interest. Comparisons of GHG emissions per kWh produced by a particular resource based on an assumed operating level may or may not translate to similar comparisons at the portfolio level.
3. Weights are used to help individuals make consistent evaluations of portfolios. However, weights are not averaged across the members of the Consultative Committee. The process of consensus building and of narrowing down a larger set of initial portfolios to a smaller number of final portfolios is achieved by structured group discussion, not by ranking of portfolios using some set of average weights.

Members of the Consultative Committee and BC Hydro Managers group were therefore never asked to rank and compare, for example, DSM with Burrard repowering.[17] Rather they were asked to rank and compare *portfolios* that contained DSM, or Burrard repowering, or both.

Model Selection: Several options were considered for the modeling work that was required to implement the Portfolio Analysis. One option was to use an American IRP model, such as IRP Manager or MIDAS.[18] However, because of time limitations, acquisition of a complex model that required significant training and installation effort was not considered practical. Moreover, it was unclear that such a model, developed primarily to address the needs of the mainly thermal private utilities of the United States, could easily be adapted to the mainly hydro system of British Columbia.

A second option was to use BC Hydro's in-house Resource Planning Model (RPM, Drinnan and Spafford, 1997). While this was obviously ideally suited to analysis of BC Hydro's own system, as a model still implemented on a mainframe it lacked the necessary flexibility to meet the demands of the schedule.

The option finally chosen was to use ENVIROPLAN, a model developed for the World Bank Environment Department as a tool to incorporate environmental considerations in power sector planning (Meier and Munasinghe, 1994). RPM and BC Hydro's Corporate Financial Model (CFM) were used to calibrate ENVIROPLAN at various stages of the modeling, and were also used to provide detailed results for the final portfolios.

8.4 PORTFOLIO ANALYSIS

It was quickly recognised that there is an almost infinite number of possible portfolios. In theory, one might be able to define an "optimal portfolio" provided all of the objectives could be monetized: even if this were theoretically possible, the Committee expressed strong objection to the proposition that a plan would be chosen by some "black box."[19] However, it *was* agreed that the portfolio building process would benefit from some "reference" portfolio, that would in turn benefit from all of the work done by BC Hydro in its earlier 1994 Plan. The reference portfolio consisted of the following resources:

Demand-side management program assumptions:

- Implementation of BC Hydro's 1995 Demand-side Management Plan. The capacity and energy savings projected in that Plan are subtracted from the energy and firm capacity requirements that must be met.

BC Hydro's in-house supply-side resources:

- Hydro: Revelstoke 5 in 1999; Stave Falls in 2000; Seven Mile 4 in 2000; Revelstoke 6 in 2000; and Mica 5 in 2007.
- Gas: All of the six Burrard units are upgraded with selective catalytic reduction (SCR). One unit was already so fitted in 1995: remaining upgrades to be completed in 1996, 97, 98, 99, and 2000).

Independent power project (IPP) resources:

- The balance of the energy and firm capacity requirements necessary to meet the load projection were assumed met from IPP resources. These were placed in the reference portfolio in the order of their levelized unit energy cost, excluding transfers, in such a way that the lowest cost resource is built first, the second lowest cost resource next, and so on (i.e., in so-called *least cost merit order).*

Agreement was then reached on a workplan for the Committee's deliberations that had the following steps:

1. Conduct a screening analysis by removing each resource in the reference portfolio, one at a time (and replacing it by the next best alternative); followed by a screening of other options not considered in the 1994 plan (the screening of supply side options is described in Annex 3.3).
2. Conduct a first round of portfolio building in which each member of the Committee perturbed the reference portfolio in a manner consistent with his other interests and preferences. ENVIROPLAN was used to calculate the attributes for these portfolios, and results were presented at the next Committee meeting.
3. Conduct a weighting exercise (described below).
4. Compare the various portfolios for each attribute by applying the weights.
5. Reduce the number of portfolios by combining common elements.
6. Interview each participant individually to discuss his/her ranking of portfolios, and resolve inconsistencies.
7. Repeat steps 3, 4 and 5 until consensus was reached, or until such time as no further reduction in portfolios was possible.

Inevitably, as in almost every IRP conducted in North America over the past decade, demand-side management was by far the most contentious matter.[20] BC Hydro had already conducted a detailed screening of individual DSM programs (BC Hydro, 1995c), using a DSM screening model called COMPASS.[21] After extensive discussions, DSM programs were placed into three groups reflecting increasing levels of DSM:

1. *A level of DSM that included only those programs that would pass the RIM test*, i.e., those programs in which the utility's cost was less than or equal to the trailing block rate for different rate classes. This level of DSM reflected the interests of several members of the Consultative Committee who were concerned about the impact of DSM on industrial rates.
2. *A level of DSM that corresponded to BC Hydro's 1995 DSM program.* These are, in the main, DSM measures that pass the TRC test: i.e., measures whose total resource costs (including those incurred by program participants) are less than the avoided supply costs, and whose implementation can be expected to lower the overall economic costs of supplying some given level of load. This level of DSM was built into the reference portfolio. However, even if total costs fall. rates may increase.
3. *A level of DSM that added to BC Hydro's current DSM program the additional DSM potential identified as achievable in the Conservation Potential Review (CPR).* This level of DSM assumes implementation of new technologies available only after 2003. It reflected the interests of

several members of the Consultative Committee who wanted to keep the number of new supply resources to a minimum, and who were concerned in particular about GHG emissions; and those interested in promoting societal behavioral changes to reflect a sustainable conservation ethic.

8.5 TRADEOFF DISPLAY (STEP 5) AND EVALUATION OF PORTFOLIOS (STEPS 8,9)

The two sets of first-round portfolios -- the first from the Consultative Committee, the second from the Managers' group -- were in many respects quite similar. Resources such as community energy planning, alternative technologies and East Kootenay coal appeared in both sets. Both sets spanned the full range of DSM, from no DSM, to exploitation of the full DSM potential identified by the Conservation Potential Review. Figure 8.3 illustrates the extent of overlap in the *impacts* for cost versus GHG emissions. The major factor in the spread of results is the extent of DSM in each portfolio.

Figure 8.3: Comparison of portfolios: cost v. GHG emissions

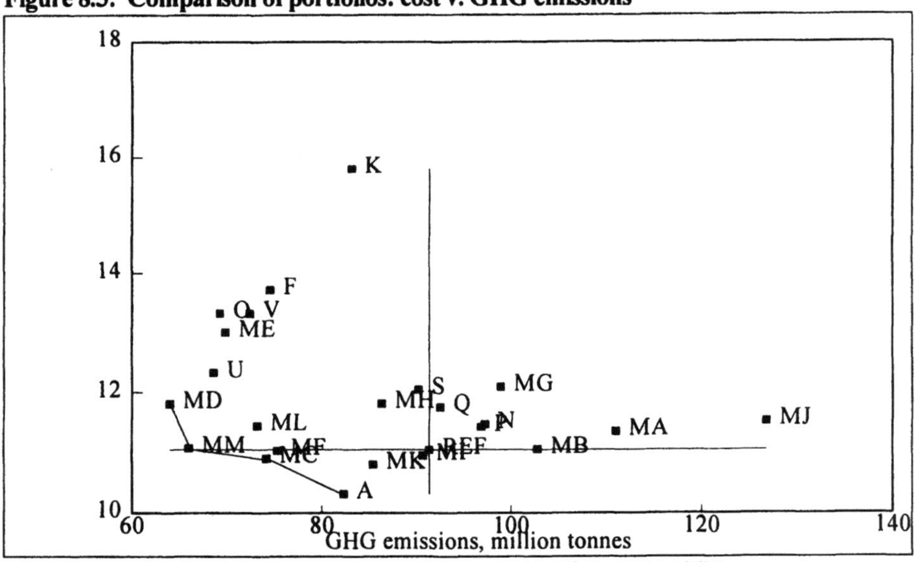

Key: Consultative Committee portfolios: A to V; Managers' portfolios MA to MM

Holistic assessment of the first-round portfolios: The results are indicated in Table 8.5.[22] The most instructive finding was that although the Reference Portfolio received no first-place scores, it had the best average rank (4.9), followed closely by Portfolio I. Portfolios such as F, M and P, which each got two first-place scores, had much lower average ranks (7.8, 7.6 and 6.4, respectively) -- because these portfolios also received very poor ratings by some. In other words, while there were sharp disagreements about the merits of some portfolios, the Reference Portfolio was the one that was *least disliked* by the Committee as a whole.

Table 8.5: Results of the holistic assessment

	REF	A	B	C	D	E	F	G	H	I	J	K	L	M	N	O	P
C	2	10	8	9	10	10	10	6	7	5	10	4	10	1	3	10	10
H	3	14	15	6	10	9	12	2	11	5	17	4	16	7	8	13	1
J	7	9	3	4	2	1	13	12	5	6	14	13	14	8	10	11	14
K	8	2	8	8	8	8	1	8	8	4	8	6	8	8	7	3	5
N	4	7	14	13	12	11	6	8	9	5	17	2	17	10	3	15	1
Q	3	5	4	6	7	8	17	9	10	1	16	11	15	12	13	14	2
R	8	3	10	9	11	7	2	17	16	4	12	5	13	15	6	1	14
U	9	2	13	12	11	6	1	13	13	5	13	7	8	13	10	3	4
V	6	1	6	6	6	6	2	6	6	3	6	5	6	6	6	4	6
W	2	10	10	5	4	10	10	10	9	6	10	7	10	3	1	10	8
avg. rank	**4.9**	**6.9**	**9.8**	**7.7**	**8.7**	**7.3**	**7.8**	**8.5**	**10**	**5**	**12.4**	**6.5**	**11.5**	**7.6**	**6.6**	**8.9**	**6.4**
best	2	1	3	4	2	1	1	2	5	1	6	2	6	1	1	1	1
worst	9	14	17	13	15	11	17	17	16	11	17	13	17	15	13	15	14

Key: **Columns** = portfolios; corresponding to the identifications shown in Figure 8.3 for Consultative Committee portfolios. **Rows** = individuals

Although the specific *numerical* results of this exercise were not further used, what the results *did* suggest was that the reference case was an appropriate starting point for consensus building. Just because it provoked no strong objections from anyone in the Committee, we deemed it an appropriate starting-point at the next consensus building stage of the process in which the 23 original portfolios were narrowed down in a series of small group discussions.[23]

The weighting exercise: In Chapter 4 we discussed in some detail the advantages and disadvantages of the various weighting methods. Unlike MATA processes used in other utilities, we consciously avoided the determination, and application, of an "average weight set". We did this for the following reasons:

- Several members of the Consultative Committee strongly objected to any averaging of weights. It was noted, correctly, that such an average would simply reflect the relative proportion of different interests which happened to be represented on the Committee, and would almost certainly not be representative of the broad public.
- In the face of such strongly held views, the use of average weights as a means of ranking portfolios was held to be spurious, and might well represent no one's interests. Rather, the Committee wished to try to seek consensus based on an informed group discussion.
- Application of an average set of weights to different portfolios tends to reduce the MATA process to one that is essentially mechanistic. As noted in the Introduction, the primary purpose of the process is to clarify and assist understanding of tradeoffs, not to produce numerical scores.

Perhaps the most important reason for maintaining weights at the individual rather than the group level, is that the determination of average weights by a Committee tends to give such weight sets a general standing that is entirely misplaced, and that will inevitably be misused:

- Weights that are appropriate for long-range planning may not be appropriate for project-level decisions. One member of the Consultative Committee noted that he might well give greater weight to environmental factors for a project-level analysis, where differences in site factors and local environmental impacts might be much greater than differences in costs, and hence merit greater weight. (This is consistent with our argument in Section 4.2 that weights should reflect the ranges of attributes involved).
- Most importantly, weights are meaningless in the abstract. Because weights are range dependent, the weights determined in this Plan had meaning only for the ranges of attribute values considered for the 1995 Electricity Plan.
- It was important for BC Hydro to understand where and why disagreements arose among stakeholders. The use of average weights hides that information.

Figure 8.4: Comparison of weights using hierarchical and non-hierarchical structures

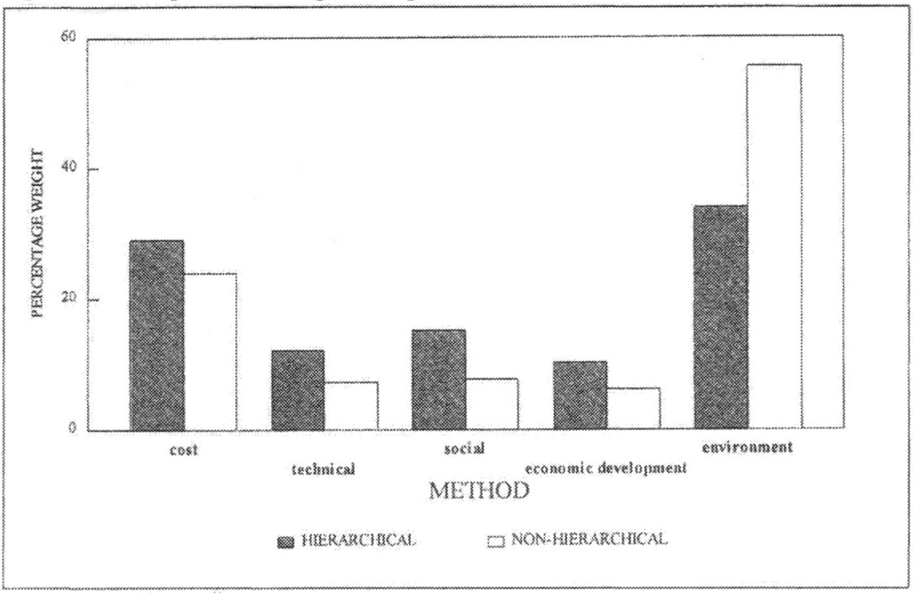

Note: "ENVIRONMENT" is the sum of six environmental attributes

Figure 8.4 shows the results of applying hierarchical and non-hierarchical weighting in a trial run conducted by the Committee. As would be expected, much higher weight was given to environmental attributes in the non-hierarchical scheme, at least in part because environmental attributes outnumbered non-environmental attributes by 9 to 3. (Chapter 10 discusses this "splitting bias" in more detail.)

Once we had demonstrated the point that how one asked weighting questions affected the result, we made no further distinction between the two methods in subsequent weighting exercises. Both types of weights for each participant were used in additive value functions to rank the portfolios. We then showed each participant their portfolio rankings resulting from both hierarchical and non-hierarchical forms at the same time, so they could use whichever weights they preferred, and could satisfy themselves that individual attribute weights properly reflected their interests and values.

8.6 GETTING TO CONSENSUS (STEP 10)

It is often assumed that the success or failure of an MCDM can be measured by the extent to which the deliberative body using the analysis comes to consensus. There is a tendency for the practitioners of MCDM to design the process in such a way that the end result is indeed a single numerical scoring and/or ranking of resources and portfolios for the entire group, in which the tradeoffs are presumed to have been incorporated.

We hold as unfortunate this presumption of a linkage between consensus and success of an MCDM. In fact, absence of consensus in a Consultative Committee is a reflection of differences in interests and values held by the participating stakeholders, and may therefore have very little to do with the validity or success of the MCDM process itself. Indeed, a successful MCDM may well highlight those differences in values and interests. Success or failure of the MCDM should therefore be gauged not by whether deliberative bodies reach consensus, but by their answers to the following questions:

1. Have the procedural objectives have been met? Did the process facilitate informed discussion? Did it help in improving the understanding of the tradeoffs that BC Hydro must make?
2. Has the process assisted the transparency and accountability of the decisions and tradeoffs BC Hydro had to make in preparing its Electricity Plan?

By the end of June 1995, the original 23 portfolios had been reduced to three (plus the reference portfolio), of which two had variations; by early August three final portfolios had been framed, which represented the limits of agreement. On some issues -- such as Burrard Repowering -- the Consultative Committee did reach consensus. In large measure this was due to the MATA, since it became self-evident that this was a part of most non-dominated portfolios. On other issues, such as the level of DSM and the extent of small/ medium hydro and alternative technologies, there remained disagreement. Table 8.6 depicts the extent to which agreement on resource portfolios was finally reached.

The F1 portfolio, advocated strongly by the representatives of industrial consumers, was designed to keep rate impacts to a minimum, incorporating a minimum level of DSM, and adding new resources solely on the basis of least cost. The F2F portfolio, advocated strongly by the environmental interests, had the objective of stabilizing GHG emissions, and minimizing air emissions: it had the highest system cost and rate impact, a consequence of the emphasis on solar photovoltaics, high levels of DSM and limitations on the amount of lower-cost thermal (gas) resources.

The final tradeoffs on the levels of DSM, alternative technologies, and low-cost gas -- points on which there was disagreement in the Committee -- were made by BC Hydro. The portfolio recommended by BC Hydro (BCH in Table 8.6) thus inevitably reflects a compromise between the two Committee portfolios.

Table 8.6: Composition of the recommended portfolios (%)[a]

	F1	BCH	F2F
Burrard (natural gas)	29	29	29
Revelstoke 5; Stave Falls; Seven Mile 4 (hydro)	18	18	18
DSM	5	5	5
Revelstoke 6 (hydro)	13	13	13
total common elements	**65**	**65**	**65**
additional DSM	0	10	12
Mica 5 (hydro)	6	6	0
wood residue	2	1	0
alternative technologies	0	0	2
other natural gas	19	10	7
Burrard (natural gas)	5	5	0
other small/medium hydro	3	3	14
	100	**100**	**100**

[a]Contribution to 2014/2015 dependable capacity (excluding reserve margin)
Source: BC Hydro (1995a)

NOTES

[1] This Chapter is based upon the work conducted by the Resource Planning Group of BC Hydro's Corporate and Strategic Planning Department for the 1995 Integrated Electricity Plan (BC Hydro, 1995a; Meier, 1995b). The writers thank Donna LeClair of BC Hydro for her contributions.

[2] As usual, the Plan ran to several large volumes. However, a 16-page overview document provides a concise summary of the main findings (BC Hydro, 1995b).

[3] This directive was to prove extremely contentious. BC Hydro appealed the directive to the Courts, not on grounds that it was opposed to planning, or IRP, or even MCDM, but that the Commission had exceeded its authority in attempting to micro-manage its planning process. In the end (early 1996), the Courts ruled in favor of BC Hydro.

[4] Because the Commission used the words "multi-attribute tradeoff analysis," the BC Hydro documents refer to "MATA."

[5] Its representation was directed by BCUC to include those of the previously established DSM collaborative. Two of the members of the Committee were also members of the group that participated in the BC Gas study, described in Chapter 9. The representatives listed in Table 8.1 were the ones who participated regularly and actively throughout the MCDM process. Some additional members of the Committee (First Nations, Canadian Homebuilders Association, and the Union of BC Municipalities) attended only occasional meetings, and were not involved in the portfolio building exercise.

[6] One member of the Committee was moved to say in his final comments on the Plan (BC Hydro, 1995a, Appendix B) *"..there were occasions on which the IPP representative blocked consensus, advocated resource portfolios which his constituents could profit from, and drew the Committee into time-consuming explorations of alleged inequities between IPP-built and utility-built projects."*

[7] In part this was a consequence of the fact that a few weeks before the work of the Committee began, BC Hydro had started legal proceedings to overturn the November 1994 BCUC decision. At the same time, BC Hydro executives were understandably irritated by the Commission's decision to levy heavy personal fines if the requested IRP was not delivered to its satisfaction on methodology, and to the unreasonable schedule demanded by the BCUC.

[8] Many of the comments (BC Hydro, 1995a, Appendix B) were begrudging: *"..the Integrated Electricity Plan is actually a much better document than we expected."* The role of the moderator was reflected in such final comments as *"...In retrospect, it was surprising that the meetings were as amicable as they were."*

[9] The time pressures imposed on the Committee's work was one of the greatest difficulties expressed by the participants, and many expressed their frustration on this matter in their final comments (BC Hydro 1995a, Appendix B).

[10] In both the Centerior and SCL studies, between 6-12 months of preparatory work preceded the first meeting of the Group that was to participate in the MCDM process.

[11] The list has numerous other definitional problems. For example, it is unclear what is meant by the word "appropriate" in the objective "promote implementation of appropriate new and existing technology." Individuals may disagree as to what is meant by "appropriate" (and surely nobody would ever admit to promoting *in*-appropriate technology).

[12] Nevertheless, this was not the end of the discussion about attributes. The decision to include the induced employment impacts of DSM in the employment attribute was subsequently discussed at some length by the Consultative Committee. Some members took strong exception

to the fact that rate-change induced employment impacts of DSM were taken into account, but that the rate-change induced employment impacts that might result from supply-side measures were *not* included. Indeed, this subject provoked some of the most heated arguments of the entire process.

[13] West Kootenay Power is a small utility in BC. It had submitted an IRP in 1994 that had largely satisfied the BCUC.

[14] In the West Kootenay Power IRP, for example, the resource DSM is given zero SO_x emissions, "low" noise impact, zero land use impact. These ratings were then used in the resource ranking process.

[15] Nevertheless, this general environmental benefit to DSM is not always the case. In some instances, if DSM results in a deferral of new, more efficient thermal plants in systems that have a significant number of older (and hence less efficient) plants, it may happen that DSM *increases* emissions in the earlier years. This occurs because older units with higher emissions get dispatched more than if the new more efficient units were built sooner.

[16] One of the problems is aggregation. The use of qualitative scales (e.g., high/medium/low) can perhaps be justified for a project-level analysis, where the options being rated and weighted are mutually exclusive (one either builds A, or B, or C). But when one has to subsequently combine ranked projects into portfolios, the validity of aggregation is an issue. Do three "lows" equal one "medium"? How many "mediums" equal a "high"? It follows that whatever plants are included in a portfolio, the portfolio itself has to be re-assessed for attributes with qualitative scales. And if the portfolio itself must be re-assessed, then the original project-level ratings may have little relevance. In short, it is unclear that the initial rating of individual resources is necessary.

[17] Burrard is an older gas-fired steam cycle plant for which one of the options was to repower and convert to combined cycle.

[18] West Kootenay Power had used MIDAS for its 1995 IRP.

[19] See also Section 7.9 for other objections to "least cost" optimization models.

[20] There were some other controversial issues, of course, but these appeared to be more in the nature of individual concerns (such as the biases against private power alleged by the IPP representative, or opposition to Alcan purchases), and the Committee was reluctant to get drawn into long discussions on these. There was no such reluctance to discuss DSM, which came up at almost every meeting.

[21] This model screens programs on the basis of widely used tests, such as the total resource cost test (TRC) and the ratepayer impact measure (RIM) test (Hobbs, 1991).

[22] Three members of the Consultative Committee were unable to attend the session at which these holistic evaluations were conducted.

[23] The extent to which the desires of members of the Committee could be accommodated is in itself an MCDM problem! Several members wanted to examine more than one portfolio (and felt quite strongly that they should be given the opportunity to do so). Against this stood the pressures of time. Fortunately the ENVIROPLAN model was well suited to quickly handling a very large number of portfolios. And so 23 portfolios from 13 Committee members, and 15 portfolios from the Manager's group, started the process. We felt that the additional work at this first stage in dealing with so many portfolios was more than offset by the goodwill it generated by acceding to the desires of the Committee; moreover, the strategy of letting the numbers speak for themselves made the process of reducing portfolios to a more manageable number in the second round a relatively easy one.

9 MULTI-METHOD MCDM AT BC GAS[1]

Agreement is made more precious by disagreement
PUBLILIUS SYRUS, Moral Sayings, 15:1 (1st C B.C.)

The most savage controversies are those about which there is no good evidence either way
BERTRAND RUSSELL, Unpopular Essays (1950)

Public involvement in energy policy serves several purposes. Some important ones are: to ensure that public values are reflected in decisions; to obtain information on impacts that might otherwise be overlooked; to inform the public; and to provide "due process" in a way that the public perceives is fair. All of these purposes can, but do not necessarily, help to build public support for the decision process and outcome.

This chapter presents a detailed example of the quantification of public values and their use in making decisions concerning utility sponsored demand-side management programs. The context was a stakeholder involvement process sponsored by BC Gas, a natural gas distribution utility located in Vancouver, BC, Canada. Details on the step-by-step elicitation and application of value judgments are provided to help the reader visualize how the steps might be executed in other contexts.

What is distinct about the methodology of this chapter is the use by stakeholders of three MCDM methods, resulting in three sets of DSM program evaluations for each stakeholder. Applying these multiple evaluations, the stakeholders each developed one set of recommendations through an interview process and several group discussions. Learning and confidence building were emphasized throughout this iterative process of developing recommendations. This multi-method approach recognizes that different MCDM methods can yield significantly different decisions, and that no one method is best for every person. In contrast, a more traditional MCDM approach would use only one method (Corner and Buchanan, 1995). Further, it would be inclined to not allow stakeholders to adjust their value judgments after reviewing the results for fear that the results would be tainted. However, our opinion is that the multiple-method approach, coupled with an interview process and structured group interaction, leads to greater understanding of technical issues and value tradeoffs by stakeholders, greater opportunity to check the data, and consequently greater credibility for the results.

In the next section (Section 9.2), we argue that no one method is best, and that applying more than one approach is worth the extra effort. Then, as an example of the multi-method MCDM approach, we explain in Section 9.3 how it can be used to screen DSM programs sponsored by a natural gas utility. The nine step MCDM procedure followed by the BC Gas stakeholder group is then outlined in Section 9.4. Section 9.5 closes the chapter with a discussion of the stakeholders' evaluation of the process. The results in Sections 9.4 and 9.5 confirm that the rankings yielded by alternative MCDM methods can significantly disagree, and that a multi-method approach builds stakeholder understanding and confidence. These results are reinforced by the experimental comparison of MCDM methods we describe in Chapter 10.

9.1 DESIGNING A MULTI-METHOD MCDM-BASED PROCESS

There has been much effort recently to try to include social costs, such as environmental effects, in economic benefit-cost analysis of DSM programs. Some of the stakeholders in the BC Gas process argued for such monetization (Keeney and McDaniels, 1999). However, as we explained in Chapter 1, we believe that MCDM methods are useful for including such considerations under any of the following circumstances:

- fundamental value conflicts exist among stakeholders;
- the decision process involves the public sector and unique local issues; or
- dollar estimates do not exist for social costs, are lacking, or are of dubious relevance.

These circumstances are common characteristics of energy planning processes, and apply to the BC Gas resource planning problem.

Although regulators and BC Gas management agreed that MCDM methods were appropriate for the BC Gas DSM management process, there remains the problem of which MCDM approach to use. Literally hundreds of

MCDM methods have been proposed. As should be clear from Chapters 3 to 5, methods differ in their general function (e.g., generation and display of tradeoffs versus elicitation of value judgments), approach to amalgamation (e.g., none versus additive versus non-linear), and how value judgments are elicited (such as direct assessment of "importance" weights on a 0 to 100 scale versus elicitation of acceptable tradeoffs). As we note elsewhere in this book, the criteria of appropriateness, ease of use, differences in results, and method validity are helpful in choosing from this bewildering variety of methods.

We have argued earlier that different methods can yield very different decisions, and document that assertion in the Seattle City Light experiment in Chapter 10.[2] Furthermore, results in Chapter 10 and in the literature show that no one method is most valid for all people: different people respond better to different methods (e.g., Corner and Buchanan, 1995). Given these results, and the fact that not all stakeholders will start the process with definite and quantifiable preferences among criteria and resources, we conclude that an MCDM stakeholder process should use more than one method and allow people to reflect upon and change their values. In particular, we suggest that the following requirements should be satisfied:

- *Stakeholder control of the process.* Stakeholders distrust processes that ask for inputs (perhaps difficult and confusing ones such as tradeoff judgments) and then process them, proclaiming results without giving people a chance to reflect and adjust their inputs. People do not like to feel out of control, especially when their inputs are tentative. The stakeholders, not a mathematical procedure, should make the ultimate recommendations.
- *Use of multiple methods to increase validity of the results.* Methods should be supported (in terms of producing valid results) by both theory and experimental evidence. In the case of weighting, theory definitely favors tradeoff judgments as a technique for choosing weights (Chapter 4; Keeney and Raiffa, 1976). This is because the tradeoff method explicitly asks how much of one criterion a person is willing to give up in order to get an improvement in another. In contrast, many other weighting methods request directly assessments of "importance", a concept that is usually vaguely defined (Hobbs, 1986; Schoemaker, 1981; Weber and Borcherding, 1993). People's direct weights have been shown, for example, to be insensitive to the ranges of criteria (Fischer, 1995; Hobbs *et al.,* 1992; and Chapter 10) even though, for instance, a doubling of the range of cost should result in a doubling or so of its weight. On the other hand, experiments also show that the difficulty of tradeoff questions can cause people to become confused and distrustful of the method, and, in practice, direct methods can sometimes better elicit people's preferences (Hobbs *et al.*, 1992). The multiple-method approach helps to capture the best features of both philosophies of weighting.
- *Opportunity for reflection and revision of value judgments.* Where people are unsure of their precise values, MCDM methods can help them better appreciate the problem, explore their feelings, form a coherent,

defensible set of values, and understand the implications of those values for the decision. Too often, MCDM methods are naively viewed as a way of obtaining a snapshot of a person's "utility function" -- as if a person's preferences were eternal and well-defined. Instead MCDM methods ought to promote learning, and allow people to adjust their judgments accordingly. Hence the inevitable inconsistencies that result from more than one method are to be welcomed (Fischoff *et al.*, 1980) -- because they facilitate learning by forcing people to think hard about what they really want. This can help build confidence in the quality of decisions.

- *Simplicity and clarity.* Applying multiple methods has a drawback: participants in experiments have found that applying more than one method, and then comparing and resolving the differences was hard work (see Chapter 10). Yet most participants in those experiments judged that this effort was worthwhile because of the learning that took place, and because they became more confident that the weights reflected their preferences. The level of effort required can be mitigated by keeping each method as simple and clear as possible. People want to understand the problem and the methods. Simple approaches (e.g., additive value functions, as opposed to multiplicative ones) are better for this reason, as long as they do not distort people's preferences.

In the remainder of this chapter, the BC Gas multi-MCDM method approach, which attempts to meet these requirements, is explained and applied.

9.2 USING MCDM METHODS TO SCREEN DSM PROGRAMS

The general approach of applying multiple MCDM methods and then resolving differences can benefit any energy policy or planning process. The three-method MCDM procedure described in the next section has been specifically developed for use in constructing a portfolio of non-exclusive alternatives; however, the general steps outlined in Section 9.4 can be easily adapted to other planning problems.

Customarily, MCDM applications assume that the alternatives are mutually exclusive: that is, just one of the options is to be chosen, and the others rejected. As explained in the last Chapter, in resource planning it is portfolios that are mutually exclusive, not individual programs. However, there are so many possible portfolios (2^{20} in the case of BC Gas, since there are 20 programs), it is desirable to simplify the process by having stakeholders examine and then accept or reject individual DSM programs. The optimal portfolio would then consist of the collection of individually acceptable programs. This simplification is possible under the following four assumptions.

First, the adoption or rejection of individual programs does not alter the marginal effect of any other program upon the portfolio's performance on the criteria. For example, we assume that individual DSM programs are small enough such that avoided (or marginal) supply costs and emissions do not depend on which programs are implemented. This assumption was judged by BC Gas to be reasonable for all criteria except for two "portfolio-level" attributes: equity and diversity. For those two criteria, whether or not a given DSM program is in the portfolio would affect the marginal contribution of other programs to those criteria. For instance, if one DSM program directed at improving energy efficiency in new buildings were already accepted, adding a second such program would not diversify the portfolio to as large an extent as did the first program.

Second, single-criterion value functions can be approximated as being linear. As explained in Chapter 4, a single-criterion value function translates a physical or economic criterion into a measure of value. In decision analysis, quadratic, exponential, and linear value functions are commonly used. Here, the assumption of linearity is reasonable because 1) the aggregate impact of all the DSM programs is not large compared to BC Gas' overall costs and environmental impact, and 2) experiments have shown that variations in weights affect decisions more than using non-linear value functions (Hobbs, 1986).

Third, an additive value function is an appropriate means of combining criteria into a single measure of resource portfolio attractiveness. An additive value function is used here because it is the most widely used multicriteria value function, and has the virtue of simplicity. Further, experiments have found its results to be robust with respect to moderate non-additivities (Stewart, 1996).

Fourth, there is no DSM budget constraint. Such a constraint would convert the problem into a "knapsack" problem which, in general, requires integer programming methods for selection of an optimal portfolio that satisfies the constraint.[3]

Under these four assumptions, it can be shown that any DSM program whose weighted sum of incremental impacts is greater than zero would increase the attractiveness of the portfolio. "Incremental impacts" are measured in relation to the marginal supply source that an efficiency DSM program would displace or that a load-building DSM program would require.[4] For instance, the adoption of energy efficiency program X might yield a cost saving of $Y due to a reduction in gas supplied by the marginal source and a reduction of Z tons of NO_x emissions.

The steps of the BC Gas MCDM procedure are summarized in Figure 9.1. Details on each step are provided in the next section. These steps do not correspond step-by-step with the general MCDM outline in Chapters 3 to 5: for instance, Steps 1 to 3 of the general procedure are compressed into one step here. The explanation that follows emphasizes the unique aspects of the BC Gas process: the use of several methods by the stakeholders, and the role of individual interviews in resolving conflicts among methods and building stakeholder confidence.

9.3 APPLICATION

The nine-step procedure was applied by a 16 member stakeholder advisory group over a six-month period in late 1994 and early 1995. The stakeholders included representatives from environmental, business, senior citizen, aboriginal, and other interests. Their role was primarily to communicate their perspectives on resource planning issues and particular resource alternatives to BC Gas, which was responsible for making the final resource choices.[5]

Figure 9.1: Steps of the three-method MCDM process

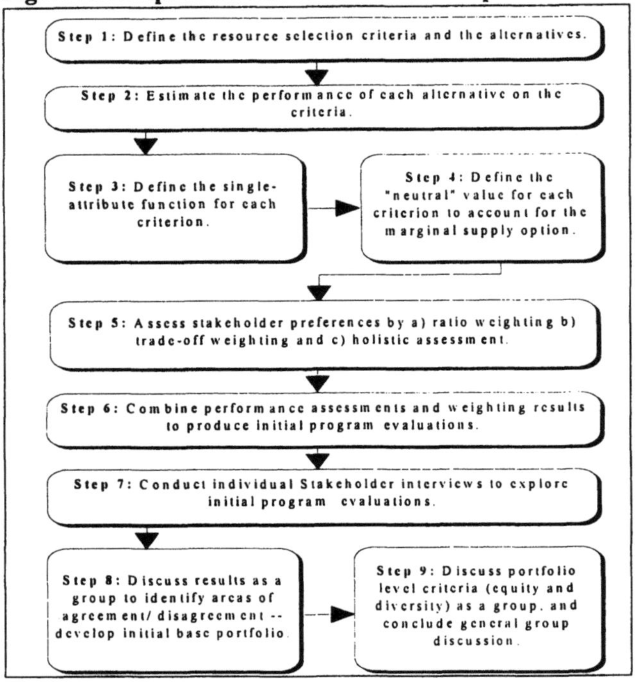

Step 1: Define criteria and alternatives

Criteria were defined by the stakeholder group during a series of meetings. The criteria definition process benefited from a multi-criteria exercise involving many of the same stakeholders held the previous year (Keeney and McDaniels, 1999). In the earlier process, emphasis was placed on understanding which fundamental objectives were important to stakeholders, and on criteria that could be used to quantify those concerns. The result was 28 criteria. The 1995 group used this previous work and other sources as a starting-point, and eventually whittled the list to nine program-level criteria, grouped under five objectives, along with two portfolio-level criteria. Program-level criteria are used in the additive value functions to evaluate programs individually; portfolio-level criteria were applied by the group late in the process (Step 9) to determine whether the portfolio was appropriately balanced. Criteria were only included if they were potentially important, measurable in some way, and were significantly affected by resource choices.

The adopted criteria are shown in Table 9.1. Because the focus of the planning exercise was on individual DSM programs, the criteria were defined as the incremental impact of a program upon the resource portfolio's overall performance. As an example, a value of zero for the criterion "net resource benefits" (designated as TRC, for "Total Resource Cost") means that the DSM program would have zero net effect upon the total cost of the resource portfolio (consistent with our discussions in the BC Hydro case study, Chapter 8).

Table 9.1: Measurable criteria and their characteristics

	criteria	units	abbrev.	range	
				worst	best
1(a)	maximize total resource benefits	$ net benefits/year	TRC	-50,000	650,000
1(b)	maximize flexibility in resource selection	flexibility/size scale	Flex	0	100
2(a)	minimize loss of environmental habitat	km linear disturbances/year	Km	2	-4
2(b)	minimize CO_2 emissions	CO_2 tonnes (or CO_2 equivalent)/yr	CO_2	5,000	-8,000
2(c)	minimize emission of other air pollutants	NO_x, tonnes/year	NOx	6	-4
3(a)	promote job creation and retention	# jobs	Jobs	-1	19
3(b)	conserve energy	change in GJ of non-renewable energy/yr	GJ	50,000	-150,000
4(a)	maximize rate-payer benefits	$/yr ratepayer impact criterion	RIM	-100,000	120,000
5(a)	minimize the risk to shareholders	risk/size scale	Risk	-100	0

As mentioned earlier, "equity" and "diversity" criteria are more appropriate when considering entire portfolios of programs. It was reasoned that while a given program may be inequitable or narrow in that it focuses on a single customer group, it may be that a combination of programs is equitable and diverse. Equity, in this context, refers to whether a portfolio of DSM programs is evenly targeted across the customer groups. Diversity concerns the mix of programs by type (e.g., energy efficiency, fuel switching, market transformation). Equity and diversity were not measured for individual programs; rather, the stakeholder group assessed the performance of portfolios on these criteria later in the process (Step 9).

The alternative DSM programs were developed by BC Gas and brought forward to the stakeholder group for review. They are shown in Table 9.2.

Step 2: Impact estimation

A_i is defined as attribute i; the specific value of that attribute, A_{ij} can be defined as the performance of DSM program j on criterion i. Table 9.3 displays their values. BC Gas used expert judgment to assess program flexibility and risk, while financial and environmental models were relied upon for the other A_{ij}.[6] Note that four commercial DSM programs (C6 through C9) have two sets of values. These are fuel switching programs in which customers would be enticed to switch from electricity to natural gas for particular end uses. Because there is considerable uncertainty concerning the long-run marginal environmental impact of electricity production in British Columbia, BC Gas felt it necessary to define two impact scenarios for each of these programs. These scenarios are labeled "M" (for "minimal electric emissions case") and "H" ("heavy emissions case") in Table 9.3.

Table 9.2: Programs proposed by BC Gas

reference number	program description	reference number	program description
R1	residential hot water savers	C4	commercial custom energy saving
R2	residential high-efficiency furnace	C5	commercial interruptible service
R3	residential energy advisory	C6	commercial gas cooling pilot
R4	residential retrofit	C7	commercial water heating
R5	residential renovation pilot	C8	commercial hot water booster
R6	residential "designed comfort" pilot	C9	commercial new construction
R7	residential off-oil conversion	NGV1.1	maintain refueling infrastructure
C1	commercial new building	NGV1.2	vehicle upgrade/cylinder purchase
C2	commercial boiler/furnace tune-ups	NGV1.3	commercial vehicle purchase
C3	commercial heating controls	NGV2.1	commercial/medium duty vehicles

Also uncertain are the prices of future gas supplies. The values of resource benefits (TRC) and rate impacts (RIM) in Table 9.3 are for a "reference avoided cost" scenario. Because the economics of DSM depend heavily on supply costs, programs were also evaluated by the stakeholders under gas prices that are lower and higher than the reference case.

Table 9.3: Program impacts

	TRC($)	Flex	Km	CO_2	NO_X	Jobs	GJ	RIM($)	Risk
R1	67,528	42.6	-1.6	-3,283	-1.5	2.2	-54,720	-82,919	-7.2
R2	88,292	15.6	-1.8	-3,600	-1.7	2.7	-60,000	-25,653	-29.1
R3	58,871	14.6	-0.8	-1,683	-0.8	2	-28,050	-23,271	-15
R4	233,072	33.3	-3.9	-7,704	-3.6	6.7	-128,400	-40,665	-23.5
R5	31,511	8.7	-0.5	-998	-0.5	1.5	-16,625	-6,467	-6
R6	7,264	3.4	-0.4	-785	-0.4	0.5	-13,080	-8,772	-16.4
R7	1,455,389	233.8	0	-3,902	-0.3	43.3	0	78,060	-26.5
C1	-23,403	4.7	-0.3	-538	-0.3	-0.4	-8,960	-6,665	-16.9
C2	21,906	22.5	-1.3	-2,592	-1.2	1.8	-43,200	-31,189	-63.5
C3	264,890	62.7	-3.6	-7,209	-3.4	8.9	-120,156	-53,298	-48.3
C4	17,905	9.3	-0.5	-1,074	-0.5	0.8	-17,892	-9,547	-10.6
C5	397,858	8	0	0	0	11.7	0	148,708	-31.7
C6-M	13,191	10.8	0.4	1,219	1.4	0.5	12,705	15,058	-11
C6-H	13,191	10.8	-0.6	-1,867	1	0.5	-22,688	15,058	-11
C7-M	72,028	30.2	1	3,421	4	2.5	32,470	73,357	-16.4
C7-H	72,028	30.2	-2	-5,238	2.8	2.5	-81,029	73,357	-16.4
C8-M	237,750	30.9	1	3,504	4.1	7.1	33,264	80,395	-11.9
C8-H	237,750	30.9	-2.1	-5,366	2.9	7.1	-83,013	80,395	-11.9
C9-M	2,762,695	119.2	3.4	13,505	15.7	80	114,852	22,118	-68.6
C9-H	2,762,695	119.2	-9.9	-20,681	11.1	80	-393,047	22,118	-68.6
NGV1.1	235,583	71.5	0	-4,109	-2.7	7.5	0	138,854	-41
NGV1.2	1,247,770	23.4	0	-2,701	-1.8	36.9	0	95,455	-49.7
NGV1.3	3,734,783	84.1	0	-9,721	-6.5	113.5	0	302,883	-211.6
NGV2.1	17,598,852	454.3	0	-52,500	-35	520.4	0	1,717,953	-985

Step 3: Value scaling of impacts

The single-criterion value function for criterion *i* is defined as $V_i(A_{ij})$. Each $V_i(A_{ij})$ is defined as a linear rescaling of the A_{ij}, in which 0 is assigned to the worst value of A_{ij} and 1 is given to the best value. A standard result of multicriteria decision theory is that lower and upper values of a rescaling are arbitrary, as long as the weights used in the additive value function are chosen appropriately.

The rescaling involved two steps. First, a base range (worst to best) was determined for each criterion to use as a denominator in the scaling calculation. This can be equal to the range over which the impacts for a given criterion vary, or it can be defined arbitrarily. Here, the base range is smaller than the actual range for most of the criteria, because a handful of the programs (involving promotion of natural gas vehicles, NGVs) are much larger than the others in terms of economic, environmental, and energy impacts. The base range was set as the range for all non-NGV programs (with some rounding). As a result, the $V_i(A_{ij})$ for *non*-NGV programs will vary between 0 and 1, while the NGV programs fall outside those bounds.[7] The reason for selecting the smaller range is to allow the variation on the smaller programs to be more visible to the stakeholders.

In the second step of the rescaling, the lower bound of the corresponding base range was subtracted from the impact and the result was divided by the base range. That is,

$$V_i(A_{ij}) = (A_{ij} - A_{i\bullet}) / (A_i^* - A_{i\bullet}) \qquad \text{Eq. [9.1]}$$

where $A_{i\bullet}$ is the worst value in the base range and A_i^* is the best value.

Step 4: Define the marginal supply option

Here A_{iN} is defined as the "neutral" value of the criterion in which the program would have no net impact upon the portfolio's performance. "Neutrality" is defined relative to the marginal gas supply option that the DSM programs would displace (or require, in the case of programs that increase gas demand). Because of its size, the source of marginal gas supplies is the same no matter how much DSM is adopted; as a result, definition of the neutral point is straightforward.[8] As an example, $A_{iN} = 0$ for total resource benefits (e.g., if the marginal gas supply saving is just offset by the expense of the program) or emissions (e.g., if the program does not change the total emissions from gas use).

The amount that a portfolio improves in terms of criterion *i* (because of the addition of a DSM efficiency program *j*) can then be represented as $V_i(A_{ij})\text{-}V(A_{iN})$. This adjustment allows a graphic representation to portray the impacts centered around the marginal supply option. Impacts better than the marginal supply option appear above the line and impacts worse than the marginal supply option lie below. For instance, Figure 9.2 shows that the *Residential Hot Water Saver Program* is better on all criteria than the marginal supply option (represented by the X axis), save for the ratepayer impact measure and risk. (That is, the program would increase gas rates, and there is some risk that its costs will not be recovered through rates.) In contrast, a program whose impacts are identical to those of the marginal supply it would displace would lie along the X axis at every point, and would have no net impact on the portfolio's performance.

Figure 9.2: Example of impacts versus the marginal supply option: residential hot water saver program

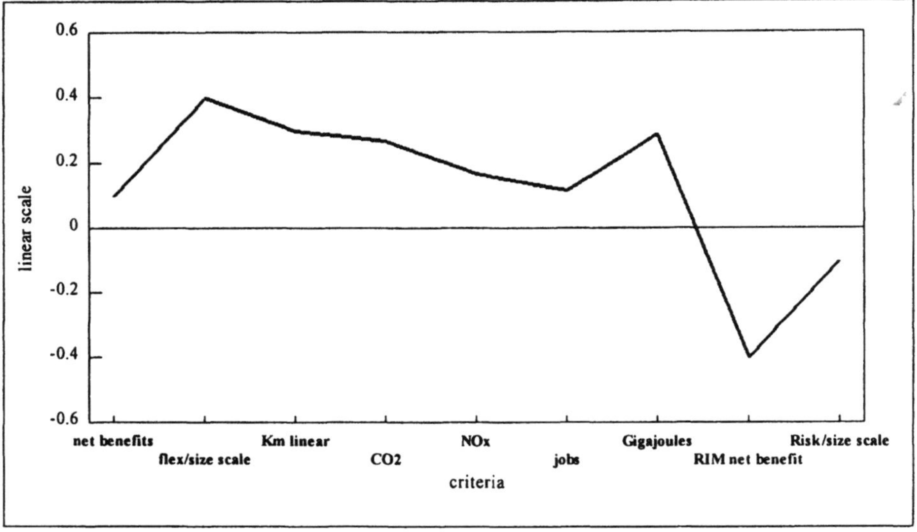

Step 5: Weighting of criteria and holistic assessments of alternatives

Steps 1 to 3, the definition of alternatives and the quantification of their performance on the criteria, are found in most multicriteria planning exercises. However this step, the application of multiple evaluation methods, is where our approach differs from traditional MCDM applications. In particular, each stakeholder applied two different weight selection methods. In addition, they made unaided, "holistic" assessments of the programs. We have argued that this multiple-method approach provides a more reliable assessment of the stakeholders' preferences. The principle could be considered analogous to triangulation. That is, in the same manner that multiple perspectives are required to identify a single point on the surface of the earth (i.e., longitudinal, latitudinal, and vertical), multiple perspectives are likely required to capture the multiple dimensions of a complex selection process that combines an individual's values systems together with a wide array of program impacts. In addition, the multi-method approach gives stakeholders a chance to reconsider and adjust their judgments as they learn.

There exist a wide range of available methods for picking criteria weights, as explained in Chapter 4. Here, we used two methods -- a ratio questioning/swing weighting hybrid and tradeoff assessment. They were adopted because they represent a good combination of reliability and ease-of-use and because they ask fundamentally different types of questions, thereby providing a useful consistency check. Both methods ask the stakeholder to state priorities among criteria, but the ratio method directly asks for ratios of weights, while the tradeoff method indirectly infers weights from tradeoff judgments.

For each weighting method, the assessment process was structured as a nominal group (Delbecq *et al.*, 1975). First, each stakeholder filled out a questionnaire. The answers to each question were then presented to the group (without identifying who gave which answers), and a round-robin discussion was then conducted. After the discussion, the stakeholders individually revised their responses based on what they learned from the discussions.

Compared to unstructured groups and DELPHI (anonymous voting without face-to-face interaction), the literature reports that such structured interactive groups are more effective and give the most satisfaction to participants (Delbecq *et al.*, 1975; Gustafson *et al.*, 1973), although decisions are not necessarily of better quality (Timmermans and Vlek, 1996). Interaction is very important, as participants in our experiments with stakeholder groups at Seattle City Light (Chapter 10) and Centerior Energy stated that they learned from each other and that this learning helped improve the quality of their weighting judgments (Chapter 9). For example, a financial analyst at Seattle City Light admitted ignorance about environmental damages caused by different types of emissions, and was grateful for information provided by others on the relative significance of different pollutants.

Figure 9.3: Hierarchy of objectives and criteria (resource level)

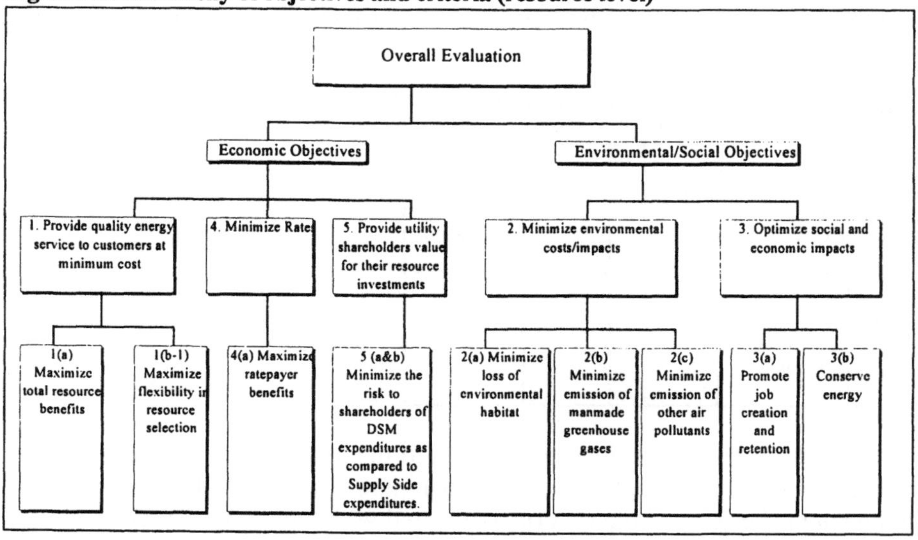

To facilitate the ratio assessments, the criteria were grouped into general objectives using the hierarchy shown in Figure 9.3. There are two purposes to the hierarchical grouping. One is to simplify assessments of weights, under the assumption that criteria within a group (e.g., two environmental criteria) are easier to compare than criteria within different groups (e.g., comparison of a socioeconomic impact with an environmental impact). A second purpose of the hierarchy is to avoid the so-called "splitting bias", in which objectives that have many criteria may be over-weighted (Stillwell *et al.*, 1987). Hierarchical weighting, in contrast, increases the relative weight of objectives that have few criteria.[9]

Ratio questions were asked at each of the three levels of the hierarchy (Figure 9.3): the individual criteria level (between pairs of criteria within a group), the objective level (among groups of criteria), and finally at the overall evaluation level (between the economic and environmental criteria groups).

Step 5a: Ratio questioning-swing weighting method

At the criteria level, the stakeholder was asked to compare two criteria, and to first state which of the two he or she would prefer to "swing up" from its least desirable value ($V_i(A_{i*}) = 0$) to its best value ($V_i(A_i^*) = 1$). It is crucial that the stakeholders carefully consider the actual criteria ranges in providing their responses. That is, the question is not (say) "Is CO_2 more or less important than NO_x?" Rather, as Figure 9.4 shows, it is "Is a *specific improvement* of 13,000 tons/yr in CO_2 (i.e. the range from 5,000 to -8000 tons/yr) more or less important than a *specific improvement* of 10 tons/yr in NO_x?" The responses would be meaningless without specifying ranges.[10] After answering this question, the stakeholders were asked to state how many times as important the range of the more important criterion is than the range of the other. For a pair of criteria A_i and A_k, the question establishes the ratio w_i/w_k of their weights w_i and w_k. For objectives that have just one criterion, no criteria-level questions were necessary.

Figure 9.4: Example of a criteria-level ratio question

Ratio	Criteria	Worst	→	Best	Units
☐	2(b) Minimize emission of manmade greenhouse gases	5,000		-8,000	CO_2 Tonnes (or CO2 equivalent)/yr
☐	2(c) Minimize emission of other air pollutants	6		-4	NO_X Tonnes/yr

The objective level questions correspond to the second level of the hierarchy in Figure 9.3. In the example shown in Figure 9.5, stakeholders were asked to consider the set of three criteria associated with Objective 2 versus the two criteria associated with Objective 3. If they felt that the set of criteria representing Objective 2 was just as important as Objective 3, then the ratio would be 1. However, if they felt, for example, that improving Objective 2 would be more important than improving Objective 3, they needed to indicate so by providing a ratio greater than 1. This was done by stating a number that indicated how many times more important a move from worst to best on Objective 2's set of criteria was than a worst to best move on Objective 3's two criteria. Then in the weight calculation procedure, the provided ratio was set equal to the ratio of the sums of the weights of the two groups.[11]

In the overall evaluation, stakeholders were asked to consider the set of criteria associated with the economic objectives versus the set of criteria associated with the environmental/social objectives. This was performed in a similar manner to the objective level questions.

Figure 9.5: Example of an objective-level ratio question

Ratio	Criteria	Worst	→	Best	Measure
	2. Minimize environmental costs/impacts				
	2(a) Minimize loss of environmental habitat	2		-4	Km Linear Disturbances/yr
	2(b) Minimize emission of manmade greenhouse gases	5,000		-8,000	CO_2 Tonnes (or CO_2 equivalent)/yr
	2(c) Minimize emission of other air pollutants	6		-4	NO_X Tonnes/yr
	3. Optimize social and economic impacts				
	3(a) Promote job creation and retention	-1		19	# jobs
	3(b) Conserve energy	50,000		-150,000	Change in GJ/yr

In total, eight ratio questions need to be asked to establish a unique set of relative weights for nine criteria.[12] Additional questions act as consistency checks, which are useful since stakeholders are generally unsure of their values and may also have difficulty understanding the questions. Redundant questions were included in the BC Gas process for this reason. Inconsistencies among the answers were resolved using the eigenvector method of the Analytical Hierarchy Process, details on which are found in Section 4.2.4, and in the many publications documenting the AHP method (Saaty, 1980). The equations were solved, yielding a set of criteria weights that sum to 1. As for all weighting methods, the sum of weights is arbitrary; the important thing is that the ratios of different weights be maintained.[13]

<u>Step 5b, Indifference Tradeoff Method</u>

The tradeoff method requires a very different response from stakeholders. With the ratio method, the stakeholder directly assesses relative importance, whereas with the tradeoff method the stakeholder states a specific amount of one criterion that they would be willing to give up to obtain a fixed improvement in another criterion.

Unlike the ratio process, the tradeoff process is generally conducted without a hierarchy. So rather than compare groups of economic criteria against groups of environmental/social criteria, stakeholders were asked to consider, for example, CO_2 versus costs directly. Eight tradeoff questions would normally be needed to identify a set of weights for the nine criteria. However time only allowed four questions to be asked, with each of the objectives involved in at least one question. Information from the ratio process was then used to complete the relationship between the lower level criteria that were not dealt with in the tradeoff process. For example, tradeoff questions were asked about CO_2 versus cost but no tradeoff questions were asked involving NO_x. Therefore, the ratio information linking CO_2 to NO_x (provided in the ratio questioning exercise) was used to determine a relative weight for NO_x.

For each question, two formats were provided as shown in Figure 9.6. Stakeholders were asked to respond to just one format, choosing whichever they were more comfortable answering. The changes in the two criteria (i.e., the value provided in the questionnaire and the stakeholder response) for a given question were then re-expressed and related to their weights as indicated in Figure 9.6.

Figure 9.6: Example of a tradeoff question

Format A

What value would you place on a 1,000 tonne reduction in CO_2 emissions?

1000 tonnes of CO_2 = __________ $ in net resource benefits

Format B

Suppose you were asked to choose between two programs

Program A reduces CO_2 emissions by zero tonnes/year and has net resource benefits of zero $/year

Program B reduces CO_2 emissions by 1,000 tonnes/yr and has net resource benefits of ________ $/yr

What would net resource benefits have to be in order for you to find the two programs equally acceptable? If you place no value on CO_2 emission reductions then you would place 0 in the space. However, if you do place a value on CO_2 emission reductions, you would place a negative value in the space to "offset" the positive value that you associate with reducing CO_2 emissions by 1,000 tons/year

For instance, if 20% of the range of TRC was judged to be equivalent to 5% of the range of CO_2, then the above equation shows that CO_2 is four times as important as TRC. (See Sections 4.2.6 and 6.6 for further explanation of the calculations.) The four tradeoff ratios were then combined with four ratios from the ratio questioning/swing weighting questionnaire to define the eight required equations. Weights for the nine criteria were then solved for and normalized so that they summed to 1.[14]

Step 5c, Holistic Program Evaluation.

The holistic process is the third of three multicriteria evaluation processes applied. It is intended to complement the weighting process by giving stakeholders a greater familiarity with the DSM programs and an additional opportunity for reflection and learning at the interview stage.

Stakeholders were given a full presentation on all the programs the day prior to the holistic evaluation. On the day of evaluation, the stakeholders were provided with a table of impacts and a chart (Figure 9.2) showing the scaled impacts. For each program they were asked to answer the question shown in Figure 9.7.

Figure 9.7: Holistic question format

Would you recommend including this resource in the portfolio?

strongly reject			neutral			strongly recommend
-3	-2	-1	0	+1	+2	+3

As expected, stakeholders generally found the holistic process simple in concept but challenging in application because of the volume of information (nine criteria, twenty alternatives) they were expected to process in a relatively short period of time. Recommendations arising from the holistic process are expected to be different from those arising from the two weighting processes for a number of reasons. These include information overload in the holistic process, lack of familiarity with the weighting processes, and differing interpretations of programs and their impacts. Psychologists have long noted that when people make unaided evaluations of alternatives under many criteria, they tend to ignore all but one or two of the most important criteria (Shepard, 1964).

Step 6: Initial evaluations by additive value function.

The two sets of weights (ratio weights and tradeoff weights) chosen by each person are used to evaluate the programs by an additive value function:

$$\textit{Net Value (Program j)} = \sum_{i=1,\ldots,9} w_i\,[V_i(A_{ij})-V_i(A_{iN})] \qquad \text{Eq.[9.2]}$$

If the net value is positive, the program would increase the attractiveness of the portfolio, assuming that the weights and criteria values are correct. The result of this procedure was two sets of quantitative DSM program evaluations for each stakeholder. These evaluations were considered tentative, and were to be reconsidered along with the holistic evaluations during the interview of Step 7.

As an example of the above calculation, the net value for program R1 using the ratio weights for stakeholder 1 is:

$$\begin{aligned}\textit{Net Value (Program R1)} &= 0.256(0.1) + 0.064(0.43) + 0.007(0.27) + 0.014(0.25)\\ &\quad + 0.057(0.15) + 0.073(0.11) + 0.007(0.27)\\ &\quad + 0.484(-0.38) + 0.037(-0.07)\\ &= -0.109\end{aligned}$$

where the coefficients are the weights and the values in parentheses are the scaled criteria values, after adjusting for the marginal supply option. Thus, because the net value is negative, this program would decrease the net value of the portfolio of programs under these weights. This process was repeated for each stakeholder, each program, and, in addition, each of the avoided supply cost scenarios mentioned in Step 2, above.

Step 7: Reconciliation of results in individual interviews: Approach

The purpose of the interview process is to explore inconsistencies among the two weighting methods and holistic results. The intent is not to favor one method over another. Rather, the goal is for the stakeholders to gain insight into their values by exploring inconsistencies across the three methods. The process also gives the stakeholder an opportunity to improve his or her understanding of individual programs by asking questions. In this regard, the three methods work in concert to cross-check each other and ultimately to produce a set of recommendations that each stakeholder felt confident in and was willing to present to the group.

The interviews were scheduled over three days with each interview lasting about 1.5 hours. The setting for the interviews was a boardroom at BC Gas. The interviews were structured as follows:

- Judgments provided by each stakeholder in the *ratio* questionnaires were reviewed and the resulting weights presented and discussed.
- Judgments provided by each stakeholder in the *tradeoff* questionnaires were reviewed and the resulting weights presented and discussed.
- The weights were contrasted and the differences (which were often significant, as we expected) were explored. Each stakeholder was then asked if they would like to revise any judgments. If they were more comfortable with one set than another, reasons for this were explored. Resolution of differences between the two weight sets was not pushed at this point because it is possible that very different sets of weights would produce the same program recommendations.
- The stakeholders were then shown the program recommendations under each of the two sets of (perhaps just revised) weights. A consultant who was knowledgeable about DSM technology and economics was present to answer any questions the stakeholder might have about individual programs. In the interest of time, the discussion focused on those programs that received conflicting evaluations under the two methods. Programs that were clearly in (or out) by both methods but just to varying degrees were not reviewed. After reflection, the stakeholder was asked to indicate which of the two sets of recommendations was most consistent with their values and why. This sometimes led to further revisions of weights and additional iterations, the result being a single set of weights that yields a set of recommendations that the person was (tentatively) most comfortable with.
- The tentative weight-based recommendations were then compared with the holistic questionnaire results and, as with the comparison of the two weighting processes, the differences were discussed. Where reconciliation was needed, the stakeholder was asked to make one recommendation per program. Where the final recommendation differed from the weighting evaluation, the stakeholder was asked to give a rationale.

Recommendations along with explanations were noted and returned to the stakeholders for use in the nominal group discussions of Step 8.

Step 7: Interviews: Examples of inconsistencies and their resolution

To illustrate the procedure, a pre-interview record of stakeholder 3's responses comparing the ratio method, the tradeoff method and the holistic method is shown in Table 9.4. The shaded boxes are used to highlight the negative scores; this makes it easy to tell at a glance where inconsistencies across methods exist. If each method produced a negative recommendation for a given program, all boxes in that program's row would be shaded. If each method produced a positive recommendation across a given program, all boxes for that program would be clear. (Note that each program has three recommendations under each weight set, one per avoided supply cost scenario. In addition, there are two sets of recommendations for programs C6 through C9 because, as mentioned in Step 2, there are two emissions scenarios associated with each.)

Table 9.4: Example of pre-interview set of weights and recommendations (for stakeholder #3)

	Ratio Method			Trade-off Method			
Program	Low Avoided-Cost Scenario	Reference Avoided-Cost Scenario	High Avoided-Cost Scenario	Low Avoided-Cost Scenario	Reference Avoided-Cost Scenario	High Avoided-Cost Scenario	Holistic Input
R1	0.088	0.097	0.102	(0.406)	(0.377)	(0.362)	2
R2	(0.047)	0.019	0.052	(0.326)	(0.117)	(0.013)	2
R3	(0.005)	0.029	0.041	(0.215)	(0.106)	(0.070)	1
R4	0.036	0.158	0.219	(0.571)	(0.185)	0.006	1
R5	0.010	0.028	0.037	(0.086)	(0.029)	(0.001)	0
R6	(0.042)	(0.025)	(0.018)	(0.095)	(0.040)	(0.019)	(2)
R7	1.649	1.325	1.197	1.379	0.355	0.153	3
C1	(0.047)	(0.039)	(0.035)	(0.056)	(0.030)	(0.018)	2
C2	(0.131)	(0.078)	(0.068)	(0.311)	(0.142)	(0.111)	(2)
C3	0.009	0.190	0.205	(0.684)	(0.242)	(0.212)	0
C4	(0.011)	0.009	0.013	(0.109)	(0.043)	(0.032)	(1)
C5	0.064	0.228	0.274	0.156	0.676	0.822	3
C6-M	0.009	0.012	0.010	0.058	0.068	0.062	(3)
C6-H	0.016	0.019	0.017	0.058	0.068	0.062	(3)
C7-M	0.110	0.100	0.096	0.368	0.333	0.319	1
C7-H	0.132	0.122	0.118	0.368	0.333	0.319	1
C8-M	0.213	0.201	0.194	0.401	0.365	0.344	(1)
C8-H	0.235	0.223	0.217	0.401	0.365	0.344	0
C9-M	1.860	1.544	1.388	0.815	0.101	(0.252)	2
C9-H	1.955	1.639	1.483	0.815	0.101	(0.252)	2
NGV1.1	0.310	0.283	0.275	0.712	0.631	0.607	0
NGV1.2	0.667	0.649	0.646	0.492	0.434	0.423	(2)
NGV1.3	1.937	1.862	1.844	1.611	1.377	1.319	2
NGV2.1	9.389	9.013	8.902	8.993	7.809	7.458	2

For this particular stakeholder, the mechanics of the second weighting method (the tradeoff process) were not clear and as a result he answered some questions in a manner that created the opposite effect of what he had intended. As a result, his ratio weights were used as his final weights and these were then compared against his holistic recommendations. While one or two stakeholders were perfectly consistent among the methods, this was not generally the case. That is, most stakeholders did *not* have all clear rows for some programs and all shaded rows for the other programs.

Table 9.5 sums up the results of the last stage of the interview, again for stakeholder 3. At this point the stakeholder had a chance to review inconsistencies and made changes as he felt appropriate.

Table 9.5: Example of post-interview results

	Results Based on Final Weights			
Program	Low Avoided-Cost Scenario	Reference Avoided-Cost Scenario	High Avoided-Cost Scenario	Stakeholder Initial Conclusions
R1	0.09	0.10	0.10	2
R2	(0.05)	0.02	0.05	2
R3	(0.01)	0.03	0.04	1
R4	0.04	0.16	0.22	1
R5	0.01	0.03	0.04	0
R6	(0.04)	(0.02)	(0.02)	(2)
R7	1.65	1.32	1.20	3
C1	(0.05)	(0.04)	(0.04)	(1)
C2	(0.13)	(0.08)	(0.07)	(2)
C3	0.01	0.19	0.21	0
C4	(0.01)	0.01	0.01	1
C5	0.06	0.23	0.27	3
C6-M	0.01	0.01	0.01	2
C6-H	0.02	0.02	0.02	2
C7-M	0.11	0.10	0.10	1
C7-H	0.13	0.12	0.12	1
C8-M	0.21	0.20	0.19	(1)
C8-H	0.23	0.22	0.22	0
C9-M	1.86	1.54	1.39	2
C9-H	1.95	1.64	1.48	2
NGV1.1	0.31	0.28	0.28	0
NGV1.2	0.67	0.65	0.65	2
NGV1.3	1.94	1.86	1.84	2
NGV2.1	9.39	9.01	8.90	2

In resolving inconsistencies, stakeholders gave a variety of reasons for altering their initial holistic recommendations. For instance, stakeholder 3 offered the following reasons for his changes:

- C1: "made holistic assessment too quickly"
- C4: "it is a low cost program so would be prepared to go with weighting results"
- C6: "thought gas was being used as a refrigerant for C6." In this case the weights helped to reveal an area where the stakeholder had an incorrect, yet plausible, interpretation of the program. The stakeholder was familiar with gas itself being a refrigerant for industrial processes and thought that *Commercial Gas Cooling* (C6) implied that type of use.

- NGV1.2: it was pointed out to the stakeholder that GJ of energy savings was incorrectly portrayed on the original chart; on the basis of a revised chart he changed his response.

The above inconsistencies emphasize the importance of using a combination of approaches and reconciling them with an interview process. The process adds clarity and provides cross-checks (for both the stakeholders and the analysts), resulting in more rigorous and defensible recommendations.

Step 7, Interviews: Analysis of inconsistencies

Below, we summarize the extent of inconsistencies between the methods and their significance. First, we review inconsistencies between the weights and evaluations resulting from the two weighting methods, and then the contradictions between the evaluations based on the final weights and the holistic evaluations from Step 5. At the close of the interview, there remained some differences between the evaluations by each stakeholder's final weights and the recommendations he or she was willing to make. These were summarized. This analysis concludes by pointing out an important advantage of additive value functions over holistic assessment: the former result in more consensus, while the latter seems to exaggerate differences of opinion.

Early in the interview, each stakeholder was asked to review both sets of weights, and to produce a final set. There were often important differences between the two sets. Several people misunderstood the tradeoff process so that their tradeoff weights could not be interpreted meaningfully, and were discarded. Of the 10 stakeholders who were able to assess meaningful tradeoff weights, the average correlation between the ratio and tradeoff weights was just 0.45, and was actually negative for three of the participants. This implies that the two types of weighting methods often elicit apparently contradictory value judgments.[15]

To summarize the results of the comparison and resolution of weights, 13 of the 16 stakeholders felt that the ratio method contributed the most to their final weights while the other 3 stakeholders stated that the tradeoff weights were the greatest influence. We calculated correlations to confirm this, and found that the final weights correlated most closely to the tradeoff weights for only 4 of the 16 stakeholders. This imbalance between the two methods reflects in part the added complexity of the tradeoff method. However, as pointed out later in this Chapter, most shareholders found that the tradeoff exercise was a valuable consistency check and that insights resulted from the group discussion of the results.

Of course, differences in weights do not matter if they do not result in differences in program evaluations. But here, as in other studies, different weighting methods often yielded different decisions (see also Chapter 10). The average correlation between each person's ratio and tradeoff weight-based evaluations was 0.72.[16] However, 40% of those correlations were 0.5 or less, consistent with our observation that the two weighting methods often yielded very different evaluations. On average, 9% of the tradeoff weight based program evaluations were negative when the ratio weight-based evaluations were positive, or vice versa.

There were even more contradictions to resolve when the stakeholders turned to comparing the final weighting-based evaluations with their initial holistic evaluations. This was indicated by the average correlation of 0.46 between holistic evaluations and the final weighting evaluations by each person (compared to the 0.72 between the ratio and tradeoff based evaluations). Another index of disagreement is the fraction of program evaluations that were negative for one method but positive for another. Fourteen percent (one of every seven) of the holistic evaluations disagreed with the final weighting evaluations, and required resolution.

The resolution of the holistic and final weighting evaluations yielded a set of recommendations. Most inconsistencies were resolved in favor of the weight-based results. This is reflected in the correlations we calculated between the final recommendations and each preliminary set of evaluations (ratio questioning, tradeoff, and holistic); they show that the weighting methods were closer to the final recommendations for all but three stakeholders. This can be interpreted as indicating that additive value functions are, in general, a more valid representation of stakeholder preferences than unaided holistic evaluations.[17] Further examination of our results indicates that ratio questioning is more valid by this correlation measure than tradeoff weighting for a large majority of our participants. However, for four of the stakeholders, the final evaluations were closer to the initial tradeoff based evaluations; since this is 25% of the group, we believe that this is sufficient justification for use of tradeoff weighting in addition to ratio questioning.

On average, each stakeholder made one program recommendation that was in contradiction with the final weights, which he or she was asked to explain. Most of these inconsistencies stem from disagreements, both positive and negative, regarding assumptions underlying the A_{ij}. Examples of these inconsistencies and their accompanying explanations are summarized below.

- Stakeholder 5 recommended C2 although the weighting results were negative. He explained that furnace tune-ups, in his experience, are done less often than people realize. Consequently, that program's benefits will be greater than expected. He also gave NGV2.1 a neutral rating despite a positive rating according to the weights. The reason given was that he felt competition from electric cars had not been adequately considered.

- Stakeholder 6 gave C2 a positive recommendation in spite of a negative weighting-based outcome, saying that "programs could be combined with other programs thus creating efficiencies that are not yet recognized in the weighting assessment". He also recommended C1 stating that it was needed so that BC Gas could learn more about this type of DSM. He gave R2 a neutral recommendation despite positive weighting results because he didn't believe the assumption that only 2% of participants would be free riders (i.e., would have adopted the technology anyway in the absence of the program).

- Stakeholder 9 did not recommend program C5 despite a positive weighting result because he felt the assumptions regarding emissions and GJ of non-renewable energy impacts were incorrect. The assumptions were subsequently re-examined and partially revised later in the process, at which point the stakeholder accepted the program.

The tentative recommendations resulting from the interviews were the input to Step 8, in which the group reviewed the results and attempted to build a consensus. An important difference between the holistic results, weighting results, and final recommendations is that use of formal MCDM methods yielded more agreement among the stakeholders than holistic evaluations. This is a familiar result from the literature, where a number of case studies have shown that holistic alternative evaluations by different people differ more than do additive value function-based evaluations (Edwards, 1977). (See also Chapter 10, below; but see Bell *et al.*, 2000, for an exception.) This phenomenon has been explained as resulting from the tendency of holistic evaluations to focus on just a few of the most important criteria, while treating the others inconsistently. This tendency can exaggerate differences of opinion between people, compared to the situation in which all criteria are considered -- as they are by additive value functions.

In the BC Gas case, the average correlation between holistic evaluations by different stakeholders was 0.18, which was significantly less than the average correlation between evaluations based on the final weights chosen by each person (0.35).[18] Meanwhile, the between-persons correlations among the post-interview recommendations averaged 0.29. We interpret this result as meaning that the multi-method approach leads to more convergence of opinions among stakeholders than if only holistic judgment was relied upon. This convergence eases the negotiations of Steps 8 and 9.

Step 8: Group review of results

The program evaluations resulting from each stakeholder interview were presented anonymously to the group at three separate one-day meetings over a one month period. The goal of the discussions was to identify agreements and disagreements, and possible bases for consensus. The evaluations were tentative, subject to modification based on what was learned during group review of the results.

In the first of the three meetings of this step, the group was presented with the individual interview results as shown on the left side of Table 9.6. The X's indicate that a stakeholder tentatively recommended that the program not be included while the O's indicate that the stakeholder would be prepared to go with the consensus of the group. Blanks indicate that the stakeholder accepted the program. Each program was discussed in turn and then voted on by the group.

Table 9.6: Summary of program recommendations

	Interview results for individual stakeholders: meeting 1 (Blank = program recommended; 0 = stakeholder would go with group consensus; X = program not recommended)																Results of Group Evaluation, Meeting 2[a]		BC Gas proposed portfolio
Program	1	2	3	4	5	6	7	8	9	10	11	12	13	14	15	16	# Yes	# No	
R1	X																12	1	Included
R2		O			O	O											13	0	Included
R3	X			O													13	0	Included
R4																	13	0	Included
R5			O		O												13	0	Included
R6	X	X	X	X	X	X				O	O		O			O	10	3	
R7																	13	0	Included
C1	X	X	X	X	X		X			O	O					O	8	5	
C2[b]	X	X	X	X				X									10	3	
C3[b]																	13	0	Included
C4[b]	X	X		O	O					O				O			10	4	
C5									X							O	12	2	Included
C6		O								O			X	X	X	X	10	4	
C7										O		O			X	X	11	3	Included
C8			X							O		O		X	X	X	11	3	
C9												O			X	X	11	3	Included
NGV1.1			O														13	0	Included
NGV1.2																	13	0	Included
NGV1.3			O														13	0	Included
NGV2.1					O												13	0	Included

[a]These columns reflect the number voting to include the program in the portfolio

[b]The discussion ended for the day with some stakeholders feeling that these programs (C2,C3,C4) should not be combined, while others felt that they should. BC Gas agreed to explore some re-design before the next meeting.

Several trends can be seen in Table 9.6. First, a number of stakeholders had concerns about programs that would increase gas rates. These stakeholders are gathered on the left of Table 9.6, and placed X's by programs with poor RIM scores (shown in Table 9.3). Second, a few stakeholders disapproved of programs involving switching of end uses from electricity to natural gas (the grouping of X's to the right). And third, all stakeholders were in favor of (or at least not against) the natural gas vehicle programs, at least tentatively.

The results of the subsequent voting process during the first meeting are also shown in Table 9.6. It can be seen that consensus was reached on R2, R3, R4, R5, R7, and all the NGV programs. The confirmation of C3, although it received a consensus vote once, was postponed until the following meeting. Note that the votes differed from the interview recommendations. This can be explained by both absenteeism and the fact that further discussion led to some individuals changing their vote.

Other important products of the first meeting were suggestions on how individual programs might be improved to address particular concerns. Before the second meeting, the design of these programs was altered, and they were then brought back to the group for further comment. Discussion during the second meeting focused on the non-NGV programs, given that consensus had been reached on the NGV programs. In light of program revisions, it was unanimously agreed that C3 should be included as part of the base portfolio.

The group then agreed to look at other portfolios of programs as a means of exploring the tradeoffs that would result from additions of further programs. This was achieved by portraying the total impact of various portfolios of programs requested by the stakeholders. The format is shown in Figure 9.8. The sub-table to the left indicates which programs are in the portfolio being compared, while the sub-table to the right indicates the comparative raw impacts of the two portfolios. The chart at the bottom shows the combined impacts of each portfolio.

The message to be drawn from Figure 9.8 is that the impact, across each criterion, of adding Program C5 to the original consensus portfolio (including C3) is largely positive. As can be seen from the chart, TRC (net resource benefits), RIM (rates), and Jobs are improved with the addition of C5 at the cost of only a slight worsening in Risk.

The discussion ensuing from this resulted in C5 being unanimously added to the base portfolio. The base portfolio now consisted of R2, R3, R4, R5, R7, C3, C5, and all the NGV programs. Further suites of programs were then explored that included switching from electricity, but it became evident that the group would not reach consensus as long as there were such fuel-switching programs in the portfolio.

Figure 9.8: Portfolio model for exploring suites of programs

Portfolio Comparison Model

DSM Progs	Base 1	Base 1 + C5
R 1		
R 2	in	in
R 3	in	in
R 4	in	in
R 5	in	in
R 6		
R 7	in	in
C 1		
C 2		
C 3	in	in
C 4		
C 5		in
C 6		
C 7		
C 8		
C 9		
NGV1.1		
NGV1.2		
NGV1.3		
NGV2.1		
NGV2.2		
NGV2.3		

	Base 1	Base 1 + C5	Base 1 + C5 - Base 1	% Difference Base 1 + C5 vs Base 1
TRC	$ 2,132,025	$ 2,529,883	$ 397,858	19%
Flexibility	368.7	376.7	8.0	2%
Km - M	(10.6)	(10.6)	-	0%
Km - H	(10.6)	(10.6)	-	0%
CO2 - M	(25,096)	(25,096)	-	0%
CO2 - H	(25,096)	(25,096)	-	0%
NOX - M	(10.2)	(10.2)	-	0%
NOX - H	(10.2)	(10.2)	-	0%
Jobs	65.0	76.8	11.7	18%
GJ NR - M	(353,231)	(353,231)	-	0%
GJ NR - H	(353,231)	(353,231)	-	0%
RIM	$ (71,294)	$ 77,415	$ 148,708	209%
Risk	(148.3)	(180.0)	(31.7)	-21%

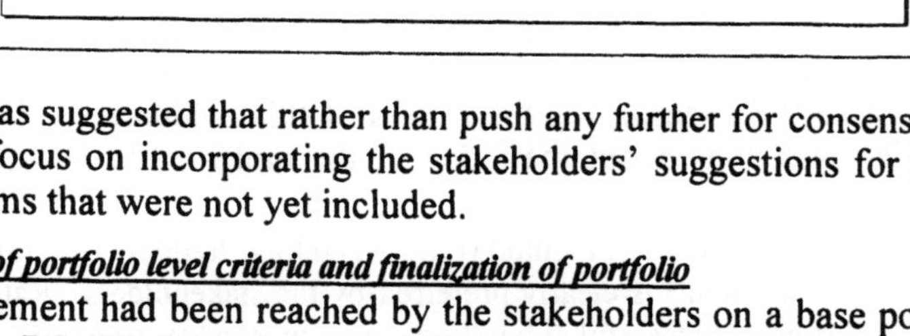

At this point it was suggested that rather than push any further for consensus, BC Gas should focus on incorporating the stakeholders' suggestions for re-designing programs that were not yet included.

Step 9: Application of portfolio level criteria and finalization of portfolio

Once initial agreement had been reached by the stakeholders on a base portfolio (R2, R3, R4, R5, R7, C3, C5 and the NGV programs), the portfolio was then evaluated by the group relative to the two portfolio-level criteria, equity and diversity. This evaluation was impressionistic in nature, with no assistance from the value functions assessed earlier. The group discussed whether addition or subtraction of any DSM programs would enhance the equity and diversity of the portfolio, and whether that improvement would justify decreases, if any, in the portfolio's performance relative to other criteria.

The base portfolio was presented as a portfolio together with the programs that were approved in the previous (1994) BC Gas resource planning exercise. No negative comments were recorded in terms of equity. However a number of stakeholders were uncomfortable that some programs (C6, C9 and R6 were mentioned) had been "left out". However it was explained that they were not left out. The base-portfolio on which the group had reached consensus was exactly that -- a base. Comments received would be reviewed and incorporated into the final portfolio recommendation from BC Gas.

The final portfolio that BC Gas recommended in their Integrated Resource Plan is shown in the right-hand column of Table 9.6. As of this writing, the final portfolio has survived the regulatory process, with the exception of the fuel switching programs C7 and C9; as in the stakeholder meetings, there was disagreement over their net environmental impact, and as a result, the company ultimately agreed to drop them.

9.4 STAKEHOLDER EVALUATION OF THE PROCESS AND CONCLUSION

The purpose of this multi-method MCDM process was to provide a vehicle for stakeholders to form and share their values and perceptions in order to help them make recommendations to BC Gas. This was accomplished by having stakeholders assess each program by three multicriteria methods, review the resulting recommendations individually (making changes as required), and finally discuss the recommendations as a group. This unique multi-method approach (with follow-up interview and structured group-discussion) was employed because it was felt to provide better opportunity for discussion, reflection and revision which would in turn lead to more defensible recommendations.

In order to evaluate the strengths and weaknesses of the process, we asked the stakeholders to fill out questionnaires in which they reported their perceptions of the appropriateness, ease of use, and validity of the methods used. They were requested to respond to a number of questions on a -3 to +3 scale. In addition, an opportunity for detailed written comments was provided.

Table 9.7 summarizes the results of this evaluation. They reinforce our earlier assertions about the value of the multi-method MCDM process. In particular:

- No one MCDM method (ratio weighting, tradeoff weighting, or holistic assessment) is clearly preferred by the stakeholders. Ratio questioning is most successful at building confidence, while tradeoff weighting has a slight advantage in focusing on the important issues. Overall, holistic assessment was most highly recommended for use in resource planning. These results, combined with the significant differences that often arose between the results of different methods, implies that a multi-method approach is desirable.

- Learning and confidence building is best accomplished by considering the results of different MCDM methods, resolving their inconsistencies in individual interviews, and reviewing results as a group. For instance, one stakeholder commented that "it is essential that one-on-one meetings take place with (a decision analysis facilitator) in order that the selection of programs reflect each stakeholder's wishes." The respondents felt strongly that this part of the process improved the consistency of the process and the ability of the stakeholders to consider the most important issues. The group discussions, especially of weight judgments, were also important in this regard.
- An explicit multicriteria approach is an appropriate methodology for considering the important trade-offs involved in energy planning.

Table 9.7: Summary of responses to the follow-up questionnaire

	Ratio questioning weighting	Tradeoff-based weighting	Holistic evaluation of programs	Resolution of methods	Nominal group discussions of weight judgments	Nominal group discussions of overall evaluations
Is the amount of effort likely to be required to apply the method in real planning exercise reasonable, given resource and time constraints?	1	1	1.15	1	1.25	1.33
Do you think that this approach would help improve the consistency of decisions (in terms of consistency of treatment of the criteria across alternatives) compared to holistic judgment?	1.54	1.17	n/a	2.09	1.66	1.43
Would use of this method increase your confidence in the rankings of alternatives (compared to holistic ranking)? (That is, would more defensible and trustworthy decisions that are more consistent with your values result?)	1.23	1	1	1.66	2.11	1.29
Does this method help you focus on the important issues and better understand the problem?	1.36	1.5	1.36	1.83	2	1.25
Would you recommend use of this method in utility resource planning?	1.46	1.5	2.07	2.15	2.25	1.5

Key: -3 = Definitely no, 0 = Not sure, +3 = Definitely yes

NOTES

[1] Graham T. F. Horn of Planit Management is co-author of this chapter. Portions of this chapter are reprinted from *Energy Policy*, Vol. 25(3), B.F. Hobbs and G.T.F. Horn, "Building Public Confidence in Energy Planning: A Multimethod MCDM Approach to Demand-Side Planning at BC Gas," pp. 357-375, with permission from Elsevier Science.

[2] Not all experiments, however, reveal such differences (e.g., Harte and Koele, 1995).

[3] A heuristic solution to the knapsack problem could be obtained in two steps. First, the DSM programs could be ranked in decreasing order of the ratio of their additive value function score to the budget dollars they require. Then, the DSM programs with the best ratios would be added to the portfolio, one at a time, until the budget is exhausted. This procedure can be suboptimal if programs cannot be scaled down in size. The reason is that the procedure assumes that the last program adopted can be scaled down in size so that the entire budget is used up, without altering that program's ratio of benefits to budget expenditures.

[4] This approach of evaluating DSM programs on an individual basis is similar in philosophy to the MCDM DSM evaluation method proposed by Vine and Harris (1990).

[5] Keeney and McDaniels (1999) provide background on the institutional context of the BC Gas study, including a frank assessment of the political obstacles to successful use of MCDM. They make several useful suggestions for value elicitation in a regulatory context, including the following. 1) If values are to be elicited from regulatory staff, keep that process separate from value judgments by the regulated entity, and be aware of legal difficulties that might ensue. 2) Carefully document all results and points of agreement in a manner appropriate for subsequent hearings, and allow participants to review and revise all results (as we do in this Chapter). Iterations are helpful to clarify information needs and to provide contextual information. 3) Thoroughly document elicitation activities and issues such as how the values will be used, as they may be subjected to regulatory scrutiny and even cross-examination.

[6] The values of flexibility and risk shown in Table 9.3 are actually the product of a subjectively assessed factor (low = 0.33, medium = 0.67, and high = 1.0) and the size of the program (GJ of energy savings, in the case of flexibility, and $ program cost, in the case of risk). "Risk" represents the possibility that BC Gas will not be allowed by the BC utility commission to recover all costs associated with the program. Under the assumption that BC Gas is (approximately) risk neutral, it is appropriate to assume that the total risk (as defined here) of a portfolio is the sum of the risks from the individual programs. This allows risk to be treated as a program-level criterion, rather than as a portfolio-level criterion in Step 9.

[7] Late in the process, some of the non-NGV programs were increased in size in response to stakeholder suggestions. Table 9.3 includes these changes; consequently, the attribute values for a few non-NGV programs also fall outside the "worst-best" range defined in Table 9.1.

[8] The choice of neutral point cannot be arbitrary because a shift would result in more or fewer DSM programs having positive net benefits. Thus, the additive function described creates a ratio scale. In contrast, in most MCDM assessments only an interval scale is needed and the zero point is arbitrary.

[9] See Chapters 8 and 10 for further explanation and demonstrations of that bias in a resource planning context.

[10] The purpose of the "swing" question is to force stakeholders to explicitly consider the range of the criterion when weighting it, a prerequisite for valid weight assessment (von Winterfeldt and Edwards, 1986). Other methods that ask directly for weights, such as the Analytical Hierarchy Process or rating on a 0-100 scale, fail to confront respondents with criteria ranges. As a result, the weights may reflect some general sense of criterion importance (such as "environment is just as important as economy"), rather than the significance of one criterion's range versus to the ranges of others (Schoemaker, 1981). For further discussion, see Section 4.2.

[11] In particular, let $RO_{2,3}$ be the ratio of importance of Objective 2 to Objective 3. Then the weights w_i of the individual criteria i would have to satisfy $(w_{2(a)}+w_{2(b)}+w_{2(c)}) / (w_{3(a)}+w_{3(b)}) = RO_{2,3}$.

[12] This is because n-1 independent linear equations involving the w_i are needed to establish a set of n w_i's that are unique up to a scaling constant. The sum of the weights can then be set to any arbitrary positive number (1.0, in the BC Gas case).

[13] That is, multiplying all weights by the same positive constant would not change the ranks of alternatives, or whether their net value is positive or negative.

[14] In particular, let RT_{ik} be the ratio of importance of criteria i and k determined by the tradeoff method, and RR_{ik} be the ratio determined by the ratio process. Eight independent equations, plus a scaling equation (e.g., $\Sigma_i w_i = 1$), are needed to determine a set of nine weights w_i. Here, four equations $w_i/w_k = RT_{ik}$ are based on the tradeoff questions and four other equations $w_i/w_k = RR_{ik}$ rely on ratios supplied from ratio questioning/swing weighting. As long as all nine criteria i are involved in at least one such equation, and no equation is linearly dependent on the other seven equations, then these eight equations can be used to derive a set of tradeoff weights (actually, hybrid tradeoff-ratio questioning weights).

[15] One measure of the significance of these differences is to compare this average "between methods" correlation with the average correlation between weights chosen by different people ("between persons" weights). The latter is 0.02, which is significantly less than the mean ratio-tradeoff weight correlation. This indicates that, first, the stakeholders disagree over priorities and, second, which stakeholder chooses the weights matters more, on average, than which weighting method is used. Nevertheless, because the weighting methods disagree by a large amount for some people, and because other studies (summarized in Hobbs, 1986; also see Chapter 10) found that between-method differences can be as large as between-person differences, we conclude that it is important to use more than one method as a consistency check.

[16] The correlations between evaluations by different methods were calculated by first converting all alternative numerical evaluations into the following scoring scheme: -1 = reject, 0 = neutral, and +1 = accept. Then Pearson correlations were calculated using the revised scores. This rescoring was necessary because the different types of scales used by holistic assessment (-3 to +3) and additive value functions (continuous scale) cannot be correlated meaningfully.

[17] Lai and Hopkins (1995) suggested that this concept is a more useful measure of method validity than others that have been proposed (see Chapter 10). They used the correlation between initial evaluations by various methods with the final evaluations resulting from an iterative process of reconciling differences between methods as a validity index.

[18] Based on a one-tailed non-parametric test similar to that used in Hobbs *et al.* (1992).

10 MCDM EXPERIMENTS AT SEATTLE CITY LIGHT[1]

Reason also is choice
MILTON, Paradise Lost, 3.108 (1667)

Ah! Don't say that you agree with me. When people agree with me I always feel that I must be wrong
OSCAR WILDE, Intentions (1891)

10.1 INTRODUCTION

We have reviewed a range of MCDM methods in previous chapters and many others have appeared in the literature. Despite the energy industry's widespread use of these techniques, too little attention has been paid to their validity and reliability. They differ in their theoretical underpinnings, in the type of questions they ask users, and in the results they yield. Therefore, users need reliable information on the performance of these methods in practice. There is a large literature in which psychologists and decision-scientists have contrasted the perceived usefulness, results, and validity of various MCDM methods (e.g., von Winterfeldt and Edwards, 1986; Hobbs, 1986; Corner and Buchanan, 1997; Zapatero *et al.*, 1997), but most of these experiments have involved students or have simplified problems.[2]

The purpose of this case study is to present the results of an experiment in which utility planners and interest group representatives apply several MCDM methods to a realistic planning problem. The focus is on methods for quantifying the preferences of users. The results are intended to help energy planners and policy makers choose MCDM methods, and to avoid the biases that method misuse can cause. Compared to Chapter 9, this chapter's experiment involves a wider range of methods and more extensive tests of method validity.

Our experiment addresses two basic questions:

1. What is the ease of use and appropriateness of the methods, as perceived by the participants in the experiment?
2. How valid are the results of the methods; that is, how well do their evaluations of options actually reflect the preferences of the users?

The question of validity is important because if different methods lead to different decisions, validity matters. But if different methods lead to the same results, then a method can be chosen considering just appropriateness and ease of use alone. Experiments have shown that the alternative rank orders yielded by different methods can differ significantly (e.g., Belton, 1986; Borcherding *et al.*, 1991). Indeed, the difference between the results of two methods applied by one person is often greater than the difference between rank orders chosen by several people using the same method (Hobbs, 1986; Hobbs *et al.*, 1992). There is evidence that disagreements between methods become larger and more important as the number of alternatives and criteria increases (Jia *et al.*, 1998); since energy planning and policy involves many options and attributes, the choice of MCDM method may well affect resource choices. Our experiment also tests this hypothesis.

In the next section of this chapter (Section 10.2) we give a brief overview of the general questions we address in this experiment. Section 10.3 provides background on Seattle City Light, a predominantly hydro utility, and the resource planning case study. The experimental design for the case study is also described in Section 10.3, followed by the results in Section 10.4 and conclusions in Section 10.5.

10.2 QUESTIONS ADDRESSED

The purpose of this experiment was to assess the appropriateness, ease of use, differences in method results, and validity of several common MCDM methods, including rating and tradeoff weighting, additive value functions, and goal programming. After applying the methods, the users assessed the methods' appropriateness and ease of use by filling out a questionnaire. Method results were compared using correlation analysis. In that analysis, we ask if differences *among methods* are as important as differences in opinion among people.

The issue of validity is addressed in three different ways. Predictive validity concerns the ability of a method to satisfactorily predict unaided, or "holistic", rank orders. A related concept, which we call estimative validity, is based on the methods' ability to predict the ultimate recommendations of the participants. All the methods are compared using predictive and estimative validity tests. Finally, construct validity concerns whether a method is sensitive to inputs in appropriate ways. Two construct validity tests (range sensitivity and splitting bias) are used to assess the validity of rating weights. Below, we summarize the approach used in each of the analyses.

Appropriateness and Ease of Use

Because the users are experienced in resource planning, we asked them to compare the appropriateness and ease of use of the various methods they applied. An appropriate method:

- would provide the information that is desired (e.g., rank orders of resource portfolios, sensitivity of ranks to different value judgments);
- would mesh with the decision making style of the organization (in the case here, quantitatively oriented but with an emphasis on openness to public input and scrutiny, implying that transparent methods are preferable);
- would be perceived as improving the consistency and quality of decisions; and
- would be compatible with the data that is available.

Questions were asked of the participants concerning these dimensions of appropriateness. To assess the ease of use of a method, users were queried about the time and training required to apply it and the comprehensibility of its questions.[3]

Differences in Results

That differences arise between the results of different methods should not be surprising for two reasons (Fischoff *et al.*, 1980). First, different methods involve very different modes of thinking. For instance, some methods ask people to focus on quantitative goals or thresholds, while other methods explicitly confront people with tradeoffs between alternatives. Just as subtle changes in phrasings of poll questions can cause people to change positions on controversial issues, differences in how value elicitations are posed can influence the judgments expressed. The second reason is that people -- even experienced decision makers -- are uncertain of precisely (or even vaguely) which tradeoffs are acceptable to them. When people are uncertain, the effect of alternative question phrasings is magnified. Further, when a decision is difficult because it involves a unique problem along with strongly held and conflicting values, most people will be even more unsure of their priorities. This certainly characterizes most energy planning and policy situations.

As an example of where we might expect differences to arise between methods, consider direct and tradeoff weighting. Direct weighting asks users to give numerical assessments of importance. As we mentioned in Section 4.2.3, the difficulty with direct methods is that "importance" is ambiguously defined, and the definition users have in mind may have little to do with the tradeoffs they are willing to make. For instance, "economy" and "environment" were assigned equal weight in one siting study by Detroit Edison because the utility thought that those criteria should receive equal priority (Rowe *et al*., 1979). However, this sort of judgment is very different from tradeoff judgments of the type required by indifference tradeoff weighting, which asks people how much of one attribute they are willing to give up for a given improvement in another. Thus, there is no reason to believe that weights based on tradeoffs that Detroit Edison is willing to make should be equal.

Since there are many methods for accomplishing the scaling, weighting, and amalgamation of attributes (Chapter 4), experiments comparing the results of each could be undertaken. Scaling methods could be compared in terms of the single-attribute value/utility functions they yield (e.g., Barron *et al.*, 1984; Krzysztofowicz and Duckstein, 1984), and whether the final rank order of alternatives differs. Similarly, the weights yielded by different weighting approaches could be directly compared, as could the resulting rank orders of options. It is desirable to compare the parameters (single-attribute value functions, weights) themselves in order to make conclusions that are independent of the alternative set being compared. A set that involves large tradeoffs and no apparent "knee" in the non-inferior set (Chapter 3)[4] is much more likely to show significant differences among methods than a set of options in which one or two alternatives come close to dominating all the others. On the other hand, the importance of differences in parameters cannot in the end be assessed unless their implications for decisions are examined. For this reason, we examine both the parameters themselves (in our case, weights from indifference tradeoff and different versions of rating methods) and the resulting decisions (ranks of resource portfolios).

How can we judge if a difference between methods is "significant"? We gauge significance by comparing correlations between results of *different methods* with correlations between results chosen by *different people*. We assume that differences of opinion among participants matter. Thus, as we argue in Section 5.1, if *which* method is adopted makes as much difference as *who* applies it, then we should conclude that differences between methods are important, too.[5]

Therefore, our comparisons of the results of different methods consist mainly of statistical tests of differences between "between-methods" and "between-persons" correlations. We also compare "between methods" correlations among different pairs of methods in order to assess which methods yield fairly similar results, and which diverge the most.

Validity

In general, the validity of a measurement instrument (whether a tape measure or MCDM method) is its ability to measure what it purports to measure (Larichev, 1992). The concept of validity is straightforward for a tape measure, since there are standard units of measurement against which it can be compared (for example, at the US National Bureau of Standards). The concept is much more ambiguous for MCDM methods. There are no objective measures of a person's values to which the results of an MCDM method (scales, weights, ranks of options) can be compared. First of all, the "values" being measured might not exist or may be very vague. Second, there are no universally accepted ways of measuring those values.

Yet this is not to say that MCDM method validity cannot be assessed, or that all methods are equally valid. There are several concepts of validity that psychologists and decision-scientists have found to be useful, even if they do not represent universal standards. The ones most relevant to this case study are discussed below.[6]

Predictive validity: For behavioral scientists, the purpose of MCDM methods is to understand and predict unaided decisions. Hence, predictive validity -- which evaluates a method on the basis of how well it replicates unaided subjective evaluations -- is the primary criterion of validity. In our case study, the first task we asked of the user group was for them to make an initial ranking of the options, given all of the data about the options and their impacts. As in Chapters 8 and 9, we term these unaided evaluations "holistic rankings". Predictive validity was then established by comparing the results of the different methods with the initial holistic rankings. Methods whose results correlate most highly with the unaided evaluations would be considered to have the highest predictive validity.

Estimative validity: The decision-scientist's rejoinder to behavioral scientists is that predictive validity is an inappropriate standard, because unaided, holistic judgments may in fact be fundamentally biased and/or inconsistent. A perhaps more useful type of validity is estimative validity: to ask whether a method produces rankings that are consistent with some standard of good decisions, such as those reached by experts.[7] However, this standard of validity presumes that there is a "right" evaluation; this may be applicable when judging, e.g., credit card applications (in which the ultimate standard is default risk), but not for ranking plans, where people are expected to differ and all points of view are to be respected.

An alternative yet practical approach to estimative validity is to have each participant repeatedly apply each method, alternating between them and contrasting their results, until their rankings converge and the user is satisfied (Lai and Hopkins, 1995). Then the most valid method is defined as the one whose initial rankings were closest to the final ones. This is, we believe, a particularly useful type of validity because a purpose of MCDM methods is to help people construct their values (Gregory *et al.*, 1993); thus, their final rankings of plans should be more trustworthy than their initial holistic judgments. This type of validity was tested in our BC Gas study of Chapter 9; an analysis encompassing a wider range of methods is presented in this chapter.

Construct (or empirical) validity: A method is said to have construct validity when its evaluations change in appropriate ways when relevant problem characteristics are altered, but do not change when irrelevant characteristics are changed (Stewart and Ely, 1984; Weber and Borcherding, 1993).[8] For this reason, some observers claim that construct validity is a much more objective test than predictive validity, because the latter depends on unaided evaluations which are themselves untrustworthy and fail to provide an objective standard (Pitz *et al.*, 1980).

For rating weights we considered two measures of construct validity: range sensitivity and an absence of splitting bias. It turns out that our groups' rating weights failed both tests.

Range sensitivity addresses whether weights are sensitive to the ranges of attributes, as they should be in theory. When the range of a criterion is expanded (say, from $100 million to $200 million for cost), its weight should also increase (i.e., W_{cost} should be larger, although not necessarily exactly twice as large).[9] Previous studies have shown that rating weights fail to consistently show such "range sensitivity", and so it is doubtful whether they represent people's willingness to trade off criteria (e.g., Fischer, 1995; Hobbs *et al.*, 1992; Stewart and Ely, 1984).[10]

An explanation for the observation that rating weights fail to show appropriate range sensitivity is that the sort of "importance" that people are assessing when they choose rating weights has little or nothing to do with how many units of one attribute they are willing to give up to obtain a unit improvement in another. Rather, "importance" may deal more with how much consideration or salience someone believes that an attribute should receive in a decision process. In this context, a statement that "environment and economics are equally important" makes sense. But such a judgment should not be interpreted as implying that, e.g., a single wild salmon is worth $1,000,000 or $1 -- either value being a possible implication of equal weights, depending on the range of the criteria.[11] Such "importance" judgments are a questionable basis for weights that are used to trade off attributes and make decisions.

The "splitting bias" test instead looks at whether weights are inappropriately sensitive to a theoretically irrelevant input: the hierarchical structure of criteria. In theory, how the hierarchy is structured should not affect the final weight assigned to each criterion. For instance, in Figure 10.1, the final weight for attribute x_5 should not depend on whether the non-hierarchical or hierarchical approach is taken.[12] However, as noted earlier, psychologists have found that hierarchical structure does affect rating weights. Non-hierarchical weights tend to be "flatter" (more equal), while hierarchical weights are "steeper" (have a greater variance) (Eppel, 1992; Stillwell *et al.*, 1987; Bell *et al.*, 2000). Criteria from groups with many criteria tend to receive lower weight per criterion than those in smaller groups.

In the experiment of this chapter, we tested whether planners and interest group representatives were subject to the splitting bias when they assessed rating weights. If they were, we would expect environmental criteria to receive more weight in Figure 10.1a than in Figure 10.1b because in Figure 10.1b, the total weight for environment would be split among many attributes (x_3 through x_{11}). Meanwhile, we would anticipate that reliability (measured as expected loss of energy ELOE, x_{12}) would receive less weight in Figure 10.1a, because it is compared to many other attributes at once. If two different hierarchies result in significantly different rating weights, then whether rating weights truly represent willingness to trade off can be questioned.[13]

Figure 10.1 (a) Non-hierarchical and (b) hierarchical weight assessment for twelve attributes in Seattle City Light case study

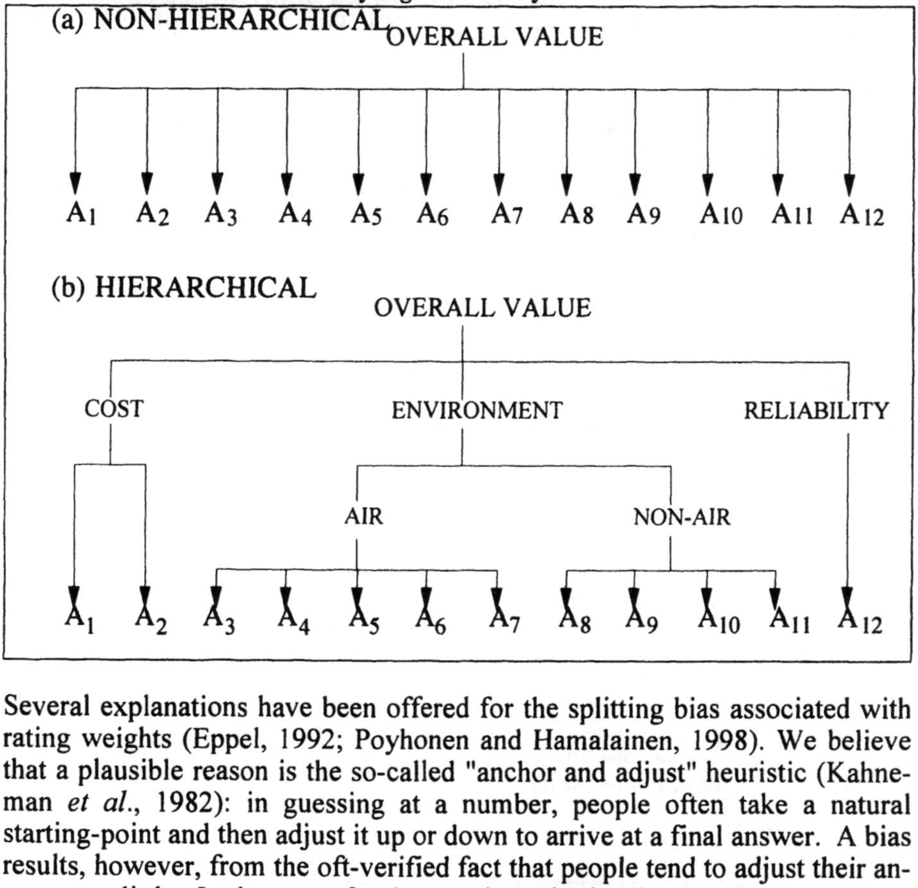

Several explanations have been offered for the splitting bias associated with rating weights (Eppel, 1992; Poyhonen and Hamalainen, 1998). We believe that a plausible reason is the so-called "anchor and adjust" heuristic (Kahneman *et al.*, 1982): in guessing at a number, people often take a natural starting-point and then adjust it up or down to arrive at a final answer. A bias results, however, from the oft-verified fact that people tend to adjust their answers too little. In the case of rating, we hypothesize that people start with an "anchor" of an equal weight for each criterion (say, 50 on a scale of 0-100), and then adjust it down for less important criteria and adjust it up for more important criteria. Because adjustments are too small, all criteria within a group receive fairly similar weights. Thus, in the hierarchy of Figure 10.1, each of the three major groups receives roughly the same weight; consequently, each environmental criterion receives less weight because that category's weight must be split among more criteria.[14]

It should be noted that since there is no objective standard for evaluating the validity of weights, it cannot be definitely stated that hierarchical weights are necessarily better than non-hierarchical weights. Experiments can only show that they are systematically different. We recommend that if hierarchical and non-hierarchical weights differ, that the user be shown the differences, and then resolve them. Generally, we have found that people are more comfortable with the hierarchical approach.

10.3 RESOURCE PLANNING AT SEATTLE CITY LIGHT

Seattle City Light (SCL) is a municipally owned utility whose service territory is the City of Seattle, Washington. SCL has a peak load of just over 2000 MW. It owns about 1800 MW of hydroelectric generating capacity and part of the Centralia coal-fired generating station. In 1992, its firm generating capacity was distributed as follows: 5% coal, 49% own hydro, and 46% purchases, mostly from the Bonneville Power Administration (BPA). SCL also sells energy to utilities in the northwestern and southwestern US, averaging 165 MW in a typical year.

SCL's load growth was expected to be quite moderate over the next decade (see Figure 10.2): of the additional 167 MW that was expected to be needed by 2010, 92 MW will come from new SCL generation, 17 MW from improvements to system efficiency, and 57.5 MW from new DSM program initiatives.

Figure 10.2: SCL load and resource projections

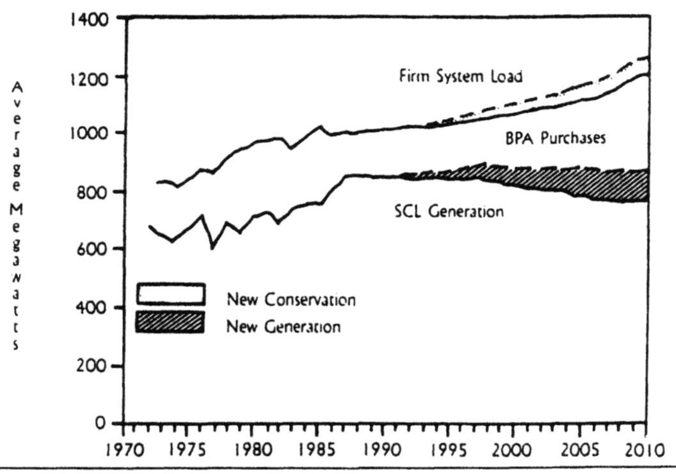

Source: SCL (1991), p. 2

In this case study, a hypothetical planning problem for SCL was analyzed by a group of 12 people. The participants included SCL staff (three environmentalists, two DSM planners, a rates analyst, two resource planners, and an operations planner), a planning consultant, an environmental group representative, and a coal company manager. The group was presented 12 hypothetical portfolios of resources for the year 2010 that were rated on 12 criteria, and each participant used several MCDM methods to rank the options.

The portfolios were created using the Annual Optimization Model (ANN) (Wilson, 1989), a linear programming capacity expansion model solved using Benders decomposition.[15] The version of ANN we applied chooses a mix of supply- and demand-side resources for a given year that minimizes the annualized capital and operating costs of the SCL system, net of revenues received from bulk power sales. The model calculates expected production cost over hundreds of different demand and hydropower scenarios, each representing a different combination of temperature and streamflow for a given time period (e.g., weekend-day). Production costs can include a penalty for CO_2 emissions; by varying that penalty, we obtained several resource mixes and levels of emissions dispatch that represent a range of tradeoffs between utility cost and CO_2 emissions. Additional portfolios were obtained by imposing constraints capturing concerns other than utility cost and CO_2. Examples of such constraints include: a prohibition on coal use during the summer in an attempt to improve visibility in the region; lower or upper bounds upon certain resources; and relaxation of the reliability constraint, allowing positive unserved energy (ELOE). In all, 36 plans were obtained, which we then trimmed to a final list of 12 to be presented to the group (see Table 10.1). Portfolios were dropped if they were dominated by or very similar to other portfolios. Table 10.2 shows the resource mixes for each of the final 12 portfolios.

Table 10.1: Definition of portfolios, year 2010

plan	description	abbrev.
1	Least utility cost mix, relying primarily on integrated coal gasification combined cycle (IGCC), gas-fired combined cycle (CC), and molten carbonate fuel cells (FC); no emissions dispatch; no DSM costing $40/MWh or more	b
2	Same as #1, except natural gas burned in Centralia during summer and 16 MW of $40/MWh DSM	cv
3	$20/ton CO_2 penalty used to optimize generation mix and dispatch; resulting in cogeneration, FC, and 27 MW of $40/MWh DSM	blo
4	Like #3, with participation in fourth NW-SW AC intertie; also, 30 MW of hydro retrofitted at existing dam	b4aci
5	$40/ton CO_2 penalty used to optimize mix and dispatch: includes cogeneration, hydro, and 54 MW of $40-$50/MWh DSM, in addition to CC and FC capacity	bmed
6	Same as #8, except adds wind and geothermal instead of CC	bcog
7	Same as #5, but less reliable (less CC and FC), and gas used in Centralia during summer	cv-rel
8	Same as #5, but with participation in fourth AC intertie and gas used in Centralia during summer	cv+4aci
9	Same as #5, but with wind and geothermal instead of CC	cv-ren
10	Same as #9, except Centralia burns coal in summer	b-ren
11	$60/ton CO_2 penalty used to optimize generation mix and dispatch, including wind, geothermal, CC, FC, 54 MW of $40/MWh-$50/MWh DSM	bnoH
12	$80/ton CO_2 used to optimize, including wind, hydro, CC, geothermal, FC, 81 MW of $40-$60/ton DSM	cvhi

Table 10.2: Resource mix for each portfolio

	plan	CO_2tax $/ton[a]	new generation resources (MW capacity): wind	IGCC	hydro	cogeneration	comb. cycle	geothermal	fuel cell	energy efficiency[b]: $40/ MWh	$50/ MWh	$60/ MWh
1	base	0		200			200		58			
2	cv	0		200			200		46	16		
3	blo	20				34	169		200	27		
4	b4aci	20				34	186		200	27		
5	bmed	40			30	34	127		200	27	27	
6	bcog	40	181		30	34	63	50	120	27	27	
7	cv-rel	40			30	34	85		200	27	27	
8	cv+4aci	40			30	34	135		200	27	27	
9	cv-ren	40	181		30	34	63		200	27	27	
10	b-ren	40	181		30	34	53		200	27	27	
11	bnoH	60	87				115		200	27	27	
12	cvhi	80	181		30		62		200	27	27	27

[a]Cost penalty on CO_2 emissions used for dispatching generation plants. See Section 7.6.

[b]These are energy efficiency (DSM) programs over and above presently planned levels. Three general programs are considered, each with a different average cost, as shown.

Table 10.3: Portfolio performance on each attribute

plan	cost, 10^6\$/y	rate increase over plan 1	CO_2 10^6 t/y	SO_2 t/y	NO_x (NW) t/y	NO_x (SW) t/y	visibility	water use ft^3/s	solid waste tons/y	remote sites, MW	new hydro, MW	ELOE, MWh/y
1	136.7	0.0%	1.51	3,720	-3	-439	40%	3.4	156,000	0	0	0
2	137.7	1.6%	1.42	3,160	-3	-445	0%	3.3	148,500	0	0	0
3	150.8	5.8%	-0.07	2,000	-31	-681	22%	1.2	21,100	0	0	0
4	153.9	6.4%	-0.22	1,960	-32	-822	22%	1.3	20,700	0	30	0
5	159.1	10.0%	-0.36	1,300	-31	-841	14%	1.1	13,600	0	30	0
6	175.9	14.3%	-0.79	1,290	-31	-1,152	14%	0.8	13,500	231	30	0
7	158.8	10.0%	-0.51	1,130	-32	-974	0%	1.2	11,800	0	30	7,200
8	162.4	10.9%	-0.51	1,130	-32	-973	0%	1.2	11,800	0	30	0
9	173.6	13.7%	-0.63	1,040	-29	-978	0%	0.8	10,800	231	30	0
10	173.1	13.6%	-0.62	1,250	-29	-983	14%	0.8	13,000	231	30	0
11	182.2	15.9%	-0.87	760	-2	-1,145	8%	1.1	8,000	137	0	0
12	193.8	21.5%	-1.09	680	-1	-1,378	0%	1.2	7,100	231	30	0

Table 10.3 shows the performance of each portfolio on the twelve attributes. "*Cost*" is defined as SCL's variable power supply cost, plus the annualized capital expense of new generation facilities, minus bulk power sales revenues. "*CO_2*" is the annual tonnage of CO_2 emissions, net of reductions that occur because of SCL's extensive power exports during the spring and early summer. These exports allow utilities in the southwest US to back off fossil-fueled stations; we felt that SCL would be justified in taking credit for the resulting CO_2 reductions. These reductions are so extensive that SCL's net CO_2 emissions are negative for many cases, as shown in Figure 10.3.

Figure 10.3: The tradeoff between CO_2 emissions and system cost for SCL

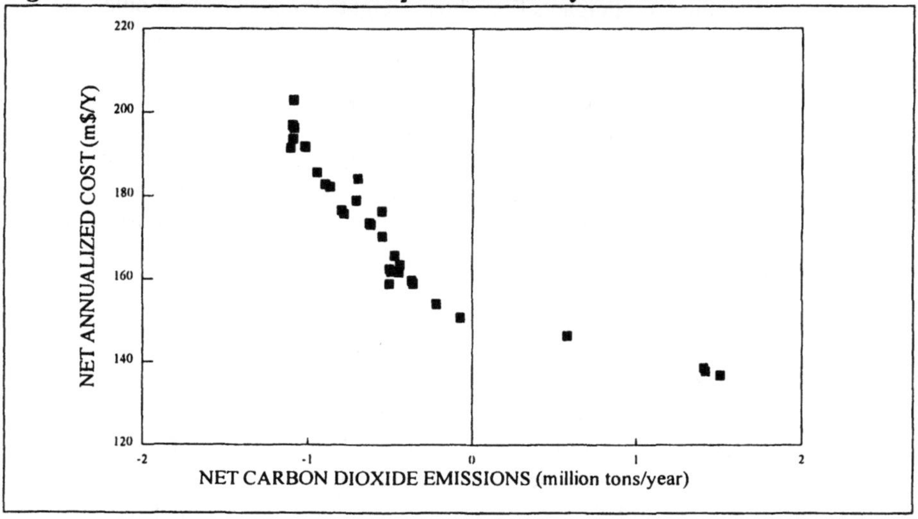

Another major air emission, NO_x, is divided into two criteria: one for emissions occurring in the northwest ("*NO_x(NW)*"), and the other for emissions in the southwest, primarily California ("*NO_x(SW)*"). This geographical division was made because smog problems are worse in the southwest, yet participants might feel more strongly about pollution in the northwest than elsewhere. Either criterion can be negative because bulk sales by SCL result in a reduction in emissions elsewhere that can exceed SCL's emissions.

The other environmental attributes also require some explanation. The attribute "*Visibility*" is defined as the Centralia station's summer capacity factor multiplied by the fraction of its fuel that is coal; higher values mean more potential visibility impact at nearby national parks. "*Water*" is the increase in consumptive water use over and above that now used at Centralia. The "*Remote sites*" criterion is the amount of wind and geothermal development, and relates to concerns about developing new transmission and generation facilities in natural settings. Similarly, the "*New hydro*" criterion is included because of opposition by some study participants to new hydroelectric plants.

The participants applied the following methods during the course of a two-day workshop:

- *Holistic assessment,* in which each participant directly rated the quality of the 12 plans on a 0-100 scale without the help of an MCDM method. They did this twice, once before the weighting exercises, and once at the end of the workshop. These assessments are referred to as "Holistic I" and "Holistic II", respectively.
- *Scaling*, in which a value function $V_i(A_{ij})$ was created for each attribute i. The purpose was to familiarize participants with the criteria, and with the principles of scaling.
- *Rating weighting*, where participants each directly chose weights on a scale of 0-100. Half the participants picked weights for all criteria at

once, while the other half used the hierarchical approach. In the hierarchical method, participants first chose weights for each overall objective in Figure 10.1(b), and then those weights were allocated among the criteria comprising each objective. In either case, we advised the participants that weights should represent the tradeoffs they were willing to make, and we gave concrete examples of weight calculations from tradeoffs as illustrations. The rating exercise was repeated later in the workshop; in the second assessment, the ranges of three of the criteria were either shrunk by 67% or expanded by 200%. The purpose there was to test the sensitivity of weights to criteria ranges as a measure of construct validity.

- *Tradeoff weighting*, in which participants traded off each criterion against cost by stating what change in cost was equivalent to a given change in the other criterion. The weights were then calculated using the method described in Section 4.2.6, assuming linear $V_i(A_{ij})$.
- *Goal setting*, wherein participants selected levels of each criterion that were acceptable, or that they wanted to obtain.

At the end of the second day, each participant filled out a questionnaire, answering questions about the appropriateness and ease of use of the methods. We then used the chosen weights to create rank orders of the plans in Table 10.1 for each person by each of the following amalgamation methods:

- **AVF**: additive value functions (Section 4.3.1) with tradeoff weights **AVF(T)** and with rating weights **AVF(R)**. The value of plan j is defined as $\Sigma_i w_i V_i (A_{ij})$; higher values are better.
- **GP(*P*,-)**: goal programming using rating weights and penalizing only bad deviations; $P = 1,2,\infty$ (Section 4.3.2). The total deviation of a plan from j the goals is defined as $\{\Sigma_i [w_i Max\ (0, G_i - V_i(A_{ij}))]^P\}^{1/P}$ (Eq. [4.8]); lower values are preferred. G_i is defined as the goal for attribute i.
- **GP(*P*,+/-)**: goal programming using rating weights and penalizing both positive and negative deviations; $P = 1,2,\infty$. The deviation of a plan from the goals is defined here as $\{\Sigma_i [w_i |G_i - V_i(A_{ij})|]^P\}^{1/P}$ (Eq. [4.7]). As in the previous case, lower values of the deviation are preferred.

In all cases, linear $V_i(A_{ij})$ were used, in part because some participants had difficulty understanding the scaling exercise. Also, we found that changes in the form of $V_i(A_{ij})$ made much less of a difference than changes in weights.

Within two weeks of the workshop we conducted individual follow-up interviews with each participant. During the interviews, they examined alternatives whose ranks shifted the most among the methods. In particular we asked them to make the following comparisons:

- **AVF(R)** vs. **AVF(T)**: for the rating weights, we calculated the tradeoffs that would be consistent with those weights. For the tradeoff weights, we deduced the rating weights that would be equivalent to the tradeoff answers. We asked which of the two sets of weights/tradeoffs was more consistent with the person's preferences, or if the person would now choose a different set of weights/tradeoffs. We then asked which of the two sets of plan ranks (**AVF(R)** or **AVF(T)**) best matched his or her preferences.

- **GP(*P*,+/-) vs. GP(*P*,-)**: for $P = 1$, we asked which set of plan ranks was more consistent with the person's preferences, explaining the differences between the mechanics of the two methods. Where the results for $P = \infty$ differed strongly from $P = 1$, we repeated this question for $P = \infty$.
- **GP(1,+/-) vs. GP(∞,+/-), or GP(1,-) vs. GP(∞,-)**: for the preferred form of goal programming (either +/- or -, depending on which method the person preferred in the above comparison), we explained how each version of GP worked, and asked which set of plan ranks best matched their preferences.
- **GP(*P*,+/- or -) vs. AVF(R or T)**: for the most favored goal programming and additive value functions, as determined by responses to the above questions, we asked which rank set best fit the participant's preferences.

Because of time limitations, three disagreements at most were resolved for each pair of methods. A "disagreement" was defined as a significant shift in the ranking of one plan. We emphasized plans that were highly ranked by at least one method, and considered shifts of three or more ranks (out of 12 plans) to be the most significant. The result of this process was selection of one method that each person felt best modeled their preferences. The purpose of this comparison was to obtain a measure of estimative validity similar in spirit to Lai and Hopkins (1995), as explained in Section 10.2, above. By this measure, the most valid method within a pair is one whose ranks are closest to the preferences stated during the final interview.

10.4 RESULTS OF METHOD COMPARISONS

We now present the results of the workshop, in which we examine, in turn: method appropriateness and ease of use; differences in results; and the various measures of validity. In footnotes to the text, we contrast our conclusions with other results published in the literature, focusing on other experiments we have undertaken.[16]

Ease of use and appropriateness

Table 10.4 summarizes the responses to the questionnaires. Some of the differences are statistically significant; for example, a sizable majority found that rating was easier to understand and use than tradeoff weighting, and that rating did a better job of helping users to understand the problem. A majority felt that each method made sense and was logically sound, improved consistency of judgment, and aided understanding of the problem. However, only a minority felt that tradeoff weighting was easy to understand, required a reasonable amount of effort, increased confidence in the results; or should be recommended for planning.[17]

Table 10.4: Mean responses to appropriateness questions

Question	Rating weighting	Tradeoff rating	Goal setting	Additive value function	Goal programming
How easy were the concepts underlying the method to understand?	1.3	0.2	1.3	0.6	0.5
How easy were the specific questions used by the method to understand?	1.5	0	1.3	0.6	0.7
Does the method make sense to you in terms of rationale and logical soundness?	1.4	0.6	1.2	1.2	1.4
Is it reasonable to expect users to acquire the skills needed to use the method?	1.7	0.2	1.4	1.1	0.9
Is the amount of effort required reasonable, given resource and time constraints?	0.8	-0.2	0.9	0.5	0.3
Does the method improve decision-consistency compared to holistic ranking?	0.8	0.5	0.6	0.7	0.6
Does the method increase confidence in the rankings compared to holistic ranking?	0.4	-0.1	0.6	0.8	0.6
Does it help you focus on important issues and better understand the problem?	1.5	0.7	1.3	0.9	1.1
Would you recommend use of this method in utility resource planning?	1.2	0	1	1	1

Key: -3 = "definitely hard" or "definitely no"; +3 = "definitely easy" or "definitely yes"; 0="neither"

These results were confirmed when participants were asked which weighting method they preferred: nine chose rating, and only two picked tradeoff weighting. Yet later, when asked to resolve differences in the follow-up interviews, several of the participants felt strongly that tradeoff weighting was helpful, because it was often more faithful to their preferences than rating weighting. This implies that although tradeoff weighting should not be relied on by itself, it is useful to ask at least some tradeoff questions as a check upon weights derived by other methods. The reader may recall that we reached similar conclusions in Chapter 9.

The questionnaires also asked which of the amalgamation methods was preferred: goal programming, additive value functions, or holistic judgment. Six participants picked goal programming, three chose holistic judgment, and only two preferred additive functions. From our discussions with the group, this seemed to be because the goal setting exercise gave additional insight.

Although people were generally positive about the various methods (except tradeoff weighting), they were less enthusiastic than the BC Gas participants in Chapter 9 (Table 9.7). Based on our interaction with the participants, we believe that this difference is due to the compressed time schedule in the Seattle City Light study, in which a wide range of methods was applied during a ten-hour workshop over two days. In contrast, the BC

Gas study allowed much more time for digestion of the methods and results, including group discussions of the results of the assessments and interviews. It is clear that adequate time must be taken to process and review results, or they will be distrusted.[18]

Comparison of results: Do different weighting methods yield the same weights?
In this subsection, we contrast the weights obtained by rating and tradeoff weighting. In the next subsection, we shall compare the effects of weighting method and amalgamation technique upon the plan ranks. We examine the weights themselves in addition to plan ranks in order to derive conclusions that are independent of the particular plans in our case study.

A necessary condition for all methods to be valid is that they produce essentially the same weights and/or alternative ranks. If their results differ significantly, then at least one method must be invalid. We use correlation statistics as a measure of closeness of two sets of weights or two sets of plan ranks. Correlations vary between -1 and +1, with +1 indicating a perfect match of the two sets, and -1 showing that the sets are in exact opposition. An r of zero indicates no correlation. Since weights are on a ratio scale, we use a parametric statistic -- Pearson's r -- to compare sets of weights.

As we explained in Section 10.2, we assess the significance of disagreements among method results by comparing "between-method" correlations with "between-persons" correlations. If method differences are similar to or larger than differences of opinion between people, we judge that choice of method matters. As an example of "between-persons" correlations, there are 66 (=12x11/2) possible pairings of the first set of rating weight sets chosen by the twelve participants. We therefore can calculate 66 "between persons" correlations (Pearson's r), one per pair: the resulting average r of those pairs is 0.32. For tradeoff weights, the mean between-persons r rises to 0.48. Thus, the participants' weight sets agree more than they disagree (i.e., the mean $r > 0$), but by no means are they unanimous about which criteria are most important.[19] Meanwhile, we also calculated Pearson's r between each person's tradeoff and first rating weights. The mean of these 12 *between-methods* correlations is 0.42, which doesn't differ significantly from the above mean between-persons correlations. For two of the participants, this between-methods correlation is even negative. Thus, in this case we conclude that *which* weighting method is used makes as much difference in this study in the weights as *who* chooses the weights.

Other studies have noted important differences between the results of direct assessment and tradeoff weighting methods.[20] Some of these differences appear to be systematic. In particular, there is ample evidence that whichever criterion is chosen to be varied in a tradeoff question tends to be overweighted, because people tend to understate the change that they are willing to make in that criterion (Eppel, 1992; Delquie, 1994). For instance, in Borcherding *et al.* (1991), tradeoff weights were obtained by asking participants to vary a health and safety criterion. Averaging across the subjects in that study, that criterion received 53% of the total weight assigned to all criteria, much more than when weights were chosen by other methods. Yet when participants were instead asked to vary a dollar amount, cost received an average of 80% of the total weight.

The possible existence here of this overweighting bias can be tested by comparing the rating and tradeoff weights to determine if the tradeoff weight for cost (the criterion varied in the tradeoff questions) is higher than its rating weight. The answer is yes. Cost received, on average, 24% of the weight under the tradeoff method, but only 14% with rating; the difference is statistically significant. Thus, we cannot rule out the presence of this *overweighting* bias here.

But neither can we conclude that this bias definitely exists, because there is another possible explanation of this result. Many of the criteria, such as NO_x(NW) and water, involve relatively trivial changes in emissions or resource consumption. Yet, as one participant explained, because people are reluctant to give zero weight to a criterion, they would give a "minimal" level during the rating exercise -- e.g., 10 on a scale of 0-100. This level is high enough to drive down the relative weight of cost, if there are enough unimportant criteria. Thus, rating may *underweigh* cost if there are several inconsequential criteria that are also being weighted.

Comparison of results: Do different methods yield the same plan ranks?

Analogous to our comparison of weights, we test whether differences in plan ranks resulting from different methods are as, or more important than, differences between ranks chosen by different people. If so, then we judge differences in methods to be significant.[21]

We use Spearman's *r* (a non-parametric correlation statistic) as a measure of closeness for two sets of plan ranks chosen by different methods. Because ranks are only on an ordinal scale, a parametric statistic (such as Pearson's *r*) would be inappropriate.

Differences of opinion between individuals are measured by the Spearman's correlations between sets of plan ranks chosen by different people. The mean of the correlations between the 66 possible pairs of **AVF(R)** plan ranks chosen by different people is 0.63. This is similar to the average between-persons correlation of 0.60 for **AVF(T)**. However, for **GP(1,+/-)**, the average falls to 0.25. There is even less agreement among the holistic ranks: there, the average between-persons correlations is *zero*. These results are broadly consistent with other studies.[22] Thus, when an MCDM method is used, there is some but not complete agreement among the participants as to which plans are best, and the use of goals results in greater disagreement. Further, MCDM methods result in more agreement among participants than does holistic judgment.

These results are displayed graphically in Figure 10.4, in which we show the agreement in plan ranks between individuals for additive value functions using rating and tradeoff weights, and one set of holistic rankings. There is considerably less agreement among holistic rankings by different people than among additive value function rankings.

Figure 10.4: Comparison of methods, plan ranks by individuals

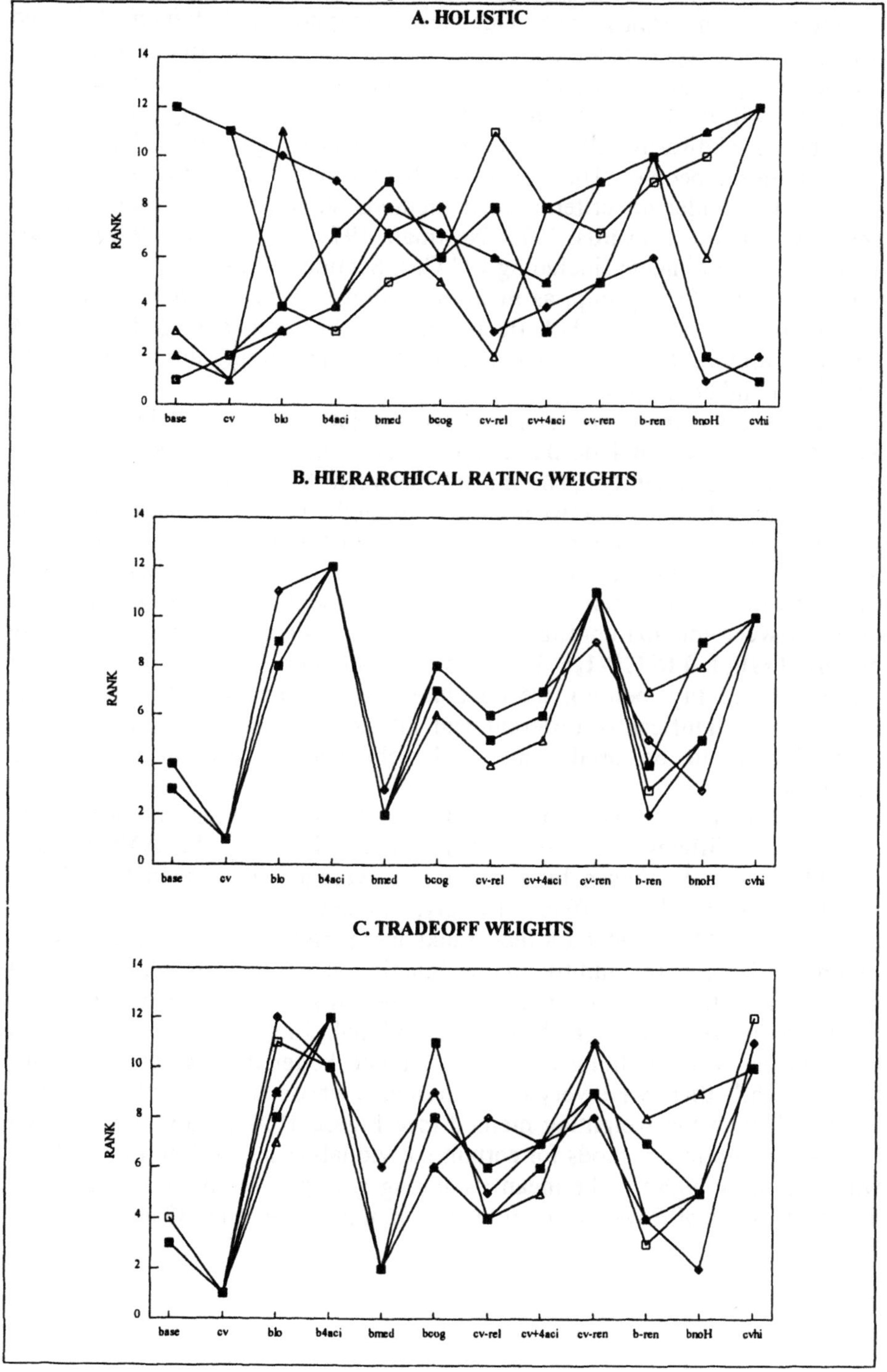

That the MCDM methods yield more consensus than holistic judgment can be interpreted in various ways. Our view is that holistic judgment is flawed because such judgments are colored by prejudices about particular resources. Further, people can consistently consider only a small subset of criteria at a time, and will tend to focus on those that most strongly engage or reinforce their biases (Edwards, 1977). These flaws can exaggerate differences of opinion among people. Thus, because MCDM methods have no such prejudices and consider all criteria, such methods should yield greater convergence of opinion (Edwards, 1977; Newman, 1975; Vlek *et al.*, 1999).[23] Just as in other experiments (including BC Gas, Section 9.4), such a convergence occurred in the SCL group, as mentioned above.[24] Several participants appeared to moderate their holistic judgments after considering the MCDM rankings, moving towards acceptance of plans representing intermediate levels of cost and emissions.

Adopting the above between-persons correlations of the MCDM methods as a standard, we conclude that additive value functions and goal programming differ significantly. The mean *r* between **AVF(R)** and **GP(1,+/-)** is 0.44; hence, differences between those two methods are comparable to those between persons (0.63 for **AVF(R)**, and 0.25 for **GP(1,+/-)**, as just stated). A major reason for this divergence of methods is **GP(1,+/-)**'s penalization of overachievement of goals. If instead **GP(1,-)** is contrasted with **AVF(R)**, the mean between-method correlation rises to 0.79. Since most people strongly preferred **GP(1,-)** to **GP(1,+/-)** in the follow-up interviews (see discussion of estimative validity, below), we conclude that penalization of overachievement yields significantly different and inferior choices. Thus, if goal programming is to be used, only underachievement of goals should be penalized.[25]

But differences between other pairs of methods appear less important than inter-person differences. For instance, the *r* between the **AVF(R)** and **AVF(T)** rank sets chosen by each person averaged 0.84, even though the weight sets themselves differed strongly, as noted above. Thus, on average, choice of weighting method affects ranks less than who chooses the weights. Figure 10.5 compares additive value function scores (Eq. [4.1], Chapter 4) yielded by tradeoff and hierarchical rating weights: we show the results averaged over all participants, and for one individual. In general, the correlations between the two weighting methods are much better than weighting versus holistic; while there were very few individuals whose holistic rankings correlated well with the weighting method (see Figure 10.8, below), consistency between weighting methods for certain individuals (such as A in Figure 10.5) was extremely high.[26] The mean *r*'s among goal programming models with different values of *P* were also quite high, ranging between 0.85 and 0.96.

Figure 10.5: Comparison of plan ranks from tradeoff weights and hierarchical rating weights

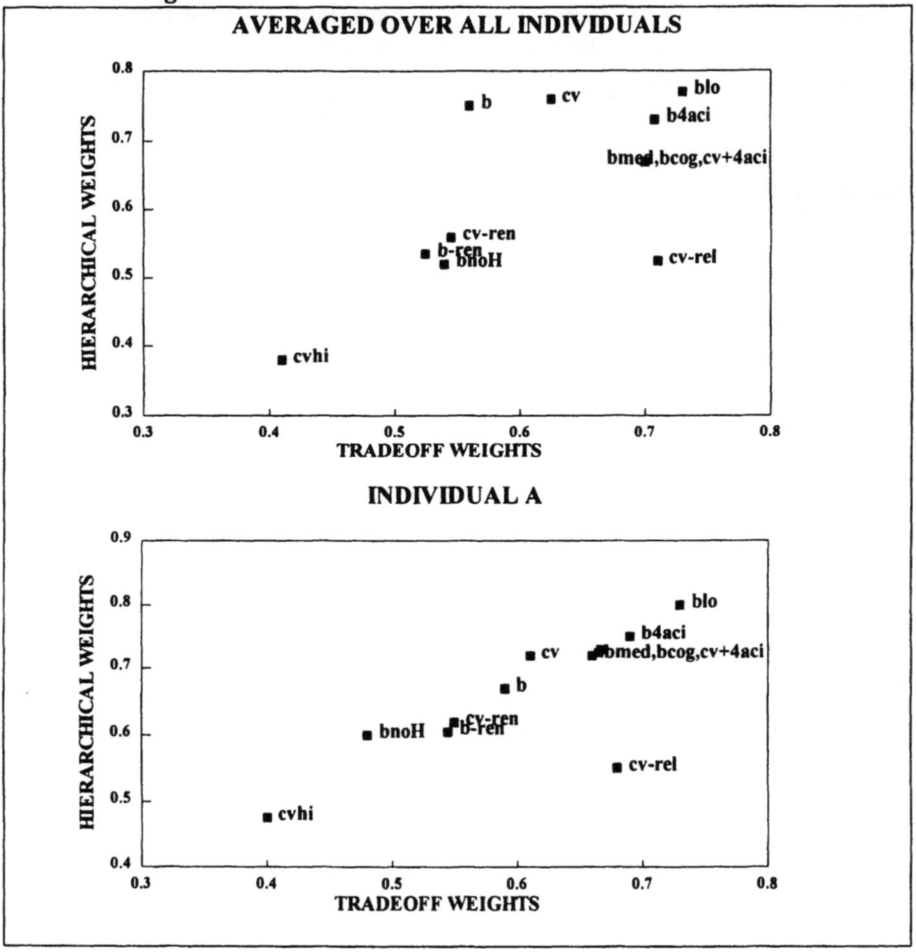

Predictive validity: How well do the methods predict holistic judgments?

The last three subsections of Section 10.4 discuss the validity test results. We begin with predictive validity.

Behavioral scientists often evaluate MCDM models by checking how well they predict holistic rankings. In this experiment, no method did well by this test. We used Spearman's r to measure how closely a method's rank set matched a holistic rank set.

Consistent with other studies,[27] the average r across the 12 participants between a given method's ranks and the holistic ranks is fairly poor. For instance, the mean r between **AVF(R)** and the holistic ranks is 0.40, while that between **AVF(T)** and the holistic ranks is 0.45. The best performing method is **GP(*P*=1,+/-)**, whose average correlation is 0.53, while the worst is

GP(*P*=∞,-), at 0.38. None of the means was statistically different from any of the others. Thus, all methods have doubtful validity by this test.

Figure 10.6 shows a comparison of holistic ranks against those obtained using the additive value function for a selection of specific individuals, as well as overall averages. Figure 10.6 also displays additive value function results based on tradeoff weights. These graphs illustrate the actual set of results implied by the low correlation coefficients noted above.

Figure 10.6: Comparison of holistic rankings with additive value functions

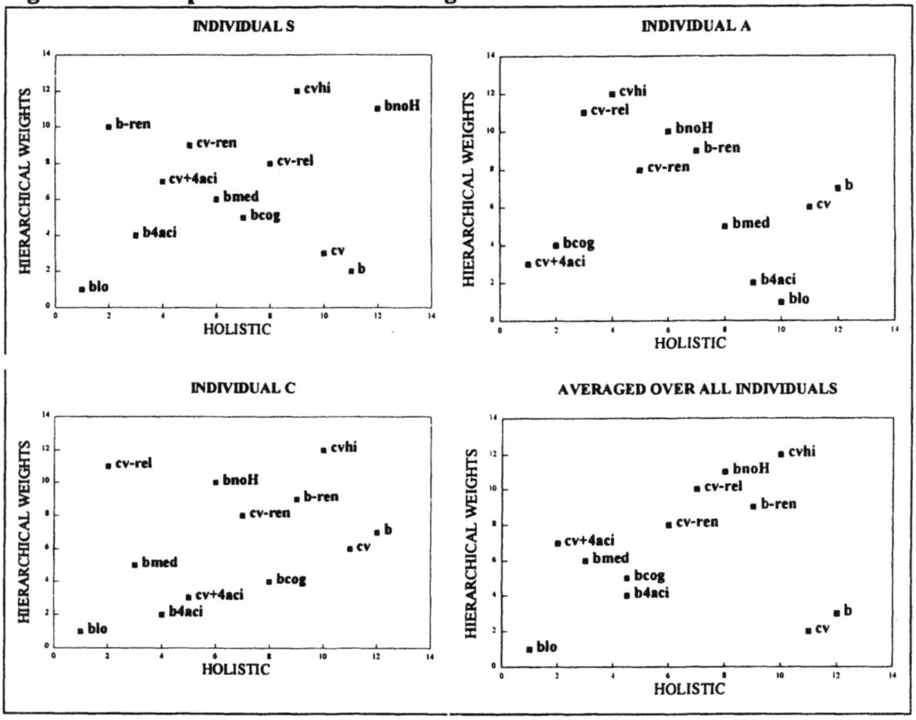

As just one example, one of the participants (Individual A in Figure 10.6) chose plan 3 ("blo") as best by both **AVF(R)** and **AVF(T)**, while his holistic ranking placed it as only the tenth best plan. In contrast, some of the plans he ranked highly using holistic judgment ("cv-rel", "cvhi") were ranked at the bottom by the additive value functions. In general, we found that holistic judgments were often more extreme, in terms of choosing alternatives that are either at one end or the other of the CO_2-cost tradeoff curve. In contrast, MCDM methods often yielded more of a compromise. There are at least two possible reasons for this. One is that people give insufficient attention to less important criteria when making unaided judgments, whereas MCDM methods consider all criteria. Alternatively, people may assign weights that are too equal, resulting in MCDM methods putting too much emphasis on unimportant criteria.[28]

Estimative validity: Which method's results better match the final preferences?

We have noted previously that holistic judgment may be fundamentally biased and/or inconsistent, and that we should not expect a valid method to predict holistic judgments properly. Estimative validity may therefore constitute a better test. Unfortunately, the approach suggested by Lai and Hopkins (1995) -- repeatedly applying methods until results converged -- takes substantially more time than was available to us in this case study. So we instead asked each person to contrast the results of the methods during the follow-up interview, reflect on the reasons for the differences, and then state which rank sets were closest to their final preferences.

The results surprised us, considering the overwhelming preference expressed in the questionnaires for rating and goal programming. In particular, a plurality (6:5, one tie) concluded that the tradeoff weights were closer to their "true" weights than rating weights, and a majority (7:3, two ties) thought that **AVF(T)**'s ranks better represented their preferences than **AVF(R)**'s. However, tradeoff weighting's advantages were not statistically significant. In the case of goal programming versus additive value functions, the methods yielded the same ranks for four persons, one person was indifferent between the two approaches, and the other six were evenly split (3:3) as to which method came closest to their preferences. In light of these results, some participants changed their mind about their earlier rejection of tradeoff weights and additive value functions.

Construct validity: Are rating weights sensitive in appropriate ways?

Construct validity tests are designed to see whether the results of a method change in appropriate ways when inputs to the method are altered. For rating weights, we considered two measures of construct validity: range sensitivity and an absence of splitting bias. Our groups' rating weights failed both tests.

As noted in Section 10.2, previous studies have shown that rating weights fail to consistently show such "range sensitivity", and so do not appropriately represent people's willingness to trade off attributes. This result is confirmed here. In the second rating exercise, some of the criteria ranges were significantly contracted or expanded, as described in Section 10.3. Only 44% of the weights chosen by the participants for those criteria changed in the appropriate direction (higher weight for wider ranges, lower weight for narrower). Thirty percent of the weights did not change, and 26% changed in the wrong direction. Thus, by this measure, rating weights fail to show construct validity most of the time.

Valid weights should also be insensitive to the hierarchical structure of criteria. As noted earlier (Sections 8.5 and 10.2), other experiments have shown that nonhierarchical rating weights are flatter (are more equal) than hierarchical weights. This splitting bias was observed in Seattle as well. The coefficient of variation[29] of rating weights for the group members who chose non-hierarchical weights was significantly lower than for those who used the hierarchy (0.83 versus 1.27).

If a splitting bias is present, then non-hierarchical weighting should give more weight to attributes that belong to categories with many attributes (e.g., in our case study, environmental attributes, of which there are nine -- see Figure 10.1b), while giving less weight to attributes in small categories (cost and reliability). To make this comparison, we normalized each set of weights so they sum to 100%. Consistent with the splitting bias, the mean normalized weight for the nine environmental attributes was significantly higher for the non-hierarchical group (7.4% of the total weight for each, vs. 3.5% for the hierarchical group), while the mean reliability weight was lower (5.8% versus 16%).[30]

In Figure 10.7 we display this result in a different way, by looking at the range of the difference between the attribute that received the lowest and the highest weights. Both tradeoff and hierarchical weights are seen to be much more "peaked" than non-hierarchical rating weights. This dramatically illustrates that by manipulating the form of the hierarchy, an analyst can increase or decrease the relative weight given to certain groups of attributes. This reinforces the need for tradeoff questions as consistency checks of the validity of rating weights.

Figure 10.7: Peakiness of hierarchical and non-hierarchical weights

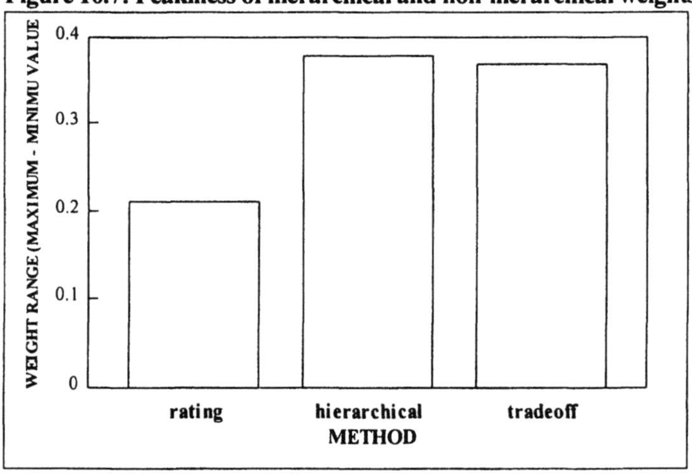

10.5 CONCLUSIONS

Most of the participants in the experiment agreed that MCDM methods help people gain insight into energy planning problems, and increase their confidence in the decisions reached. However, the experiment also shows that different methods can often yield significantly different decisions: thus, users of MCDM methods in utility planning should beware of invalid methods that could result in biased or undependable results.

Unhappily, no single MCDM method is unambiguously more valid than the others. Each has potential flaws. For example, rating weights are often too flat and fail to represent tradeoffs that people are willing to make.

However, if rating weights are going to be used, we strongly recommend the use of hierarchical structures, because users tend to feel more comfortable with that approach. Meanwhile, tradeoff questions are difficult and may overweigh some criteria. For some people, additive value functions and tradeoff weights best represent their preferences; for others, goal programming and rating are superior. Therefore, as we argued in Chapter 5, more than one approach should be used in any application of MCDM methods. For instance, some tradeoff questions should be asked as consistency checks upon weights that are assessed by direct methods such as rating or the Analytic Hierarchy Process. Both goal programming and additive value functions can be used, and the results presented to decision-makers who can then mull over the differences. In evaluating the results of different methods, the potential for the biases that we and others have uncovered should be kept in mind. The extra effort in applying more than one method is not large, but the potential benefits, in terms of enhanced confidence and a more reliable process, are great.

It must be recognized that the extent to which *any* MCDM method will provide consistent plan rankings among individuals is a function of the diversity of values held by those individuals: the greater the differences, the less likely is it that individuals' rankings will agree with each other. This was illustrated in Figure 5.4, which shows the average weights for the different individuals in the SCL participant group, with a comparison of the weights in a study we conducted at Centerior in Ohio. Clearly, the SCL group shows much greater differences in weights, which reflects the much greater diversity of backgrounds -- and hence also values -- than in the Centerior case. In this instance, the group consisted entirely of electric utility planners.

Yet uncovering disagreements among participants' values, and making explicit how different values and weights may lead to different results, is precisely the point of applying MCDM methods. Indeed, despite the need to reconcile the results that reflect very different perspectives, we would still argue for a diversity of viewpoints in a planning group that plans to use MCDM methods. In the SCL exercise, the presence of outside representatives from an environmental group, and from a business group, provided valuable perspectives. It is critical that decision-makers understand and consider these perspectives early in the process.

NOTES

[1] Portions of this chapter are based on "Multicriteria Methods for Resource Planning: An Experimental Comparison" by B.F. Hobbs and P. Meier which appeared in *IEEE Transactions on Power Systems*, Vol. 9(4), pp. 1811-1817, Nov. 1994, Copyright 1994 IEEE.

[2] Other researchers have used Monte Carlo simulation to compare how the weights of different methods might diverge under certain assumptions (Jia *et al.*, 1998; Stewart, 1996; Zanakis *et al.*, 1998).

[3] A number of studies (Adelman *et al., 1984*; Eckenrode, 1965; Nutt, 1980; Sarin *et al.*, 1980) have contrasted the time and effort required of decision-makers to use a variety of weighting and value-scaling techniques.

[4] For an example, see Figure 10.3, below.

[5] We could also compare variations between the results of different methods with variations between results from the same method applied by the same person at two separate times (test-retest validity). If method differences are less than differences in a person's values at two different times, we might judge method differences to be unimportant compared to random variations due to uncertainty and a person's mood. However, such a comparison is often difficult for two reasons. One is the "carry-over" bias, in which a person will unconsciously or consciously attempt to make a second assessment consistent with the first, which understates the actual randomness inherent in his/her judgments. A second reason might instead exaggerate the importance of a person's uncertainty: learning and reflection so that values change over time, which could exaggerate differences in test-retest results.

[6] Others include convergent validity, inter-expert validity, inter-temporal validity, and normative validity (Hobbs, 1986):

Convergent validity: If two methods lead to similar parameters or evaluations, then confidence in their validity is increased. Important differences suggest that the methods cannot both be valid. But convergence does not guarantee validity, since both methods could have the same systematic bias. Convergent validity is examined in both of the case studies: and we do indeed find that some methods give quite different results than others.

Inter-expert validity: The assumptions of this criterion are that evaluations by fully knowledgeable and rational people should agree. Invalid methods are presumed to unduly exaggerate differences of opinion or value. For example, Newman (1975) discovered more convergence among evaluations by different judges' weighting summation rules than among their unaided judgments, and therefore concluded that the latter were less valid. However, the inadequacy of the inter-expert convergence criterion is evident when one realizes that under this criterion, equal weighting would be the most valid weighting method. Moreover, since differences in values lie at the heart of many utility planning problems, forced uniformity would only bury disagreements that should be made explicit.

Inter-temporal validity: A method has inter-temporal validity if, when applied by the same person at different times, its evaluations do not change. Yet differences between a first and a second application may have as much to do with learning as with the validity of the method itself. Or differences may signal that the application of formal MCDM techniques has indeed achieved one of its aims, namely that it has forced the user to think harder about how he or she feels about the issues. Balzer *et al.* (1983) is one of the few studies to have looked at this question in a carefully controlled experiment: their results suggest the unsurprising conclusion that direct weighting is more trustworthy if the users have been allowed to examine the alternatives and become very familiar with the problem.

Normative validity: A method is said to have normative validity when it is based on the best available information, and is the result of careful thought (Stewart and Ely, 1984). The method must be understood, and judged reliable by the decision-maker -- who would then be committed to the method's results and willing and able to defend them. This is self-evidently a purely empirical question, because it can be tested only by asking the decision-makers for their opinions. It may be difficult to treat this question as purely one of validity: the decision-maker's overall comfort with a particular method, and his willingness to defend decisions reached by that method, are likely to be as much questions of appropriateness and ease of use.

7 A good example of such a study is Adelman *et al.* (1984), who compared the accuracy of five weighting techniques with two group discussion methods, adopting as a standard a model actually used by the US Marines to evaluate combat readiness.

8 Perhaps one of the most obvious tests of construct validity is to change a decision-problem by adding a significantly dominated alternative: rankings ought not to change in a method that meets construct validity.

9 This assumes that the single attribute value functions $V_i(A_i)$ are defined so that the worst level in an attribute's range receives a value $V_i(A_{i*}) = 0$, while the best level in the range receives value $V_i(A_i^*) = 1$ (or some other fixed value).

10 von Nitzsch and Weber (1993) show that ratio and multiple regression weights suffer from the same problem. In our case study at Centerior Energy, we found that another method, swing weighting, also lacks range sensitivity (Hobbs and Meier, 1994a). This is a particularly disturbing result because, as we explain in Chapter 4, swing weighting questions are designed to explicitly confront assessors with the ranges of attributes.

11 A value of $1,000,000/salmon could occur if the range of the cost attribute were [$0, $$10^{10}$] and that of salmon [0, 10^4], if salmon were the only environmental attribute and it received the same weight as cost. But if each attribute had a range of [0, 10^6], then a $1/salmon value would instead be implied if the two attributes were equally weighted.

12 More precisely, each attribute's weight, as a percentage of the sum of the weights assigned to all attributes, should not change. Only the ratios of weights matter when ranking alternatives. For instance, if there are only two attributes, say cost and CO_2, then the weight sets (w_{cost}=10, w_{CO2}=20) and (w_{cost}=2, w_{CO2}=4) would yield the same rank ordering of plans, and so would be considered the same.

13 We mentioned in Chapter 8 that this bias was found for rating weights in the BC Hydro study. The Seattle study was the first in which we found this bias, and so we discuss it at more length here.

14 Another possible explanation for the splitting phenomenon is the double counting that can occur in the absence of a hierarchy because respondents are inclined to accord a weighting to a single criterion as though it were a proxy for a higher level objective. For example, one individual may, in the absence of a hierarchy, be more inclined to treat CO_2 as a proxy for "environment" and thereby assign a disproportionate weight to it.

15 This model is based on the linear programming models in Turvey and Anderson (1977). The Benders decomposition methodology is used to break up the problem into a resource choice model (the Benders "master problem") and resource operations models ("subproblems"), one for each operating scenario. The algorithm iterates between the master problem and subproblems until a convergence criterion is satisfied. The Benders method is described in Bloom *et al.* (1984).

[16] A thorough review of the earlier literature may be found in von Winterfeldt and Edwards (1986).

[17] In other studies, users have also found trade-off weighting to be more confusing and difficult than direct weighting methods, such as rating (Hobbs, 1980; Hobbs *et al.* 1992; but see Bell *et al.*, 2000, in which the tradeoff method was liked just as much as direct methods). Borcherding *et al.* (1991) give evidence that trade-off weighting can be made easier and more consistent by phrasing questions in such a way that respondents are never asked to say how much of a deterioration they would accept in an adverse attribute, such as lives lost or environmental impact, since their participants could not answer such questions consistently. Rather, trade-off questions should be phrased in terms of how much *improvement* they would accept in such attributes, because improvements are easier to accept psychologically.

[18] Lack of time was the major reason why the numerical results of Keeney and McDaniels's (1999) utility function exercise at BC Gas were not accepted by the stakeholder group.

[19] These between-person correlations are comparable to other studies. Hobbs (1986) reports that two panels of power plant siting experts chose weights by several direct methods, along with trade-off weighting and multiple regression. The direct methods yielded mean inter-person correlations of weights of between 0.42 and 0.72. However under indifference trade-off weighting, the two panels had mean correlations of 0.08 and 0.81, respectively. Another siting study (Hobbs, 1980) found a value of 0.18 for the between-persons correlation for rating weights. Other siting studies have found values of inter-person correlation of rating weights in the range of 0.2 to 0.4 (Dobson, 1976; Otway and Edwards, 1977). Meanwhile, a value of 0.49 for rating weights resulted in a water resources planning study (Hobbs *et al.*, 1982). Inter-person weight correlations of 0.02-0.06 were found for direct weighting in Chapter 9 and Bell *et al.*(2000). A higher value (0.46) resulted from tradeoff weighting in the latter study because nearly everyone chose a high weight for cost in that method.

[20] For example, in a power plant siting study, Hobbs (1980) found that the mean between-method correlation of weights (0.31) was not significantly different from the between-persons correlation (0.18). In another siting study, Hobbs (1986) reports that indifference tradeoff's correlation with directly chosen weights averages 0.32 to 0.56, which were similar to between-persons correlations. Borcherding *et al.* (1991) report that two direct methods (ratio questioning and swing weights) had mean correlations of -0.05 to 0.41 with two versions of tradeoff weighting. But in other studies, differences between tradeoff and rating weights were less important. In Chapter 9, the average between-method correlation was 0.45, well above the average between-person correlations, which were approximately zero. In a water resources study involving 20 US Army Corps of Engineers planners, tradeoff and rating weights were even more highly correlated, averaging 0.71 over the 20 participants; this mean was significantly higher than the mean between-persons correlation of 0.19 (Hobbs *et al.*, 1992). Tradeoff weights had an average correlation of 0.39-0.51 with direct weights in Bell *et al.* (2000), well above the mean inter-person correlations for direct weights (0.03-0.06). Thus, in general, the difference between tradeoff weights and rating weights can be as large as the differences of opinion among people -- but not necessarily so.

[21] As noted in Section 10.2, an alternative measure of "significance" would be to compare between-method correlations to correlations of ranks resulting from the same person using the same method at two different *times*. If such test-retest differences arise only because of random error, then any between-method correlations that are not significantly less than test-retest correlations would be judged insignificant. In this study, the holistic method was applied twice, and the mean *r* between a person's two holistic rank sets was 0.71. This is much less than test-retest correlations involving MCDM methods in other studies. For example, in Hobbs *et al.* (1992), involving water resource plans for the US Corps of Engineers, the test-

retest correlation for **AVF(R)** was 0.86. We believe that the low correlation for the holistic ranks reflects not so much error as the learning that took place during the course of the workshop. Therefore, we do not use the test-retest correlation of 0.71 as a threshold of significance for differences between methods.

[22] Calculations of between-person correlation of alternative ranks have been made in other studies, or can be obtained from reported ranks. Section 9.4 reports values for the BC Gas study. Hobbs (1986) reported the following results from two power plant siting panels: 0.91 and 0.38 for **AVF(R)**, for panels I and II, respectively; and 0.28 and 0.73 for their holistic evaluations. The first panel consisted of scientists and engineers from Brookhaven National Laboratory, while the second was a diverse group of scientists, a public interest group representative, and a town planner. The diversity of the second panel might account for the greater differences between their **AVF(R)** rankings. In another siting study, Hobbs (1980) gives a value of 0.77 for between-person correlations of **AVF(R)** scores. In the greenhouse gas policy study of Bell *et al.* (2000), AVF rankings had between-person correlations of 0.17 (swing weights/AHP) to 0.74 (trade-off weights). Finally, in a water resources planning context, Hobbs *et al.* (1992) and Goicoechea *et al.* (1992) found between-person correlations of **AVF(R)** rankings to be about 0.50. Like this SCL study, goal programming resulted in more disagreement than AVF in Hobbs *et al.* (1992) and Bell *et al.* (2000), yielding mean correlations of 0.13-0.26.

[23] Newman (1975) concludes that because MCDM methods result in more agreement, their evaluations are more trustworthy than unaided holistic evaluations. His assumption is that evaluations by fully knowledgeable and rational individuals should agree, and that invalid approaches exaggerate differences of opinion. However, we believe that value differences lie at the heart of energy-environment controversies, so artificial uniformity would only bury disagreements that should be made explicit.

[24] Bell *et al.* (2000) is a rare exception; inter-person correlations for holistic judgment averaged 0.32, while all MCDM methods except fuzzy sets and tradeoff weights produced values of that magnitude or lower.

[25] Goal programming has been found to yield very different results from additive value functions in other studies when users are asked to set goals equal to "satisfactory levels" of achievement. Hobbs *et al.* (1992) found that **GP(1,+/-)** has a mean correlation of only 0.19 with **AVF(R)** ranks for 20 participants, which was less than the between persons **AVF(R)** correlation of 0.5. However, if each goal is set equal to the very best level of each attribute (the "ideal"), then differences are minor. (In that case, only underachievement is possible). In Bell *et al.*'s (2000) greenhouse gas policy study, **GP(2,+/-)** and **GP(infinity,+/-)** had mean correlations of 0.17-0.48 with **AVF(R)**. In contrast, Hobbs (1986) used "ideal" goals, in which **GP(1,+/-)** and **GP(2,+/-)** had mean correlations of 0.95 and above with **AVF(R)**. However, **GP(infinity,+/-)**'s correlations with **AVF(R)** were somewhat less, averaging 0.91 and 0.73, respectively for the two panels in that study. They were still no smaller than the between person correlations, however. Other amalgamation methods have also been compared to additive value functions in Hobbs (1986) and Hobbs *et al.* (1992), with mixed results. Multiplicative utility functions were highly correlated with additive value functions (see also Fischer, 1977). (However, Keefer and Pollock (1980) found that the use of the wrong form could result in large utility losses.) But ELECTRE yielded lower correlations. In Bell *et al.* (2000), a fuzzy set MCDM method's ranks were essentially uncorrelated with AVF ranks. See also Goicoechea *et al.* (1992), Gershon and Duckstein (1983, discussed in Hobbs, 1984), and Massam and Askew (1982).

[26] Some other studies have also shown that indifference tradeoff and rating yield similar rankings, but others disagree. In Hobbs *et al.* (1992), **AVF(R)** and **AVF(T)** had a mean

correlation of 0.77, which was significantly higher than the between-persons correlation of 0.58 for **AVF(R)**. However, Hobbs (1986) found mixed results. For the first panel in that latter study, there was high convergence: both between-weighting method and between-person correlations of site ranks were very high (0.77 to 0.91) and could not be distinguished. But for his second panel, the mean between-method correlation of **AVF(R)** and **AVF(T)** was 0.55. This exceeded the between-persons correlation for **AVF(R)** of 0.38, but was less than the **AVF(T)** between-person correlation of 0.85. Meanwhile, Hobbs (1980) found correlations between **AVF(R)** and **AVF(T)** to be, on average, as low as between-persons correlations for **AVF(R)** (0.78 v. 0.77, respectively).

[27] Karni (1992) reports a mean Spearman's *r* of 0.55 of 19 individuals' holistic rankings of electric rate policies and rankings constructed using additive value functions and direct weighting. Goicoechea *et al.* (1992) found values of Kendall's Tau (another non-parametric correlation measure) between 0.45 and 0.66 for several MCDM methods in a water resources planning problem. In Hobbs (1986), there was a somewhat better predictive validity for two power plant siting problems. There, the mean Spearman's *r* was between 0.53 (for AVF(T)), and 0.88 (for AVF(R)) for a range of weighting methods and amalgamation techniques. In Bell *et al.* (2000), AVF(T) had a significantly lower correlation (0.14) with holistic than direct AVF with direct weights. However, this poorer performance of tradeoff weights was not found in our Seattle case study. Numerous studies have also examined the predictive validity of other amalgamation methods other than AVF (Huber, 1974; Goicoechea *et al.*, 1992; Hobbs *et al., 1992*; Bell *et al.*, 2000): generally they do not perform significantly better than additive value functions.

[28] Too equal, or "flat" weights, can result for several reasons. One is that non-hierarchical weighting yields weights that are too flat (Section 8.5). As we noted in Section 10.2, people also are reluctant to give zero weight even to negligibly important attributes, which tends to make weights too equal. Another possible explanation for flat weights is the phenomenon of logarithmic perception: people report magnitudes of physical phenomena (e.g., loudness, or brightness) on a log scale (Stevens, 1966). Some decision analysts believe that value judgments are also expressed on log scales (e.g. Lootsma *et al.*, 1990). For instance, if *X* is three times as important as *Y*, the stated rating weights would instead have a ratio of log (3/1). Such weights will be flatter because the weights of the most important criteria will be depressed.

[29] The ratio of the standard deviation to the mean.

[30] The significance was assessed by applying a non-parametric two-sample paired comparison test. Our BC Hydro case study (Chapter 8) showed similar results: non-hierarchical weights there were much flatter than hierarchical weights.

11 CLOSING REMARKS

Great is the art of beginning, but greater is the art of ending.
LONGFELLOW, Elegiac Verse (1881)

Better is the end of a thing that the beginning thereof
OLD TESTAMENT, Ecclesiastes 7:8

As we hope the many examples in this book have shown, multicriteria analysis can be tremendously useful. Light can be shed on what tradeoffs, uncertainties, and value judgments are crucial to the decision, and what issues do not matter. MCDM provides a logical way of structuring thinking and discussions about what is important. By better understanding what is at stake, and by exploring the implications of tentative value judgments, you can form priorities and make recommendations with more confidence.

MCDM analysis can usefully play at least three types of roles in environmental decision making and other public issues (Stewart, 2000). One role is initial impact assessment and screening. In this task, technical staff attempt to identify alternatives, and then narrow them down to a "short list" of options to present to a broader audience, including management and external stakeholders. Accompanying the list would be an initial characterization of the tradeoffs and uncertainties involved. MCDM is useful at this stage for structuring comparisons of alternatives and for filtering out options that are obviously inefficient or fail exclusionary criteria. In Chapters 3 and 4, we described several methods that can fill this role. MCDM studies can also provide an "audit trail" which allows external parties to examine the value judgments that have been made in the course of developing and screening options. This role is often legally mandated, and was a major motivation for the case studies of Chapters 8 and 9.

The second role is "within interest" structuring and evaluation. The point of this task is for individual stakeholders or groups to develop coherent positions on the alternatives, which will then become an input to the political process. Because an interest group will share certain values, it can make sense to talk about a "group's" evaluation. However, although group members may share general concerns and have some sense of what their priorities are, they are unlikely to have specific positions on questions such as the rate they are willing to tradeoff different quantified criteria. MCDM analysis can help people to develop and articulate values judgments in a systematic way that can be used to rank alternatives; to gain insight on the implications of different judgments and ways of viewing the problem; and to identify consensus positions or disagreements within the group. In our detailed case studies (Chapters 8-10), we show how asking stakeholders to apply more than one MCDM method and then to think hard about resolving differences between the methods' results can improve understanding and confidence.

The third role identified by Stewart (1998) is to facilitate "between interest" negotiation and decision-making. Once stakeholders have developed an understanding of the tradeoffs and an initial position on the alternatives, the next phase is communication among the stakeholders and (if applicable) the people ultimately responsible for the decision (e.g., agency heads, corporate management, legislators, governing boards). The purpose of this phase is to identify possible compromises and consensus and to understand the reasons for any remaining disagreements. MCDM can help with these tasks in at least three ways. First, it attempts to get participants to focus on ultimate objectives: as a result, it is often possible to identify new alternatives that creatively address fundamental concerns (Keeney, 1992). MCDM also provides a framework for systematic discussion of tradeoff and uncertainty issues by participants in the process. Second, as we emphasize in Chapters 5 and 9, MCDM can then identify options (if any) that are found attractive by most or all groups, and help explain the reasons for disagreements over other alternatives (see also, e.g., Brown, 1984). Finally, MCDM can be used in a game theory mode to explore how possible strategies that might be adopted by various groups could interact to resolve the decision problem (e.g., Fraser and Hipel, 1984; Hipel *et al.*, 1997).

Hammond, Keeney, and Raiffa's (1999) book on "Smart Choices" inspires some of our final thoughts. When you are faced with a tough problem, first clarify your objectives -- what are you trying to achieve, and is your analysis solving the right problem? Then work hard at creating better alternatives from which to choose. Focus on tradeoffs, being aware that tough compromises will almost always be needed in energy-environmental planning and policy. Be aware that any one method for considering these tradeoffs is probably the "wrong" method -- because there will almost always be one participant in your decision process who will misunderstand it or whose preferences will be distorted as a result of the many difficulties and biases we document in Chapter 4 and in our case studies.

Since no one method is "right" for everyone, use consistency checks and allow time to apply more than one method carefully; comparing the results of different methods gives valuable opportunities for reflection and insight. Give ample opportunity for participants to share their opinions and results. Design the process so people can learn and easily change their minds.

Finally, read more. In a book such as ours that emphasizes applications, we can only introduce the basic mechanics and a few of the assumptions of the more popular methods. Luckily, there are many excellent textbooks and articles that will help you gain a more sophisticated understanding of the theory of MCDM, and which convey many valuable lessons from other applications. A fuller appreciation of theory and what has been accomplished by practitioners cannot but help you design better processes. It will also leave you better prepared to deal with the inevitable surprises and difficulties that occur in any real decision-making process.

REFERENCES

Adelman, L., P. J. Sticha, and M. L. Donnell, 1984. The Role of Task Properties in Determining Relative Effectiveness of Multiattribute Weighting Techniques, Organizational Behavior and Human Performance, 33, 243-262.

Analytical Solutions, 1997. A/S Plan User's Manual. 197 Seaman Avenue, Bayport, New York, 11705.

Anandalingam, G., 1987. A Multiple Criteria Decision Analytic Approach for Evaluating Acid Rain Policy Choices, European Journal of Operational Research, 29, 336-352.

Andrews, C. J., 1992a. Spurring Inventiveness by Analyzing Tradeoffs: A Public Look at New England's Electricity Alternatives, Environmental Impact Assessment Review, 12, 209-234.

Andrews, C. J., 1992b. The Marginality of Regulating Marginal Investments: Why We Need a Systemic Perspective on Environmental Externality Adders, Energy Policy, 20, 451-475.

Andrews, C. J., and S. Govil, 1995. Becoming Proactive About Environmental Risks: Regulatory Reform and Risk Management in the US Electricity Sector, Energy Policy, 23(10), 885-892.

Arthur D. Little of Canada, Ltd., 1995. Northern Telecom Environmental Performance Index, User's Manual, Centre for Environmental Assurance, prepared for NORTEL.

Atanackovic, D., D. T. McGillis, and F. D. Galina, 1998. The Application of Multicriteria Analysis to Substation Design, IEEE Transactions on Power Systems, 13(3), 1172-1178.

Athanassopoulos, A., N. Lambroukos, and L. Seiford, 1999. Data Envelopment Scenario Analysis for Setting Targets to Electricity Generating Plants, European Journal of Operational Research, 115, 413-428.

Balestieri, J. A. P., and P. de B. Correia, 1997. Multiobjective Linear Model for Prefeasibility Design of Cogeneration Systems, Energy, 22(5), 537-548.

Balzer, W. K., J. Rohrbaugh, and K. R. Murphy, 1983. Reliability of Actual and Predicted Judgments Across Time. Organizational Behavior and Human Performance, 32, 109-123.

Baron, J., 1997. Biases in the Quantitative Measurement of Values for Public Decisions, Psychological Bulletin, 122(1), 72-88.

Barron, F. H., and B. E. Barrett, 1996. Decision Quality Using Ranked Attribute Weights, Management Science, 42(11), 1515-1523.

Barron, F.H., D. von Winterfeldt, and G.W. Fischer, 1984. Theoretical and Empirical Relationships Between Risky and Riskless Utility Functions, Acta Psychologica, 56, 233-244.

BC Hydro, 1994. Resource Acquisition Policy, Resource Planning, Corporate and Strategic Planning, Vancouver.

BC Hydro, 1995a. Integrated Electricity Plan: Main Report, Vancouver

BC Hydro, 1995b. Integrated Electricity Plan: Overview, Vancouver.

BC Hydro, 1995c. 1995 Demand-side Management Plan, Market Development Group, Vancouver.

Beccali, M., M. Cellura, and D. Ardente, 1998. Decision Making in Energy Planning: The ELECTRE Multicriteria Analysis Approach Compared to a Fuzzy-Sets Methodology, Energy Conversion & Management, 39(16-18), 1869-1881.

Bell, M. L., B. F. Hobbs, E. M. Elliott, H. Ellis, and Z. Robinson, 2000. An Evaluation of Multicriteria Methods in Integrated Assessment of Climate Change, in Haimes and Steuer (2000).

Belton, V., 1986. A Comparison of the Analytic Hierarchy Process and a Simple Multiattribute Value Function, European Journal of Operations Research, 26(1), 7-21.

Belton, V., and T. Gear, 1997. On the Meaning of Relative Importance, Journal of Multicriteria Decision Analysis, 6(6), 335-338.

Bernow, S., B. Biewald, and D. Marron, 1991. Full-Cost Dispatch: Incorporating External Costs in Power System Operations, The Electricity Journal, 4(2), 20-33.

Biewald, B., 1996. Position Paper of the Vermont Department of Public Service in Docket 5854 (Restructuring), Appendix 4, Preserving the Environment in a Restructured Electric Industry, Burlington, VT, March 25.

Bishop, A. B., 1974. An Approach to Evaluating Environmental, Social and Economic Factors in Water Resources, Water Resources Bulletin, 8(4), 724-734.

Black and Veatch International, 1988a. Trincomalee Thermal Power Project: Social and Environmental Assessment. Report to Ceylon Electricity Board, Colombo, Sri Lanka.

Black and Veatch International, 1988b. Thermal Generation Options Study, Report to the Ceylon Electricity Board, Colombo, Sri Lanka.

Bloemer, J. A., and D. C. Knue, 1993. Integrated Resource Planning/CAAA Compliance Planning at the Cincinnati Gas & Electric Company, presented at the PROSCREEN/ PROVIEW User's Forum, Atlanta, GA, September.

Bloom, J. A., M. Caramanis, and L. Charny, 1984. Long-Range Generation Planning: Using Generalized Benders Decomposition: Implementation and Experience, Operations Research, 32, 314-342.

Bonneville Power Administration (BPA), 1986. Environmental Costs and Benefits Associated with Hydropower Development in the Pacific Northwest, Report DE-AC79-83BP11546, Portland, Oregon.

Bonneville Power Administration (BPA), 1992. Puget Sound Area Reliability Plan: Environmental Impact Statement, Portland, Oregon.

Borcherding, K., T. Eppel, and D. von Winterfeldt, 1991. Comparison of Weighting Judgments in Multiattribute Utility Measurement, Management Science, 37, 1603-19.

Brans, J. P., and B. Maréschal, 1986. Geometrical Representation for MCDA, Internal Report STOOT/213, Vrije Universiteit, Brussels.

Briggs, T., P. Kunsch and B. Maréschal, 1990. Nuclear Waste Management. An Application of the Multicriteria PROMETHEE Methods, European Journal of Operational Research, 44(1), 1-10.

BC Utilities Commission (BCUC), 1994. British Columbia Hydro and Power Authority: 1994/95 Revenue Requirements Application: Decision November 24, 1994, Vancouver.

Brown, C.A., 1984. The Central Arizona Water Control Study: A Success Story for Multicriteria Planning and Public Involvement, Water Resources Bulletin, 20(3), 331-338.

Brown, C.A., 1990. Managing Conflict Over a Dam Safety Problem, in W. Veissman, Jr. and E.T. Smerdon, eds., *Managing Water-Related Conflicts; The Engineer's Role*, American Society of Civil Engineers, New York, 223-237.

Burtraw, D., W. Harrington, and A. J. Krupnick, 1995. Optimal "Adders" for Environmental Damages by Public Utilities, Journal of Environmental Economics and Management, 29(Part 2), S1-S19.

Cavanagh, R., 1996. Restructuring for Sustainability: Toward New Electric Services Industries, The Electricity Journal, 9(6), 70-79.

Chang, C. S., and W. Fu, 1998. Stochastic Multiobjective Generation Dispatch of Combined Heat and Power Systems, Proc. IEEE, Generation, Transmission & Distribution, 145(5), 583-591.

Chankong, V. and Y. Haimes, 1983. *Multiobjective Decision Making: Theory and Methods*, North- Holland, Amsterdam.

Chattopadhyah, D., and J. Momoh, 1999. A Multiobjective Operations Planning Model with Unit Commitment and Transmission Constraints, IEEE Transactions on Power Systems, 14(3), 1078-1084.

Chedid, R., T. Mezher, and C. Jarrouche, 1999. A Fuzzy Programming Approach to Energy

Resource Allocation, International Journal of Energy Research, 23(4), 303-317.

Chen, Y. L., 1998. Weighted-Norm Approach For Multiobjective VAR Planning, IEE Proceedings -- Generation Transmission & Distribution, 145(4), 369-374.

Chen, J. F., and S. D. Chen, 1997. Multiobjective Power Dispatch with Line Flow Constraints Using the Fast Newton-Raphson Method, IEEE Transactions on Energy Conversion, 12(1), 86-93.

Clemen, R. T., 1996. *Making Hard Decisions: An Introduction to Decision Analysis*, 2nd Edition, Duxbury Press, Belmont, CA.

Climaco, J., C. H. Antunes, A. G. Martins, and A. T. Almeida, 1995. A Multiple Objective Linear Programming Model for Power Generation Expansion Planning, International Journal of Energy Research, 19, 419-432.

Cohon, J. L., 1978. *Multiobjective Programming and Planning*, Academic Press, New York.

Cohon, J. L., *et al.*, 1980. Application of a Multiobjective Facility Location Model to Power Plant Siting in a Six-State Region of the US., Computers and Operations Research, 7, 107-123.

Cook, W. D., M. Kress, and L. M. Seiford, 1996. A General Framework for Distance-Based Consensus in Ordinal Ranking Models, European Journal of Operational Research, 96, 392-397.

Corner, J. L., and J. T. Buchanan, 1995. Experimental Consideration of Preference in Decision Making under Certainty, Journal of Multicriteria Decision Analysis, 4, 107-121.

Corner, J. L., and J. T. Buchanan, 1997. Capturing Decision-Maker Preference: Experimental Comparison of Decision Analysis and MCDM Techniques, European Journal of Operational Research, 98 (1), 85-97.

Crousillat, E., 1989. Incorporating Risk and Uncertainty in Power System Planning World Bank, Industry and Energy Department, Energy Series Paper #17, Washington, DC.

Crousillat, E. and H. Merrill, 1992. The Tradeoff/Risk Method: A Strategic Approach to Power Planning, World Bank, Industry and Energy Department, Washington, DC.

Crousillat, E., P. Dorfner, P. Alvarado, and H. M. Merrill, 1993. Conflicting Objectives and Risk in Power System Planning, IEEE Transactions on Power Systems, 8, 897-893.

d'Avignon, G. R., and M. Saubageau, 1996. L'Aide Multicritere a la Decision: Un Cas D'Integration de Critere Techniques, Economiques et Environmentaux a Hydro-Quebec, Recherche Operationnelle, 30(3), 317-332.

Daniels, S. E., R. L. Lawrence, and R. J. Alig, 1996. Decision-Making and Ecosystem-Based Management: Applying the Vroom-Yetton Model to Public Participation Strategy, Environmental Impact Assessment Review, 16(1), 13-30.

Davey, A., and D. Olson, 1998. Multiple Criteria Decision Making Models in Group Decision Support, Group Decision and Negotiation, 7(1), 55-75.

De, M. and K. W. Hipel, 1987. A Fuzzy Multicriteria Model for Comparing Energy Projects, Energy, 12(7), 599-613.

DeCicco, J. M., S. S. Bernow, and J. Beyea, 1992. Environmental Concerns Regarding Electric Power Transmission in North America, Energy Policy, 20(1), 30-39.

Dee, N., *et al.*, 1973. An Environmental Evaluation System for Water Resource Planning, Water Resources Research, 9(3), 523 - 535.

Delbecq, A. L., A. H. Van de Ven, and D. H. Gustafson, 1975. *Group Techniques for Program Planning - A Guide to Nominal Group and Delphi Processes*, Scott Foresman and Company, Glenview, IL.

Delhaye, C., J. Teghem and P. Kunsch, 1991. Application of the ORESTE Method to a Nuclear Waste Management Problem, International Journal of Production Economics, 24, 29-39.

Delquié, P., 1994. Inconsistent Trade-offs Between Attributes: New Evidence in Preference Assessment Biases, Management Science, 39(11), 1382-1395.

Ditz, D. and J. Ranganathan, 1997. Measuring Up: Toward a Common Framework for Tracking Corporate Environmental Performance, World Resources Institute, Washington, DC.

Dobson, J., 1976. The Maryland Power Plant Siting Project: An Application of the ORNL-Land Use Screening Procedure, ORNL/NUREG/TM-79, Oak Ridge National Laboratory, Oak Ridge, TN.

Dodd, D. E. and J. A. Lesser, 1994. Can Utility Commissions Improve on Environmental Regulations?, Land Economics, 70, 63-76.

Doyle, J. R., R. H. Green, and P.A. Bottomley. 1997. Judging Relative Importance: Direct Ratings and Point Allocation Are Not Equivalent, Organizational Behavior & Human Decision Processes, 70(1), 65-72.

Drinnan, J. H., and K. R. Spafford, 1997. A Stochastic Resource Planning Model, Paper presented at the Canadian Energy Association Spring Meeting, Vancouver.

Duckstein, L., and S. Opricovic, 1980. Multiobjective Optimization in River Basin Development, Water Resources Research, 16(1), 14-20.

Dyer, J. S., 1990. Remarks on the Analytic Hierarchy Process, Management Science, 36(3), 247-275.

Dyer, J. S. and R. K. Sarin, 1979. Measurable Multiattribute Value Functions, Operations Research, 27(4), 810-822.

Earl, G., and R. Clift, 1998. How Important is Environmental Performance? A Case Study Measuring the Environmental Preferences of Business-to-Business Consumers, Journal of Sustainable Product Design, Issue 6, 18-27.

Eckenrode, R. T., 1965. Weighing Multiple Criteria, Management Science, 12, 180-192.

Economides, S., and M. Sharifi, 1978. Environmental Optimization of Power Lines, Journal of the Environmental Engineering Division, ASCE, 104(EE4), 675-684.

Edwards, W., 1977. How to Use Multiattribute Utility Measurement for Social Decision-making, IEEE Transactions on Systems, Management and Cybernetics, SMC- 7(5), 326-340.

Ekel, P. Y., L. D. B. Terra, M. F. D. Junges, V. A. Popov, M. A. de Moraes, A. V. Prakhovnik, and O. V. Razymovsky, 1998. Multicriteria Power and Energy Shortage Allocation Using Fuzzy Set Theory, in *Proceedings, Bulk Power System Dynamics and Controls IV*, Santorini, Greece, Aug. 24-28.

Energy Management Associates, 1993. PROSCREEN Overview, Version 14, Release 1.3, Atlanta, Georgia.

Eppel, T., 1992. Description and Procedure Invariance in Multiattribute Utility Measurement, School of Management, Purdue University, West Lafayette, IN.

Epstein, M. J., 1996. *Measuring Corporate Environmental Performance, Best Practices for Costing and Managing an Effective Environmental Strategy*, Irwin Professional Publishing, Chicago.

Eto, J., E. Vine, L. Shown, R. Sonnenblick and C. Payne, 1996. The Total Cost and Measured Performance of Utility Sponsored Energy Efficiency Programs, Energy Journal, 7(1), 31-51.

Eyre, N., 1997. External Costs-What Do They Mean for Energy Policy?, Energy Policy, 25(1), 85-95.

Farber, M., E. Brusger, and M. Gerber, 1988. MIDAS, Multiobjective Decision Analysis System, Electric Power Research Institute, Palo Alto, CA.

Farrell, A., and M. Hart, 1998. What Does Sustainability Really Mean?: The Search for Useful Indicators, Environment, 40(9), 4.

Ferrell, G. C., 1978. Coal, Economics, and the Environment: Tradeoffs in the Coal-Electric Cycle. Paper PP-78-2, International Institute for Applied Systems Analysis, Laxenburg, Austria.

Fischer, G. W., 1977. Experimental Validation of Decomposed Scaling Procedures, Organizational Behavior and Human Performance, 18, 295-315.

Fischer, G. W., 1979. Utility Models for Multiple Objective Decisions: Do they Accurately Represent Human Preferences?, Decision Sciences, 10, 451-479.

Fischer, G. W., 1995. Range Sensitivity of Attribute Weights in Multiattribute Value Models, Organizational Behavior and Human Decision Processes, 62(3), 252-266.

Fischoff, B., P. Slovic, and S. Lichtenstein, 1980. Know What You Want: Measuring Labile Values, in T. Wallsten, Ed., *Cognitive Processes in Choice and Decision Behavior*, Erlbaum, Hillsdale, NJ.

Fishburn, P.C., 1967. Methods of Estimating Additive Utilities, Management Science, 13(7), 435-453.

Florig, H, 1992. Responding to Potential Health Effects of Electric and Magnetic Fields, Resources, 109, Resources for the Future, Washington, DC.

Forman, E. H., *et al.*, 1996. Expert Choice Pro 9.0, Decision Support Software, Inc., Washington, DC.

Fraser, N. M. and K. W. Hipel, 1984. *Conflict Analysis: Models and Resolutions*, North-Holland, New York.

Freeman III, A.M., 1993. *The Measurement of Environmental and Resource Values: Theory and Methods,* Resources for the Future, Washington, DC.

Gandibleux, X., 1999. Interactive Multicriteria Procedure Exploiting a Knowledge-Based Module to Select Electricity Production Alternatives: The CASTART System, European Journal of Operational Research, 113(2), 355-373.

Gas Research Institute, 1994. Results of Appraisal of GRI 1995-1999 R&D, Appendix A, Project Appraisal Methodology Description, Chicago, IL.

Georgopoulou, E., D. Lals, and L. Papagiannakis, 1997. A Multicriteria Decision Aid Approach for Energy Planning Problems: The Case of Renewable Energy Options, European Journal of Operational Research, 103(1), 38-54.

Georgopoulou, E., Y. Sarafidis, and D. Diakoulaki, 1998. Design and Implementation of a Group DSS For Sustaining Renewable Energies Exploitation, European Journal of Operational Research, 109(2), 483-500.

Gershon, M. E., and L. Duckstein, 1983. Multiobjective Approaches to River Basin Planning, J. Water Resources Planning and Management, 109, 13-28.

Goicoechea, A., E. Z. Stakhiv, and F. Li, 1992. Experimental Evaluation of Multiple Criteria Decision Models for Application to Water Resources Planning, Water Resources Bulletin, 28(1), 89-102.

Goicoechea, A., D. R. Hansen, and L. Duckstein, 1982. *Multiobjective Decision Analysis with Engineering and Business Applications*, J. Wiley, NY.

Golabi, K., C. W. Kirkwood and A. Sicherman, 1981. Selecting a Portfolio of Solar Energy Projects using Multiattribute Preference Theory, Management Science, 27 (2)

Goldman, C. A., J. F. Busch, E. P. Kahn, S. S. Stoft and S. Cohen, 1992. Review of Integrated Resource Bidding at Niagara Mohawk, Lawrence Berkeley Laboratory, LBL 31667 UC350, Berkeley, CA.

Goldstein, W. M., 1990. Judgments of Relative Importance in Decision Making: Global vs. Local Interpretations of Subjective Weights, Organizational Behavior and Human Decision Processes, 47, 313-336.

Goldstein, W. M., and H. C. Mitzel, 1992. The Relative Importance of Relative Importance--Inferring Other Peoples' Preferences from Relative Importance Rankings and Previous Decisions, Organizational Behavior & Human Decision Processes, 51(3), 382-415.

Goto, N. and T. Sawa, 1993. An Analysis of the Macroeconomic Costs of Various CO_2 Emission Control Policies in Japan, Energy Journal, 14(1), 83-110.

Gould, S. J., 1996. *The Mismeasure of Man.* Norton, New York.

Goumas, M. G., V. A. Lygerou, and L. E. Papayannakis, 1999. Computational Methods for Planning and Evaluating Geothermal Energy Projects, Energy Policy, 27(3), 147-154.

Gregory, R. and R. L. Keeney, 1994. Creating Policy Alternatives Using Stakeholder

Values, Management Science, 40, 1035-1048.

Gregory, R., S. Lichtenstein, and P. Slovic, 1993. Valuing Environmental Resources: A Constructive Approach, Journal of Risk and Uncertainty, 7, 177-197.

Guidinger, J., 1987. Right-of-Way Selection Using Link-Node Methodology and Some Recent Experiences in Its Use, presented at the Fourth Symposium on Environmental Concerns in Right-of-Way Management, Indianapolis.

Gulliver, J. W., and C. C. Vito, 1993. EMF and Transmission Line Siting: The Emerging State Regulatory Framework and Implications for Utilities, Natural Resources and Environment, 7(3).

Gum, R. L., R. G. Roefs, and D. B. Kimball, 1976. Quantifying Social Goals: Development of a Weighting Methodology, Water Resources Research, 12(4), 617-622.

Gustafson, D. H., R. K. Shukla and G. W. Walsten, 1973. A Comparative Study of Differences in Subjective Likelihood Estimates made by individuals, Interacting Groups, Delphi groups, and Nominal Groups, Organizational Behavior and Human Performance, 9, 280-291.

Haimes Y. Y., and V. Chankong, Eds., 1985. *Decision Making with Multiple Objectives*, Springer-Verlag, New York.

Haimes, Y.Y., and R. Steuer, Eds., 2000. *Research and Practice in Multiple Criteria Decision Making*, Springer-Verlag, in press.

Hamalainen, R. P., 1988. Computer Assisted Energy Policy Analysis in the Parliament of Finland, Interfaces, 18,12-23.

Hamalainen, R. P., 1991. Facts or Values-How Do Parliamentarians and Experts See Nuclear Power?, Energy Policy, 19(5), 464-472.

Hamalainen, R. P., and R. Karjalainen, 1992. Decision Support for Risk Analysis in Energy Policy, European Journal of Operational Research, 56,172-183.

Hamalainen, R. P., E. Kettunen, M. Marttunen, and H. Ehtamo, 1999. Towards Decision and Negotiation Support in Multi-Stakeholder Development of Lake Regulation Policy, in *Proceedings of the 32nd Annual Hawaii Intl. Conf. on System Sciences*, IEEE Press, Piscataway, NJ.

Hammond, J. S., R. L. Keeney, and H. Raiffa, 1999. *Smart Choices: A Practical Guide to Making Better Decisions*, Harvard Business School Press, Boston.

Hammond, K. R., J. L. Mumpower, and T. H. Smith, 1977. Linking Environmental Models with Models of Human Judgment-Symmetrical Decision Aid, IEEE Transactions on Systems, Man, & Cybernetics, 7(5), 358-367.

Hanson, M., *et al.*, 1991. Electric Utility Least-Cost Planning: Making It Work Within a Multiattribute Decision-Making Framework, Journal of the American Planning Association, 34(1), 34-43.

Hansen, U., 1998. Technological Options for Power Generation, Energy Journal, 19(2), 63-87.

Harker, P. T., 1987. Incomplete Pairwise Comparisons in the Analytic Hierarchy Process, Mathematical Modeling, 9(11), 837-848.

Harte, J. M., and P. Koele P., 1995. A Comparison of Different Methods for the Elicitation of Attribute Weights - Structural Modeling, Process Tracing, and Self-Reports, Organizational Behavior & Human Decision Processes, 64(1), 49-64.

Hersh, M. A., 1999. Sustainable Decision Making: The Role of Decision Support Systems, IEEE Transactions on Systems Man & Cybernetics Part C–Applications & Reviews, 29(3), 395-408.

Heslin, J. S., and B.F. Hobbs, 1989. A Multiobjective Production Costing System for Emissions Dispatch and Fuel Switching, IEEE Transactions on Power Systems, 4(3), 536-542.

Hipel, K. W., D. M. Kilgour, L. Fang, and X. Peng, 1997. The Decision Support System GMCR in Environmental Conflict Management, Applied Mathematics and Computation, 83(2/3), 117-152.

Hirst, E., 1992. Price and Cost Impacts of Utility DSM Programs, Energy Journal, 13 (4), 75-90.

Hirst, E., 1996. Is There A Future for Electric-Industry IRP?, in *Proceedings, ACEEE 1996 Summer Study*, American Council for an Energy Efficient Economy, Washington, DC.

Hirst, E. and C. Goldman, 1991. Creating the Future: Integrated Resource Planning for Electric Utilities, Annual Review of Energy, 16, 91-121.

Hirst, E. and M. Schweitzer, 1989. Uncertainty: A Critical Element of Integrated Resource Planning, Electricity Journal, 2 (6), 16-27.

Hirst, E., and B. Tonn, 1996. Social Goals and Electric-Utility Deregulation, Issues in Science & Technology, 12(3), 43-47.

Hobbs, B.F., 1979. Analytical Multiobjective Decision Methods for Power Plant Siting: A Review of Theory and Applications, BNL-NUREG-51204, NUREG/CR-1687, Prepared for the US Nuclear Regulatory Commission, Brookhaven National Laboratory, Upton, NY.

Hobbs, B.F., 1980. A Comparison of Weighting Methods in Power Plant Siting, Decision Sciences, 11, 725-37.

Hobbs, B. F., 1984. Discussion of Multiobjective Approaches to River Basin Planning, Journal of Water Resources Planning and Management, 110(WR1), 123-124.

Hobbs, B.F., 1985. Choosing How to Choose: Comparing Amalgamation Methods for Environmental Impact Assessment, Environmental Impact Assessment Review, 5, 301-319.

Hobbs, B. F., 1986. What Can We Learn From Experiments in Multiobjective Analysis?, IEEE Transactions on Systems, Man, and Cybernetics, SMC-16 (3), 384-394.

Hobbs, B. F., 1991. The 'Most Value' Test: Economic Evaluation of Electricity Demand-Side Management Considering Customer Value, The Energy Journal, 12(2), 67-91.

Hobbs, B. F., 1992. What Do SO_2 Emissions Cost? External Costs and Allowance Prices Journal of Energy Engineering, 120 (3), 122-132.

Hobbs, B.F., 1997. Environmental Planning for Electric Utilities, in C. S. ReVelle and A. E. McGarity (eds.), *Design and Operation of Civil and Environmental Engineering Systems*, J. Wiley, New York.

Hobbs, B. F., V. Chankong, W. Hamadeh, and E. Z. Stakhiv, 1992. Does Choice of Multicriteria Method Matter? An Experiment in Water Resources Planning, Water Resources Research, 28(7), 1767-1780.

Hobbs, B.F., and G. T. F. Horn, 1997. Building Public Confidence in Energy Planning: A Multi-Method MCDM Approach to Demand-Side Planning at BC Gas, Energy Policy, 25(3), 357-375.

Hobbs, B. F. and P. Maheshwari, 1990. A Decision Analysis of the Effect of Uncertainty Upon Electric Utility Planning, Energy, 15(9), 785-802.

Hobbs, B. F., and P. M. Meier, 1994a. Multicriteria Methods for Resource Planning: A Case Study of Centerior Energy, Report by IDEA, Inc. to Oak Ridge National Laboratory, Washington, DC.

Hobbs, B. F., and P. Meier, 1994b. Integrated Resource Planning and the Environment: A Guide to the Use of Multicriteria Decision Methods, Report by IDEA, Inc. to Oak Ridge National Laboratory, Washington, DC.

Hobbs, B.F. and P.M. Meier, 1994c. Multicriteria Methods for Resource Planning: An Experimental Comparison, IEEE Transactions on Power Systems, 9(4), 1811-1817.

Hobbs, B.F., and M. D. Rowe, 1979. A Comparison of Regional Screening Methodologies, BNL/NUREG-51205, Brookhaven National Laboratory, Upton, NY.

Hokkanen, J., and P. Salminen, 1997. ELECTRE III and IV Decision Aids in an Environmental Problem, Journal of Multicriteria Decision Analysis, 6, 215-226.

Hope, C., J. Parker, and S. Peake, 1991. A Pilot Environmental Index for the United Kingdom: Results for the Last Decade, Statistical Journal of the United Nations ECE, 8, 85-107.

Hota, P.K., R. Chakrabarti, and P. K. Chattopadhyay, 1999. An Integrated Approach to Economic Emission Load Dispatching Using Neural Network and Goal-Attainment Methods, Electric Machines & Power Systems, 27(10), 1085-1096.

Huang, J. P., K. L. Poh, and B. W. Ang, 1995. Decision Analysis in Energy and Environmental Modeling, Energy, 20(9), 843-855.

Huber, G. P., 1974. Multiattribute Utility Models: A Review of Field and Field-Like Studies, Management Science, 20, 1393-1402.

Intergovernmental Panel on Climate Change, 1996. *Climate Change 1995: Economic and Social Dimensions of Climate Change*, Cambridge University Press, Cambridge.

International Energy Agency, 1996. Implementing Agreement on Demand-Side Management Technologies and Programmes, 1995 Annual Report, Swedish National Board for Industrial and Technical Development, Stockholm, January.

Jaccard, M., 1995. Oscillating Currents: The Changing Rationale for Government Intervention in the Electricity Industry, Energy Policy, 23(7), 579-592.

Jenkins, R. T., and D. S. Joy, 1974. Wien Automatic System Planning Package (WASP) -- An Electric Utility Optimal Generation Expansion Planning Computer Code, ORNL-4945, Oak Ridge National Laboratory, Oak Ridge, TN, 1974.

Jia, J., G. W. Fischer, and J. S. Dyer, 1998. Attribute Weighting Methods and Decision Quality in the Presence of Response Error: A Simulation Study, Journal of Behavioral Decision Making, 11(2), 95-105.

Jones, M., C. Hope, and R. Hughes, 1990. A Multiattribute Value Model for the Study of UK Energy Policy, Journal of the Operational Research Society, 41(10), 919-929.

Jones, M. R., 1989. The Potential of Decision Analysis as a Means of Integrating the Environment into Energy Policy Decision-making, Environment and Planning 21, 1315-1327.

Joskow, P. L., 1992. Weighing Environmental Externalities: Let's Do It Right!, The Electricity Journal 5(4), 53-67.

Joskow, P. L. and D. Marron, 1992. What Does a Negawatt Really Cost? Evidence from Utility Conservation Programs. Energy Journal, 13(4), 41-74.

Jungermann, H., 1985. The Social Compatibility of Energy Technologies--Theoretical and Empirical Approaches in West Germany, in P. J. M. Stallen and V. R. R. Uppuluri, Eds., *Environmental Impact Assessment, Technology Assessment, and Risk Analysis*, NATO ASI Series G; Ecologic Sciences, Vol. 4, Springer-Verlag, 789-810.

Kagazyo, T., K. Kaneko, M. Akai, and K. Hijikata, 1997. Methodology and Evaluation of Priorities for Energy and Environmental Research Projects, Energy, 22(2/3), 121-129.

Kahn, E. P., *et al.*, 1989. Evaluation Methods in Competitive Bidding for Power, Lawrence Berkeley Laboratory, LBL-26924/UC-101, Berkeley, CA.

Kahneman, D., P. Slovic, and A. Tversky, 1982. *Judgment under Uncertainty: Heuristics and Biases*, Cambridge University Press, New York, NY.

Kalika, V., and S. Frant, 1999. Environmental Aspects of Power Generation, Energy Sources, 21(8), 687-704.

Karni, R., P. Feigin, and A. Breiner, 1992. Multicriterion Issues in Energy Policy-Making, European Journal of Operational Research, 56, 30-40.

Karni, E., and E. Werczberger, 1995. The Compromise Criterion in MCDM: Interpretation and Sensitivity to the Parameters, Environment & Planning B: Planning & Design, 22, 407-418.

Kavrakoglu, I., and G. Kiziltan, 1983. Multiobjective Strategies in Power Systems

Planning, European Journal of Operational Research, 12, 159-170.

Keefer, D. L. and S. M. Pollack, 1980. Approximations and Sensitivity in Multiobjective Resource Allocation, Operations Research, 28(1), 114-128.

Keeney, R. L., 1980. *Siting Energy Facilities*, J. Wiley, NY.

Keeney, R. L., 1987. An Analysis of the Portfolio of Sites to Characterize for Selecting a Nuclear Repository, Risk Analysis, 7(2), 195-218.

Keeney, R. L., 1992. *Value-Focused Thinking*, Harvard University Press, Boston.

Keeney, R. L., and T. L. McDaniels, 1999. Identifying and Structuring Values to Guide Integrated Resource Planning at BC Gas, Operations Research, 47(5), 651-662.

Keeney, R. L., and K. Nair, 1977. Nuclear Siting Using Decision Analysis, Energy Policy, 5(3), 223-231.

Keeney, R. L. and H. Raiffa, 1976. *Decisions with Multiple Objectives: Preferences and Value Trade-offs*, New York: Wiley.

Keeney, R. L., and H. Raiffa, 1991. Structuring and Analyzing Values for Multiple-Issue Negotiations, in P. Young, Ed., *Negotiation Analysis*, U. Michigan Press, Ann Arbor, MI, 131-151.

Keeney R. L. and A. Sicherman, 1983. Illustrative Comparison of One Utility's Coal and Nuclear Choices, Operations Research, 31(1), 50-83.

Keeney, R. L., O. Renn, and D. von Winterfeldt, 1987. Structuring West Germany's Energy Objectives, Energy Policy, 15, 352-362.

Keeney, R. L., and D. von Winterfeldt, 1987. Optimal Procedures to Evaluate Decisions with Multiple Objectives, EPRI EA-5433, Electric Power Research Institute, Palo Alto, CA.

Keeney, R. L., and E. Wood, 1977. An Illustrative Example of the use of Multiattribute Utility Theory for Water Resources Planning, Water Resources Research, 13(4), 705-712.

Keeney, R. L., *et al.*, 1981. Decision Framework for Technology Choice - Volume 1: A Case Study of One Utility's Coal-Nuclear Choice, EPRI EA-2153, Electric Power Research Institute, Palo Alto, CA.

Khare, R., and R. D. Christie, 1997. Prioritizing Emergency Control Problems with Fuzzy Set Theory, IEEE Transactions on Power Systems, 12(3), 1237-1242.

Kirkwood, C. and R. K. Sarin, 1985. Ranking with Partial Information: A Method and an Application, Operations Research, 33, 38-48.

Koundinya, S., D. Chattopadhyay, and R. Ramanathan, 1995. Incorporating Qualitative Objectives in Integrated Resource Planning --Application of Analytic Hierarchy Process and Compromise Programming, Energy Sources, 17(5), 565-581.

Krupnick, A. J., and D. Burtraw, 1996. The Social Costs of Electricity-Do the Numbers Add Up?, Resource and Energy Economics, 18(4), 423-466.

Krzysztofowicz, R., and L. Duckstein, 1984. Assessment Errors in Multiattribute Utility Functions, Organizational Behavior and Human Performance, 26, 326-348.

Lai, S. K. and L. D. Hopkins, 1995. Can Decision-Makers Express Multiattribute Preferences using AHP and MAUT? An Experiment, Environment and Planning, B, 22, 21-34.

Larichev, O. I., 1992. Cognitive Validity in Design of Decision-Aiding Techniques, Journal of Multicriteria Decision Analysis, 1, 127-138.

Lee, R., 1995a. The US-European Community Fuel Cycle Externalities Study: The US Research Team's Methodology, Results and Conclusions, Oak Ridge National Laboratory, Oak Ridge, TN, prepared for the EC, IEA and OECD Workshop on the External Costs of Energy, Brussels, Jan. 30-31.

Lee, R., 1995b. Externalities and Electric Power: An Integrated Assessment Approach, Oak Ridge National Laboratory, Oak Ridge, TN, presented at the Malama Aina (Preserve the Land) Conference, Honolulu, IH.

Lenssen, N., 1996. Local IRP: A New Tool, The Electricity Journal, 9(6), 26-36.

Leon, O. G., 1997. On the Death of SMART and the Birth of GRAPA, Organizational Behavior & Human Decision Processes, 71(3), 249-262.

Liu, Z. P., and Y. L. Huang, 1998. Fuzzy Model-Based Optimal Dispatching for NO_x Reduction in Power Plants, International Journal of Electrical Power & Energy Systems, 20(3), 169-176.

Lootsma, F. A., *et al.*, 1990. Choice of a Long-Term Strategy for the National Electricity Supply via Scenario Analysis and Multicriteria Analysis, European Journal of Operational Research, 48, 189-203.

Loucks, D. P. and L. Somlyody, 1988. *Assessment of Multiple Objective Water Resources Projects: Approaches for Developing Countries*, United Nations, NY.

Maekawa, T., K. Yashuda, and R. Yokoyama, 1992. Fuzzy Coordination of Multiobjectives Reflecting the Operator's Intention for Dynamic Generation Rescheduling, Electrical Power and Energy Systems, 14(5), 314-320.

Maguire, L. A. and L. G. Boiney, 1994. Resolving Environmental Disputes: A Framework Incorporating Decision Analysis and Dispute Resolution Techniques. Journal of Environmental Management, 42, 31-48.

Maloney, J., 1996. Multiple Objective Trade-Off Analysis: Theory Meets Practice in a Public Involvement Process at Eugene Water & Electric Board, unpublished manuscript, Eugene, OR.

Manne, A. S. and R. Richels, 1991. Global CO_2 Emission Reductions: The Impacts of Rising Energy Cost, Energy Journal, 12(1), 87-107.

Martunnen, M., and R. P. Hamalainen, 1995. Decision Analysis Interviews in Environmental Impact Assessment, European Journal of Operational Research, 87(3), 551-563.

Massam, B. H., and I. D. Askew, 1982. Methods for Comparing Policies using Multiple Criteria: An Urban Example, Omega, 10(2), 195-204.

Matos, M. A., 1999. A Fuzzy Filtering Method Applied to Power Distribution Planning, Fuzzy Sets & Systems, 102(1), 53-58.

McDaniels, T. L., 1994. Sustainability, Value Tradeoffs, and Electric Utility Planning, Energy Policy, 22(12) 1045-1054.

McDaniels, T. L., 1995. A Multiattribute Index for Evaluating Environmental Impacts of Electric Utilities, Journal of Environmental Management, 46, 57-66.

Medawar, P., 1977. Unnatural Science, New York Review of Books, 3, 13-18.

Megdahl, L. M. 1994. Hard to Quantify Benefits and Costs Scoping Study, Cambridge Systematics, Inc., Prepared by for the Low Income Evaluation Task Force, New York Public Service Commission, Cambridge, MA.

Meier, P., 1972. Stochastic Population Dynamics and Projection at the Design Level, Journal of the Sanitary Engineering Division, American Society of Civil Engineers, 8(SA6), 883 - 896.

Meier, P.M., 1979. Long-Range Regional Power Plant Siting Model, J. Energy Division, Proceedings of the ASCE, 105(EY1), 117-135.

Meier, P., 1995a. Integrated Resource Plan for Andhra Pradesh. United States Agency for International Development, New Delhi.

Meier, P., 1995b. Resource Trade-off Decision Analysis for BC Hydro's 1995 Integrated Electricity Plan. Appendix F, BC Hydro Integrated Electricity Plan, Vancouver.

Meier, P., 1999. ENVIROPLAN: A Model for Electric Utility Strategic Planning and Environmental Policy Assessment, Technical Paper 14, International Development and Energy Associates, Old Headington, Oxford, UK.

Meier, P. and M. Munasinghe, 1994. Incorporating Environmental Concerns into Power Sector Decision-making: A Case Study of Sri Lanka, Environment Department Paper 6, World Bank; Washington DC.

Meier, P., and M. Munasinghe, 1999. Sri Lanka Global Overlay, Environment Department, World Bank, Washington, DC.

Meier, P. and M. Munasinghe, 2000. Sri Lanka Power Technology Assessment, Environment Department, World Bank, Washington, DC.

Meier, P., M. Munasinghe, and T. Siyambalapitiya, 1993. Energy Sector Policy and the Environment: A Case Study of Sri Lanka. Environment Department, World Bank, Washington, DC.

Meier, P. and L. Ruff, 1979. The Spatial Dimension of Regulatory Impact Assessment, Report to the National Commission on Air Quality, Washington, DC.

Merrill, H. M. and A. J. Wood, 1991. Risk and Uncertainty in Power System Planning, Electrical Power and Energy Systems, 13 (2), 81-90.

Merrill, H. M., F. C. Schweppe, and D. C. White, 1982. Energy Strategy Planning for Electric Utilities: SMARTE Methodology, IEEE Transactions on Power Systems, 101(2), 340-355.

Miakisz, J., 1995. Measuring Environmental Performance at Niagara Mohawk Power, Total Quality Environmental Management, 47-56.

Mills, D., L. Vlacic, and I. Lowe, 1996. Improving Electricity Planning: Use of a Multicriteria Decision Making Model, International Transactions in Operational Research, 3(3/4), 293-304.

Mintzer, I., *et al.*, 1990. Environmental Externality Data For Energy Technologies, Center for Global Change, University of Maryland.

Mirasgedis, S., and D. Diakoulaki, 1997. Multicriteria Analysis vs. Externalities Assessment for the Comparative Evaluation of Electricity Generation Systems, European Journal of Operational Research, 102(2), 364-379.

Mollaghesemi, M., and J. Pet-Edwards, 1997. *Technical Briefing: Making Multiple Objective Decisions*, IEEE Computer Society Press, Los Alamitos, CA.

Montana Power Co., 1991. Environmental Externality Considerations For Use in Integrated Least Cost Planning, Unpublished Manuscript.

Morgan, M. G., M. Kandlikar, J. Risbey, and H. Dowlatabadi, 1999. Why Conventional Tools for Policy Analysis Are Often Inadequate for Problems of Global Change, Climatic Change, 41, 271-281.

Moro, L. M., and A. Ramos, 1999. Goal Programming Approach to Maintenance Scheduling of Generating Units in Large Scale Power Systems, IEEE Transactions on Power Systems, 14(3), 1021-1027.

Morse, J. N., 1979. A Theory of Naive Weights, in *Multiple Criteria Problem Solving* (S. Zionts, Ed.), Springer-Verlag, NY.

Moskovitz, D., 1993. Green Pricing: Why Not Customer Choice?, The Electricity Journal, 6, 42-50.

Munasinghe, M., 1993. Environmental Economics and Sustainable Development. World Bank Environment Paper 3, Washington, DC.

Muslu, M., and M. D. Anderson, 1989. A Multiobjective Optimization Method for Electrical Power Distribution Systems with Load Control Alternatives, Electric Power Systems Research, 16(1), 1-9.

Nadel, S., V. Kothari, and S. Gopinath, 1991. Opportunities for Improving End-Use Electricity Efficiency in India, ACEEE, Washington, DC.

Nangia, U., N. K. Jain, and C. L. Wadhwa, 1998. Optimal Weight Assessment Based on a Range of Objectives in a Multiobjective Optimal Load Flow Study, IEE Proceedings--Generation Transmission & Distribution, 145(1), 65-69.

New York Department of Environmental Conservation, 1980. Hazardous Waste: The Siting Board, New York State Environment, 9 (Dec.18), 3-6.

Newman, J. R., 1975. Assessing the Reliability and Validity of Multiple Attribute Utility Assessment Procedures, Social Science Research Institute, University of Southern California, Los Angeles.

NORTEL, 1996. Northern Telecom Environmental Performance Index, prepared by Arthur D. Little of Canada, Ltd., Centre for Environmental Assurance, http://www.nortel.com/cool/environ/epi/epi.html.

Nutt, P.C., 1980. Comparing Methods for Weighting Decision Criteria. Omega, 8, 163-172.

Otway, H., and W. Edwards, 1977. Application for Simple Multiattribute Rating Techniques to the Evaluation of Nuclear Waste Disposal Sites: A Demonstration, RM-77-31, IIASA, Laxenberg, Austria.

Pan, J. P, and S. Rahman, 1998. Multiattribute Utility Analysis With Imprecise Information: An Enhanced Decision Support Technique for the Evaluation of Electric Generation Expansion Strategies, Electric Power Systems Research, 46(2), 101-109.

Pastijn, H. and J. Leysen, 1989. Constructing an Outranking Relation with ORESTE, Mathematical and Computer Modelling, 12(10/11), 1255-1268.

Payne, J. W., J. R. Bettman, and D. A. Schkade, 1999. Measuring Constructed Preferences: Towards a Building Code, Journal of Risk and Uncertainty, 19(1-3), 243-270.

Pennsylvania Power & Light, 1977. DELPHI Assessment, PP&L Coal-Fired Power Plant Site Selection, Allentown, PA.

Perlack, R. D., G. K. Beim, D. C. Bowman, and B. F. Hobbs, 1995. Multicriteria Decision Frameworks for Technology Ranking Problems, ORNL-6883, Draft, Oak Ridge National Laboratory, Oak Ridge, TN.

Pirlot, M., 1997. A Common Framework for Describing Some Outranking Methods, Journal of Multicriteria Decision Analysis, 6(1), 86-92.

Pitz, C., J. Heerboth and N. Sachs, 1980. Assessing the Utility of Multiattribute Utility Assessments, Organizational Behavior and Human Performance, 26, 65-80.

Poyhonen, M. A., 1998. On Attribute Weighting in Value Trees, Ph.D. Dissertation, Helsinki University of Technology, Helsinki, Finland.

Poyhonen, M. A., R. P. Hamalainen, and A. A. Salo, 1997. An Experiment on the Numerical Modelling of Verbal Ratio Statements, Journal of Multicriteria Decision Analysis, 6(1), 1-10.

Poyhonen, M., and R. P. Hamalainen, 1998. Notes on the Weighting Biases in Value Trees, Journal of Behavioral Decision Making, 11(1),139-150.

Putta, S., 1990. Weighing Externalities in New York State, Electricity Journal, 3(6), 42-47.

Putta, S., 1991. Methods for Valuing and Incorporating Environmental Costs in Electric Resource Planning and Acquisition, in O. Hohmeyer and R. L. Ottinger, Eds., *External Environmental Costs of Electric Power*, Springer-Verlag, NY.

Quaddus, M. A., and T. N. Goh, 1985. Electric Power Generation Expansion: Planning with Multiple Objectives, Applied Energy, 19, 301-319.

Raab, J. 1994. Using Consensus Building to Improve Utility Regulation, American Council for an Energy-Efficient Economy, Washington, DC.

Rader, N. A., and R. B. Norgaard, 1996. The Renewables Portfolio Standard for Efficiency and Sustainability in Restructured Markets, The Electricity Journal, 9(6), July, 37-49.

Rahman, S. and G. Shrestha, 1992. A Technique to Incorporate New Information in Evaluating Generation Alternatives, IEEE Transactions in Power Systems, 7(2), 900-906.

Raiffa, H., 1982. *The Art and Science of Negotiation*, Belnap Press, Cambridge, Mass.

Ramanathan, R., 1999. Selection of Appropriate Greenhouse Gas Mitigation Options, Global Environmental Change-Human and Policy Dimensions, 9(3), 203-210.

Ramanathan, R., and L. S. Ganesh, 1995. Energy Resource Allocation Incorporating Qualitative and Quantitative Criteria: An Integrated Model Using Goal Programming and AHP, Socio-Economic Planning Sciences, 29(3), 197-218.

Ramirez-Rosado, I. J., and R. N. Adams, 1991. Multiobjective Planning of the Optimal Voltage Profile in Electric Power

Distribution Systems, COMPEL, 10(2), 115-128.

Rao, G. R., 1998. Demand-Side Management, Energy, Economy and Environmental Consultants, Bangalore, India.

Rau, J., and D. Wooten, 1980. *Environmental Impact Analysis Handbook*, McGraw-Hill, NY.

Rogers, M., and M. Bruen, 1998. Choosing Realistic Values of Indifference, Preference and Veto Thresholds for Use with Environmental Criteria Within ELECTRE, European Journal of Operational Research, 107(3), 542-551.

Rogers, P., 1999. Environmental Indicators, Journal of Water Resources Planning & Management, 125(2).74-75.

Rowe, M. D., B. F. Hobbs, B. L. Pierce, and P. M. Meier, 1979. Evaluation of Nuclear Power Plant Siting Methodologies, Report NUREG/CR-1689, Brookhaven National Laboratory, Upton, NY.

Russell, B., 1945. *A History of Western Philosophy*, Simon & Schuster, London.

Saaty, T. L., 1980. *The Analytical Hierarchy Process: Planning, Priority Setting, Resource Allocation*. McGraw-Hill, New York, NY.

Sage, A., and E. White, 1980. Methodologies for Risk and Hazard Assessment: A Survey and Status Report, IEEE Transactions on Systems, Man, and Cybernetics, SMC-10, 425-446.

Salleh, K. B. O., 1996. Route Evaluation and Selection for a 500 kV Electric Transmission Line in Malaysia: An Environmental Assessment, Asia Journal of Environmental Management, 4(1), 19-30.

Salminen, P., J. Hokkanen, and R. Lahdelma, 1998. Comparing Multicriteria Methods in the Context of Environmental Problems, European Journal of Operational Research, 104, 485-496.

Salo, A. A., and R. P. Hamalainen, 1997a. PRIME-Preference Ratios in Multiattribute Evaluation, Helsinki University of Technology, System Analysis Laboratory Research Reports A43.

Salo, A. A., and R. P. Hamalainen, 1997b. On the Measurement of Preferences in the Analytic Hierarchy Process, Journal of Multicriteria Decision Analysis, 6, 309-319.

Samouilidis, J. E., P. Capros, and S. Papathanassiou, 1988. Multicriteria Analysis of Energy Supply Decisions in an Uncertain Future, OMEGA, 16, 107-115.

Sarin, R. K., J. S. Dyer and K. Nair, 1980. A Comparative Evaluation of Three Approaches for Preference Function Assessment. ORSA/TIMS Joint National Meeting.

Sarin, R. K., 1977a. Interactive Evaluation and Bound Procedure for Selecting Multiattributed Alternatives, in M. Starr and M. Zeleny, eds., *Multiple Criteria Decision Making*, TIMS Studies in the Management Sciences, Vol. 6, North-Holland, Amsterdam.

Sarin, R. K., 1977b. Screening of Multiattribute Alternatives, OMEGA, 5(4), 481-489.

Schoemaker, P. J. H., 1981. Behavioral Issues in Multiattribute Utility Modeling and Decision Analysis, in J.N. Morse, Ed., *Organizations: Multiple Agents with Multiple Criteria*, Springer-Verlag, NY.

Schoemaker, P. J. H., 1991. When and How to Use Scenario Planning: A Heuristic Approach with Illustrations, Journal of Forecasting, 10, 549-564.

Schweitzer, M., M. English, and J. Altman, 1994. DSM/IRP Collaboratives: What Have They Accomplished? Energy Sources, 16, 161-176.

Seattle City Light, 1991. Strategic Corporate Plan 1990-1991, Second Edition, City of Seattle.

Shepard, R.,1964. On Subjectively Optimal Selections Among Multiattribute Alternatives, in M. Shelley and G. Bryan, Eds., *Human Judgments and Optimality*, Wiley, New York, 257-281.

Sicherman, A. 1979. A General Methodology and Computer Tool for Environmental Impact Assessment with Two Case Study Examples, in *Proceedings of the International Conference on Cybernetics and Society*, Tokyo, Japan, Nov. 3-7.

Slovic, P., and S. Lichtenstein, 1971. Comparison of Bayesian and Regression Approaches to the Study of Information Processing in Judgment, Organizational Behavior and Human Performance, 6, 649-744.

Sone Command Area Development Authority (SCADA), 1998. Environmental Issues in the Power Sector: A Case Study of Bihar. Patna, India.

Srinivasan, D., and A. G. B. Tettamanzi, 1997. A Revolutionary Algorithm for Evaluation of Emission Compliance Options in View of the Clean Air Act Amendments, IEEE Transactions on Power Systems, 12 (1), 336-341.

Stam, A., M. Kuula and H. Cesar, 1992. Transboundary Air Pollution in Europe: An Interactive Multicriteria Tradeoff Analysis, European Journal of Operational Research, 56, 263-277.

Stanton, T., 1990. Unpublished report, Michigan Public Service Commission, E. Lansing, MI.

Starr, C., and C. Whipple, 1980. Risks of Risk Decisions, Science, 208, 1114-1119.

Staschus, K., *et al.*, 1991. A Multiattribute Evaluation Framework for Electric Resource Acquisition in California, Electrical Power & Energy Systems, 13(2), 73-80.

Steuer, R. E., 1986. *Multiple Criteria Optimization: Theory, Computation, and Application*, J. Wiley, New York.

Stevens, S. S., 1966. On the Operation Known as Judgment, American Scientist, 54(4), 385-401.

Stewart, T. J., 1991. A Multicriteria Decision Support System for R&D Project Selection, Journal of the Operational Research Society, 42(1), 17-26.

Stewart, T. J., 1992. A Critical Survey on the Status of Multiple Criteria Decision Making Theory and Practice, OMEGA, 20(5/6), 569-586.

Stewart, T. J., 1996. Robustness of Additive Value Function Methods in MCDM, Journal of Multicriteria Decision Analysis, 5, 301-309.

Stewart, T. J., 2000. Policy Decisions in the Public Sector: Can MCDA Make a Difference?, in Haimes and Steuer (2000).

Stewart, T. R. and D. W. Ely, 1984. Range Sensitivity: A Necessary Condition and a Test for the Validity of Weights. National Center for Atmospheric Research, Technical Report NCAR 3141-84/14. Boulder, CO.

Stewart, W. R., Jr., and E. R. Horowitz, 1991. Environmental Factor Weighting at the Federal Energy Regulatory Commission, Socio-Economic Planning Sciences, 25(2), 123-132.

Stillwell, W. G., D. von Winterfeldt, and R. S. John. 1987. Comparing Hierarchical and Nonhierarchical Weighting Methods for Eliciting Multiattribute Value Models, Management Science, 33(4), 442-450.

Stirling, A., 1998. Valuing the Environmental Impacts of Electricity Production: A Critical Review of Some "First Generation" Studies, Energy Sources, 20(4-5), 267-300.

Susskind, L., and J. Cruikshank, 1987. *Breaking the Impasse: Consensual Approaches to Resolving Public Disputes*, Basic Books, New York.

Talaq, J. H., F. El-Hawary, and M. E. El-Hawary, 1994. A Summary of Environmental/ Economic Dispatch Algorithms, IEEE Transactions on Power Systems, 9(3), 1508-1516.

Tennessee Valley Authority, 1995. Energy Vision 2020, Volume I: Integrated Resource Plan Environmental Impact Statement, Knoxville, TN.

Teo, C. Y., and W. Jiang, 1997. Adaptive Real-Time Knowledge Based Approach For Bulk Power System Restoration, Electric Machines & Power Systems, 25(10), 1079-1088.

Thomas, H., and D. Samson, 1986. Subjective Aspects of the Art of Decision Analysis: Exploring the Role of Decision Analysis in Decision Structuring, Decision Support, and Policy Dialogue, Journal of the Operational Research Society, 37, 249-265.

Timmermans, D., and C. Vlek, 1996. Effects on Decision Quality of Supporting

Multiattribute Evaluation in Groups, Organizational Behavior and Human Decision Processes, 68(2), 158-170.

Tollefson, G., R. Billinton, and G. Wacker, 1994. Comprehensive Bibliography on Reliability Worth and Electric Service Customer Interruption Costs: 1980-1990, IEEE Transactions on Power Systems, 6(4), 1508-1514.

Tonn, B. E., and M. Schweitzer, 1996. Public Policy Responsibilities in a Restructured Electric Industry: An Analysis of Values, Objectives, and Approaches, ORNL/CON-428, Oak Ridge National Laboratory, Oak Ridge, TN, March.

Turvey, R. and D. Anderson, 1977. *Electricity Economics: Essays and Case Studies*, The Johns Hopkins University Press, Baltimore, MD.

Tzeng, G.-H., T. Shiau and C.-Y. Lin, 1992. Application of Multicriteria Decision-Making to the Evaluation of New Energy System Development in Taiwan, Energy, 17(10), 983-992.

United Nations Environment Programme, 1988. TCD/SEM.88/3, INT/84/R36, New York.

US Office of Technology Assessment, 1994. Studies of the Environmental Costs of Electricity, OTA-ETI-134, US Govt. Printing Office, Washington, DC.

US Agency for International Development, 1995. Integrated Resource Plan for Andhra Pradesh, Office of Energy, Environment and Technology, Washington DC.

US Army Corps of Engineers and Montgomery Watson Americas, Inc., 1996. Strategic Decision Making Process and Results, Columbia River Fish Mitigation Program, Walla Walla District, Walla Walla, WA.

US Nuclear Regulatory Commission, 1977. Nuclear Energy Centre Site Survey, Washington, DC.

Venkatesh, B., G. Sadasivam, and M. A. Khan, 1999. Optimal Reactive Power Planning Against Voltage Collapse Using the Successive Multiobjective Fuzzy LP Technique, IEE Proceedings--Generation Transmission & Distribution, 146(4), 343-348.

Vetschera, R., 1994. Visualization Techniques in Multicriteria Decision-Making, OR Spektrum, 16(4), 227-241.

Vine E. L. and J. P. Harris, 1990. Evaluating Energy and Non-Energy Impacts of Energy Conservation Programs: A Supply Curve Framework of Analysis, Energy, 15(1), 11-21.

Vlek, C., J. Mesken, and L. Steg, 1999. Future-Sketching and Multiattribute Evaluation May Affect Your Preference Order of Complex Policy Scenarios, Journal of Behavioral Decision Making, 12(2), 107-122.

von Nitzsch, R., and M. Weber, 1993. The Effect of Attribute Ranges on Weights in Multiattribute Utility Measurements, Management Science, 39(8), 937-943.

von Winterfeldt, D., and W. Edwards, 1986. *Decision Analysis and Behavioral Research*, Cambridge University Press, New York.

Vroom, V., and P. Yetton, 1973. *Leadership and Decision-making*, Univ. Pittsburgh Press, Pittsburgh, PA.

Watson, S. R., 1981. Decision Analysis as a Replacement for Cost/Benefit Analysis, European Journal of Operational Research, 7, 242-248.

Weber, M. and K. Borcherding, 1993. Behavioral Influences on Weight Judgments in Multiattribute Decision-Making, European Journal of Operational Research, 67, 1-12.

Weedy, B. M., 1989. Environmental Aspects of Route Selection for Overhead Lines in the US, Electric Power Systems Research, 16(3), 217-226.

Weinberg, C. J., Iannucci, J. J., and Reading, M. M., 1993. The Distributed Utility: Technology, Customer, and Public Policy Changes Shaping the Electrical Utility of Tomorrow, Energy Systems and Policy, 15, 307-322.

West Kootenay Power, 1995. Integrated Resource Plan March 1995.

White, C. C., A. P. Sage, S. Dozono, and W. T. Scherer, 1984. Performance Evaluation of a Decision Support System, Large Scale Systems, 6, 39-48.

White, D. J., 1990. A Bibliography on the Applications of Mathematical Programming Multiobjective Methods, J. Operational Research Society, 41(8), 669-691.

Wilson, A. F., 1989. Value of Energy Study, Seattle City Light, Seattle, WA.

Woo, C. K., B. F. Hobbs, R. Orans, R. Pupp, and B. Horii, 1994. Emission Costs, Consumer Bypass, and Efficient Pricing of Electricity, The Energy Journal, 15(3), 43-54.

Wood, C., 1995. *Environmental Impact Assessment: A Comparative Review*, J. Wiley, New York.

World Bank, 1989. *Striking a Balance: The Environmental Challenge of Development*, Washington DC.

World Bank, 1996. China: Issues and Options in Greenhouse Gas Emissions Control. Discussion Paper 330, Washington, DC

World Bank, 1998a. India: Environmental Issues in the Power Sector, Industry and Energy Department, World Bank, Washington, DC.

World Bank, 1998b. Economic Appraisal of the Andhra Pradesh Power Sector Reform and Restructuring Programme, New Delhi.

Xie, Z. J., and M. Kuby, 1997. Supply-Side/ Demand-Side Optimization and Cost-Environment Tradeoffs for China's Coal and Electricity System, Energy Policy, 25(3), 313-326.

Yang, H. T., and S. L. Chen, 1989. Incorporating a Multicriteria Decision Procedure into the Combined Dynamic Programming/ Production Simulation Algorithm for Generation Expansion Planning, IEEE Transactions on Power Systems, 4(1), 165-175.

Yokoyama, R., *et al.*, 1997. Multiobjective Optimal Unit Sizing of A Grid-Connected Photovoltaic System in Consideration of Its Probabilistic Characteristics, Journal of Solar Energy Engineering-Transactions of the ASME, 119(2), 134-140.

Yoon, K. P., and C. L. Hwang, 1995. *Multiple Attribute Decision Making: An Introduction*, Sage Publications, Thousand Oaks, CA.

Yosie, T. F., and T. D. Herbst, 1998. Using Stakeholder Processes in Environmental Decision-Making: An Evaluation of Lessons Learned, Key Issues, and Future Challenges, ICF Inc., McLean, VA.

Zanakis, S. H., A. Solomon, N. Wishart, and S. Dublish, 1998. Multiattribute Decision Making: A Simulation Comparison of Select Methods, European Journal of Operational Research, 107, 507-529.

Zapatero, E.G., C. H. Smith, and H.R. Weistroffer, 1997. Evaluating Multiple Attribute Decision Support Systems, Journal of Multicriteria Decision Analysis, 6, 201-214.

Zeleny, M., 1982. *Multiple Criteria Decision Making*, McGraw-Hill, NY, 1982.

INDEX

MIX
Papier aus verantwortungsvollen Quellen
Paper from responsible sources
FSC® C105338

If you have any concerns about our products,
you can contact us on
ProductSafety@springernature.com

In case Publisher is established outside the EU,
the EU authorized representative is:
Springer Nature Customer Service Center GmbH
Europaplatz 3, 69115 Heidelberg, Germany

Printed by Libri Plureos GmbH
in Hamburg, Germany